Affinity
Chromatography

CHEMICAL ANALYSIS

A SERIES OF MONOGRAPHS ON
ANALYTICAL CHEMISTRY AND ITS APPLICATIONS

VOLUME 59

A WILEY-INTERSCIENCE PUBLICATION

JOHN WILEY & SONS

New York / Chichester / Brisbane / Toronto

Affinity Chromatography
Bioselective Adsorption on Inert Matrices

WILLIAM H. SCOUTEN

Chemistry Department
Bucknell University
Lewisburg, Pennsylvania

A WILEY-INTERSCIENCE PUBLICATION

JOHN WILEY & SONS

New York / Chichester / Brisbane / Toronto

Library of Congress Cataloging in Publication Data

Scouten, William H   1942–
  Affinity chromatography.

  (Chemical analysis; v. 59 ISSN 0069–2883)
  "A Wiley-Interscience publication."
  Includes bibliographical references and index.
  1. Affinity chromatography. 2. Biological
chemistry–Technique. I. Title. II. Series.
QP519.9.A35S36   574.19′25′028   80-20129
ISBN 0-471-02649-2

Printed in the United States of America

10 9 8 7 6 5 4 3 2 1

To Nancy and the children

PREFACE

Affinity chromatography has grown to be such a significant aspect of biochemical research that almost every issue of any biochemical journal has one or more reports dealing in some fashion with this technique, and there is now one journal (*Solid Phase Biochemistry*) that deals only with this and several closely related subjects. It has become so significant that no biochemist can go long before encountering a need to utilize this technique.

The primary purpose of this particular book is to provide *a general introduction* to the field, guiding the researcher through the major aspects of the field, describing its potential and some of its applications. I have not attempted to give either a complete review of all the nearly 2000 papers that have appeared in this general area, nor to compile a detailed book of methods. Rather, I hope to combine a general introduction with a detailed description of some of the most important techniques, such that the novice to the field can determine what application can be made to his or her own research and, in general, how to begin using affinity chromatography in any desired enzyme purification. At the same time, the latter chapters provide descriptions of some of the newer advances in this area that may be of value to the more experienced researcher.

Affinity chromatography isn't limited to enzymes. Several chapters have been devoted to the purification of other biomolecules using affinity chromatography. Chromatography on insoluble polynucleotides is the major method of purifying messenger RNA (mRNA), and thus an entire chapter has been devoted to bioselective adsorption of nucleic acids. Shorter sections have been devoted to the purification of coenzymes, peptides, saccharides, and other nonenzymatic compounds of biological importance. Particular emphasis has been given to the use of antibodies in affinity chromatography.

Affinity chromatography has led, directly or indirectly, to the development of other aspects of solid-phase biochemistry. Of particular importance is hydrophobic chromatography (or "reverse-phase" chromatography), and this is discussed throughout the book, with Chapter 9 devoted solely to this subject. Similarly, brief discussions are included of such related techniques as covalent chromatography, charge-transfer chromatography, and immobilized reagents for protein modification.

I would like to express my appreciation to Janet Zimmerman for her patient typing and coordinating of the manuscript and to Hans Tramper, Bennett Willeford and Vanessa Morenzi for their critical reading of portions of the manuscript. Finally, thanks are due to Cees Veeger, Landbouwhogeschool, Wageningen, The

Netherlands and his colleagues for their hospitality during my stay in The Netherlands, where this work was begun. Thanks are also due to the Dreyfus Foundation for a Teacher-Scholar Grant and to the Research Corporation and the Petroleum Research Fund for their financial support of the author's contributions in this area of research.

WILLIAM H. SCOUTEN

Lewisburg, Pennsylvania
January 1981

CONTENTS

ix

Affinity Chromatography

INTRODUCTION TO THE BASIC CONCEPTS
OF AFFINITY CHROMATOGRAPHY

Almost any protein can be purified to homogeneity with the use of classical purification procedures (provided enough starting material is available), but very often only at considerable cost both in supplies and hours of work time. With the development of affinity chromatography, these frequently low yield, tedious, multistep procedures can often be replaced with much higher yield procedures that are relatively simple, contain far fewer steps (possibly only two or three), and save many hours of work. It is for this reason that affinity chromatography—or bioselective adsorption, as it is more appropriately termed—is developing into one of the greatest technical advances in enzymology.

The basis for this technique, as shown in Figure 1.1, is relatively simple. Each enzyme or other biomolecule that one wishes to purify has a biologically significant selective attraction for another, usually smaller, bioligand. If the bioligand is immobilized and if a crude cell extract, freed of endogenous substrate, is passed through a column of the immobilized bioligand, the enzyme to be purified will be retained on the column as an enzyme–bioligand complex. Unwanted proteins without any affinity for the bioligands will be eluted with buffer, and subsequently the desired protein can be eluted by a high concentration of soluble bioligand or by changing the conditions (pH, ionic strength, temperature, etc.) such that the enzyme–bioligand complex is no longer stable and the desired enzyme can be eluted in a much purified state.

One of the most striking illustrations of the value of affinity chromatography is its use in the isolation of the vitamin B_{12} binding protein of human serum, transcobalamin II. There is much evidence of the importance of this protein in cellular processes, even though only one of every 2 million protein molecules in human serum is transcobalamin II . . . that is, it composes only 0.00005% of the serum protein. Its purification is further complicated by the fact that transcobalamin II is similar to other serum proteins in many of its chemical and physical properties, which are the very factors used in most classical enzyme purification schemes, such as size, shape, and ionic charge. How, then, can we purify transcobalamin with a reasonable yield, with acceptable cost and with sufficient speed that the protein doesn't change or denature during the purification process?

The answer is found in the fact that, in one crucial respect, transcobalamin II

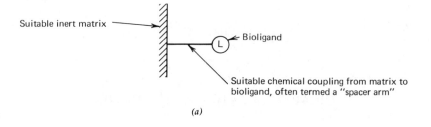

Suitable inert matrix

L

Bioligand

Suitable chemical coupling from matrix to bioligand, often termed a "spacer arm"

(a)

Crude cell extract, plasma, etc.

Exchange buffer

(b)

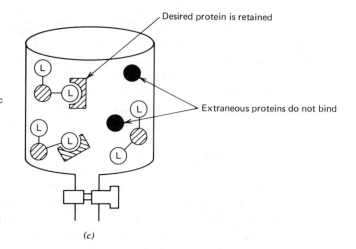

Desired protein is retained

Chromatographic packing of immobilized bioligand from step 1

Extraneous proteins do not bind

(c)

Figure 1.1. The steps of affinity chromatography. (*a*) Step 1: Immobilize a bioligand. (*b*) Step 2: Prepare crude extract and free it from any endogenous substrate. (*c*) Step 3: Apply the substrate free extract to a column of the bioselective adsorbent. (*d*) Step 4: Wash away unwanted proteins. (*e*) Step 5: The desired protein is eluted, possibly with a soluble bioligand.

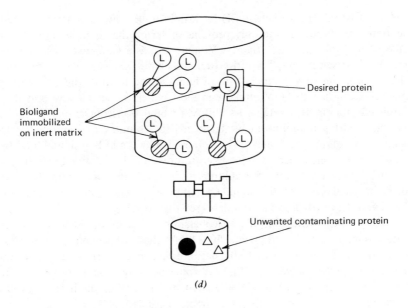

(d)

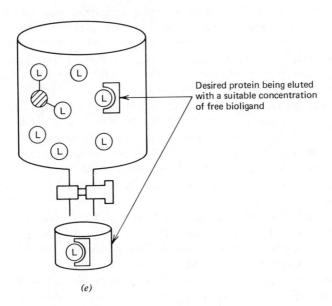

(e)

Figure 1.1. (*Continued*)

(and its "sister," transcobalamin I) is entirely unlike all other serum proteins—the transcobalamins are the only proteins of serum that bind vitamin B_{12}, and some of its analogues, very tightly. Vitamin B_{12} is a *bioligand,* and transcobalamin binds it selectively. To purify transcobalamin, one can make use of its bioselective binding to vitamin B_{12} if an immobilized form of the vitamin is prepared by coupling the bioligand to an inert matrix in such a way that the immobilized vitamin B_{12} retains its ability to bind transcobalamin. We can then use the resulting immobilized bioligand–matrix complex as a chromatographic medium. A column fashioned of the immobilized vitamin B_{12} should bind transcobalamin, but all other proteins from serum should pass directly through the column. Next, if we elute the column with a solution of vitamin B_{12}, transcobalamin will bind to the mobile vitamin B_{12}. The transcobalamin is thus eluted and the purification is complete (see Figure 1.1).

Actually, the purification of transcobalamin by Allen and Majerus (1) wasn't quite as simple as described previously. The synthesis of the immobilized vitamin B_{12}, shown in Figure 1.2, was itself a definite challenge (2). However, this example does illustrate the power of affinity chromatography. These investigators took the Cohn III fraction of serum, prepared by classical methods from 1400 l of outdated human plasma. This serum fraction totaled 34 kg of protein and was dissolved in 360 l of buffer divided equally in four 100-l trash pails. To each pail was added 63 g of ion-exchange resin. The transcobalamin that was ionically absorbed to the ion-exchange resin was subsequently eluted from the ion-exchange beads with 30 l of buffer containing 1-M NaCl. The purpose of these steps, preparatory to bioselective adsorption, was to reduce the 1400 l of viscous serum to 20 l of relatively dilute protein solution. This partially purified protein solution was applied to a 2.5- $\times$ 20-cm column of vitamin B_{12}–Sepharose. The column was subsequently washed with buffer to remove all inert proteins, and then the vitamin B_{12} binding proteins were eluted with a vitamin B_{12}-containing buffer. From this very small column 24.7 mg of homogeneous transcobalamin II was obtained. This constituted a 17.8% yield based on vitamin B_{12} binding capacity. Moreover, the entire purification was performed in less than a week! (See Figure 1.2 and Table 1.1.)

As you might expect, this is one of the most spectacular purifications effected with the use of affinity chromatography. Countless other enzymes have been purified by using bioselective adsorption. Such purifications are far less dramatic in their results and in the purification of some particular enzymes and proteins, affinity chromatography has proved to be of little or no value. However, because of general usefulness of affinity chromatography, the modern biochemist must be acquainted with the technique and its many applications and should also be familiar with its limitations and with other aspects of the related and rapidly growing areas of solid-phase biochemistry, since solid-phase biochemistry (i.e., affinity chromatography, immobilized enzymes, solid-phase peptide synthesis,

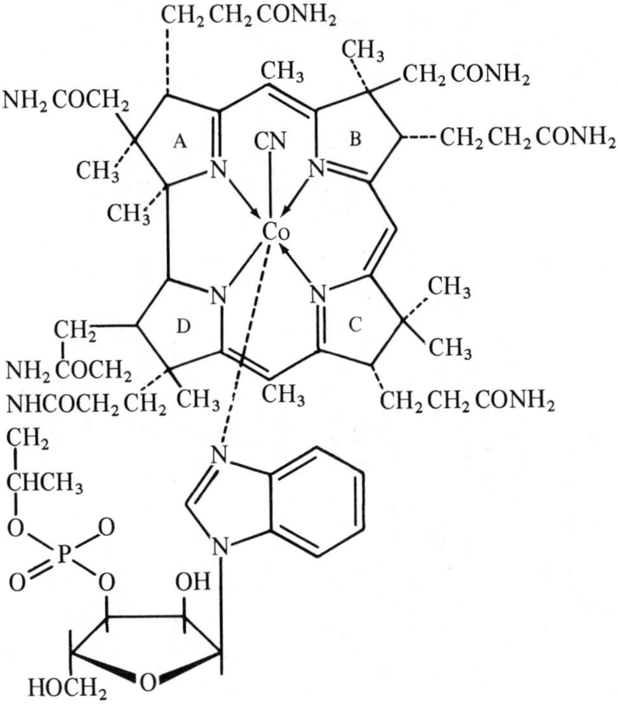

CH$_2$CH$_2$CONH$_2$

Figure 1.2. Vitamin B$_{12}$–cellulose.

etc.) constitutes one of the most important technical developments in enzymology in the past decade.

1.1. NOMENCLATURE

The term "affinity chromatography," first coined by Cuatrecasas et al. (3), is the term most often applied to the type of purification. Unfortunately, it is not very descriptive of the process, and the term is also used in other areas of chemistry with different connotations (4). For this reason several other terms have been employed. For example, the term "biospecific adsorption" was introduced (5) to emphasize that affinity chromatography utilizes a very special type of *biological* affinity between enzyme and bioligand (usually at the active site or allosteric site of the enzyme), rather than such general affinities as ion-exchange or hydrophobic interactions. The term "biospecific adsorption" is, however, misleading—the enzyme-substrate interaction is actually selective rather than specific. That is to say, no enzyme is absolutely specific in the sense that it

TABLE 1.1. Purification of Transcobalamin II

Step	Volume (ml)	Transcobalamin II–Vitamin B$_{12}$-Binding Activity (Total ng)	Protein (Total mg)	Specific Activity, (ng Vitamin B$_{12}$ Bound/mg Protein)	A$_{280}$/A$_{361}$	Fold Purified	Yield (%)
Human plasma	1,400,000	140,000,000[a]	98,000,000	0.0143		1	100
Cohn Fraction III	372,000	573,000	5,280,000	0.109		7.6	40.9
Cm-Sephadex eluate	19,700	447,000	760,000	0.588		41	31.9
Affinity chromatography on vitamin B$_{12}$-Sepharose	62.0	353,000[b]	24.7	14,000	3.56	1,000,000	25.2
DEAE–cellulose	61.5	276,000[b]	11.2	24,600	2.46	1,720,000	19.7
3,3'-Diaminodipropylamine Sepharose	82.0	258,000[b]	10.2	25,400	2.38	1,780,000	18.4
Sepharose	9.1	179,000[b]	6.26	28,600	2.04	2,000,000	12.8

[a]Values for protein content and transcobalamin II–Vitamin B$_{12}$ binding activity were not determined on the plasma used in this purification. The values given are averages obtained from analysis of other human plasma samples.

[b]Based on vitamin B$_{12}$ content.

Source: Reprinted from Allen and Majerus (2).

interacts with only *one* substrate or bioligand. And no bioligand is specific in the sense that it interacts with *only* one enzyme. Thus the term "bioselective adsorption" (6), which is much more descriptive of the actual process, is to be preferred. Affinity chromatography is the consequence of a selective interaction between an enzyme and an immobilized bioligand, and in this monograph the term "affinity chromatography," the most widely used term, is used interchangeably with the more descriptive term "bioselective adsorption."

1.2. HISTORY OF BIOSELECTIVE ADSORPTION

The potential for the purification of enzymes that utilize some aspect of their bioselective affinity for substrates and/or other bioligands has been investigated for many years. As early as 1919, Starkenstein (7) noted that amylase binds very tightly to insoluble starch. In 1933 Holmberg separated α-amylase from β-amylase by chromatography on starch gels (8). Somewhat later Thayer (9) separated microbial amylases by bioselective adsorption on insolubilized starch, followed by bioselective elution with a buffer containing soluble starch.

Other investigators have utilized the isolation of naturally occurring enzyme-insoluble bioligand complexes as the first step in enzyme purification. For example, glycogen synthetase was purified by Leloir and Goldenberg (10) by first isolating the particulate glycogen synthetase–glycogen complex from liver cells. Mere isolation of these particles effected a 30-fold purification over the cell extract. Washing of the particles with buffer containing glucose-6-phosphate and soluble starch effected a further 300-fold purification. Finally, the glycogen synthetase could be eluted from a column of insoluble glycogen particles, by using soluble glycogen as a biospecific eluant, as a homogeneous enzyme.

Perhaps the first enzyme to be purified with the use of a synthetic immobilized bioligand was mushroom tyrosinase, which was purified by Lerman (11), using azodyes immobilized on cellulose. The azodyes, known inhibitors of mushroom tyrosinase, were coupled to cellulose by way of ether linkages. Several of the immobilized dyes, (see Figure 1.3) specifically retained the enzyme, whereas most of the contaminating proteins were not adsorbed. Mushroom tyrosinase was eluted, with a concomitant 60-fold purification, by changing the pH of the eluant buffer. Therefore, to purify tyrosinase, the enzyme was adsorbed at acid pH levels, where the inhibitors normally were tightly bound and was eluted at higher pH levels, where the inhibitor-enzyme interaction was weak.

In subsequent work, a series of flavin requiring apoflavoenzymes were purified by Arsenis and McCormick (12) by chromatography on flavincellulose chromatographic packings. Liver flavokinase was purified to a specific activity of 1500 mμ moles of riboflavin-5-phosphate[flavomononucleotide (FMN)] synthesized per hour per milligram of protein by chromatography on 6-cellulose acetamido-9-

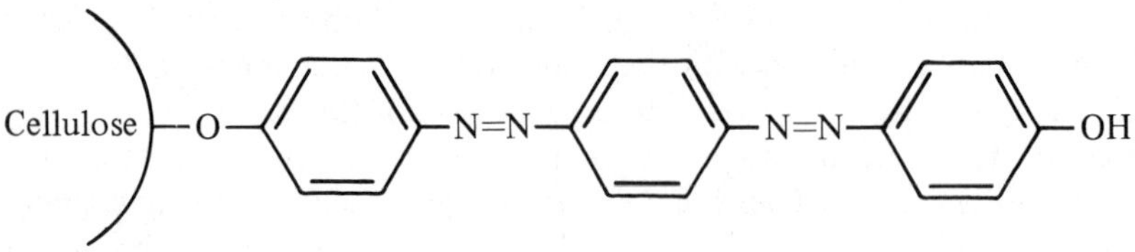

Figure 1.3. Cellulosic dye for the adsorption of mushroom tyrosinase.

(1'-D-ribityl) isoalloxazine (Figure 1.4), using a linear gradient of increasing ionic strength as the eluant. The highest previously reported activity of apparently homogeneous preparations of this enzyme was 100 mμ moles of FMN synthesized per hour per milligram of protein. The substantially higher activity was due to the elimination of trace amounts of FMN phosphatase by way of bioselective adsorption.

The same authors also purified glycolate apooxidase (13) nearly 300 fold by chromatography on FMN immobilized to cellulose by way of a phosphoester bond. Flavomononucleotide esters of diethylaminoethyl (DEAE) cellulose and phosphocellulose were also shown to be useful in the purification of the apo-enzyme forms of several other flavoproteins.

Despite the success of these and other workers, the concept of affinity chromatography did not gain widespread use until after the introduction of activated agarose (14,15) as a potential matrix for the immobilization of bioligands.

Purification of enzymes by use of their ability to complex with biologically active substrates, inhibitors, antibodies, and so on is not limited to chromatography. Many enzymes have been purified by using procedures based on adsorption to ion-exchange resins or other chromatographic resins followed by elution with a bioselective ligand. For example, fructose-1,6-diphosphatase (16) was purified 10 fold by applying a partially purified enzyme fraction to a carboxymethyl cellulose column followed by extensive washing with buffer and sub-

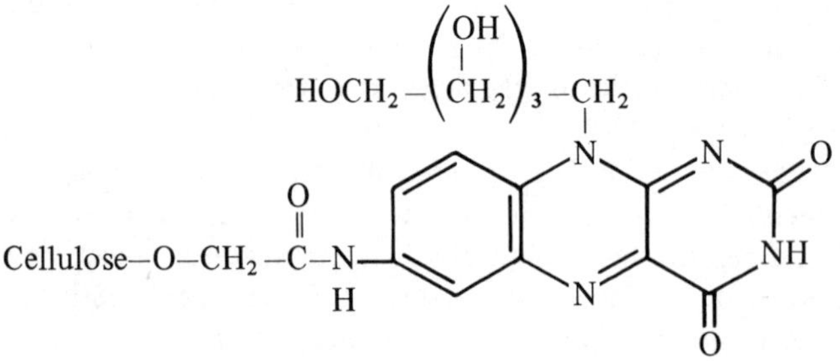

Figure 1.4. 6-Cellulose acetamido-9-(1-D-ribityl) isoalloxatine.

sequent elution of the enzyme with fructose-1,6-diphosphate malonate buffer, pH 6.8.

In a similar fashion, aldolase (17) was purified by elution of the adsorbed enzyme from a carboxymethylcellulose column by also employing fructose 1,6-diphosphate as the bioselective eluant. A 90% recovery of the enzyme with a 29-fold purification was effected. In contrast, when aldolase was eluted from carboxymethylcellulose using the classical elution procedure, by way of a buffer gradient of increasing ionic strength, a much lower yield and a highly diluted enzyme resulted. Less purification was also effected.

Note that in the purification of both aldolase and fructose-1,6-diphosphatase, the same bioselective eluant, fructose-1,6-diphosphate, was employed, giving a good example of the selectivity, rather than specificity, of substrates and other small bioligands for enzymes and similar biopolymers. When many enzymes recognize a common bioligand (NAD^+, ATP, coenzyme A, etc.) the ligand is termed a *group-specific bioligand.* (Further discussion of bioselective elution is given in Chapter 7.)

1.3. SOME GENERAL PRINCIPLES

As shown previously, the basic principle in affinity chromatography is the purification of an enzyme, antibody, cell-surface-receptor protein, or some similar biomolecule by chromatography on a bioligand that has been immobilized on an inert matrix. Much time, obviously, may be required for the synthesis of the bioselective adsorbent, but the result—a rapid purification of the protein, in good yields and high purity sometimes involving as few as one or two steps—is well worth the effort. Moreover, the synthesis of the bioselective adsorbent can often be simplified by employing one of the wide range of commercially available preformed bioselective adsorbents or, in the absence of a commercially available prefabricated adsorbant, synthesizing one's own adsorbant from the numerous partially prepared or pretreated matrices. Also, whenever an investigator applies affinity chromatography for the first time to a given protein system, it is often possible to design the proper adsorbent system based on the numerous examples that have appeared in the literature since 1970. A few basic principles need to be mastered, however, before one begins using bioselective adsorption since, like any other technique, there are numerous pitfalls in which the novice may easily become trapped.

First, the investigator must decide what the final goal will be—that is, is rapid, high-yield purification of an enzyme or some similar macromolecule to be used later in other experiments desired? Or is a bioselective adsorbent needed for the research that is demonstrably dependent on physiological interaction between bioligand and the macromolecule to be purified? If only purification is desired, the investigator may well skip the many controls needed to establish the bioselectivity of the system but will also have to temper any claims as to the existence of bioselectivity in the system. On the other hand, if bioselectivity must be demonstrated, the investigator will spend much of the time establishing the fact of bioselectivity before beginning those experiments on which it rests.

Second, the investigator must choose the ligand to be immobilized. It will, of course, be fortunate if ligands can be employed that are commercially available in preimmobilized form. If not, the investigator will need to find a ligand—usually a substrate, inhibitor or allosteric effector—that will retain "good" affinity for the desired enzyme after immobilization. "Good" affinity usually means that the K_m must be low enough (10^{-3} M or lower) such that the enzymes affinity for the ligand exceeds all nonspecific adsorption factors (ion exchange, etc.) that might be present in the system. At the same time, the affinity should be sufficiently loose that the adsorbed enzyme may be removed from the column without denaturation.

Finally, the selection of an "inert" matrix is extremely important. Most affinity separations have been made by using agarose, polyacrylamide, cellulose, or glass as the matrix. Each has its advantages and disadvantages—agarose, for example, is very porous, has few ionic groups, and is easily derivatized. On the other hand, it is also fragile, both mechanically and chemically, and many of the methods used to derivatize it create ionic groups on the surface.

Fortunately, the literature is replete with numerous examples that will serve to guide the novice in the field. There are several excellent reviews (4,5) and recent texts (18,19) in the general area of affinity chromatography. In this monograph we first briefly outline a few general examples, and then in subsequent chapters detail each step involved in designing an affinity chromatographic packing. (See also 30, 31).

1.4. SOME GENERAL EXAMPLES

1.4.1. DNA-Binding Proteins

Probably the single most widely used bioselective adsorbent is DNA–cellulose. Although DNA–cellulose is unusual, in that it is one of the few cases in which cellulose has been successfully employed as the matrix, it does provide a rapid, simple, high-yield chromatographic medium for the purification of one class of proteins, namely, DNA-binding proteins. Proteins that bind DNA are of interest to researchers in various aspects of biochemistry and molecular biology; therefore, DNA–cellulose chromatography is an extremely important technique.

Bruce Alberts (20,21) and Rose Litman (22) independently first synthesized DNA–cellulose by quite similar methods. Alberts procedure involved the codrying of fibers of DNA and cellulose. This codrying procedure causes the fibers of DNA and cellulose to become intertwined, physically trapping the soluble DNA within the fibrous insoluble cellulose matrix. Litman similarly trapped DNA within cellulose fibers, using alcoholic solvents to effect the "drying" process, while simultaneously irradiating the DNA–cellulose mixture with ultraviolet light. This causes not only physical entrapment of the DNA, but also cross-linking of the DNA molecules is photolytically induced between adjacent DNA chains or between different parts of the same DNA chain. These inter- and intramolecular cross-links cause the DNA to become far more immobile than the DNA that is only physically trapped as in the method of Alberts. Thus the DNA–cellulose chromatographic medium prepared by using ultraviolet irradiation has little, if any, free DNA leaking from it during chromatography, whereas DNA–cellulose prepared solely by codrying has a very small, but noticeable, leakage that in some circumstances may be extremely important. Nevertheless, the codrying approach of Alberts is the one most commonly used because (1) the preparation is easier and (2) the DNA-cellulose that is ultraviolet treated does not resemble physiological DNA as closely as does codried DNA. Further, the half-life of codried DNA–cellulose is 400 hours at 37°C and even several orders of magnitude longer at 0°C, the temperature at which it is usually employed. This degree of instability is sufficiently low as to be acceptable in most applications.

One of the most noteworthy of the many DNA-binding proteins isolated with the use of DNA-cellulose affinity chromatography is the protein coded for by gene 32 of T_4 phage (Figure 1.5). Mutations in gene 32 are noted by both inability to synthesize DNA and absence of DNA recombination. These observations suggest that gene 32 has a DNA-related function; therefore, the protein coded for by gene 32 could have an affinity for DNA. In an elegant series of experiments, Alberts and Frey (23) grew T_4 wild type in ^{3}H leucine-containing media. They grew T_4 gene 32 temperature-sensitive mutants by using ^{35}SO$_4^=$-containing media. The host cells were then lysed and the endogenous DNA was removed by treating the extracts with deoxyribonuclease I. (Any unremoved DNA might tend to compete with the immobilized DNA for the gene 32 protein, thus lowering the amount of gene 32 protein bound to the DNA-cellulose.) The nuclease was subsequently inactivated by addition of ethylenediaminetetraacetic acid (EDTA), and the extract was chromatographed on DNA-cellulose, using step gradients of increasing ionic strength (Figure 1.6). The eluted proteins were electrophoresed on polyacrylamide gels. The proteins from the T_4 wild type that eluted from the DNA-cellulose with 0.6-M NaCl were notably different from the proteins that were found by using the T_4 gene 32 mutants. One very prominent electrophoretic band, present in the proteins isolated from wild type, was noticeably missing from the fraction isolated from the gene 32 minus mutant. This protein peak contained the protein coded for by the T_4 gene 32.

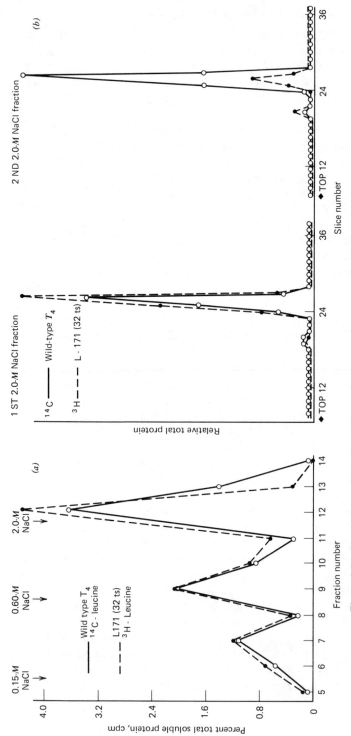

Figure 1.5. Comparison of binding ability of normal and temperature-sensitive T_4-32 protein for single-stranded DNA-cellulose. At 35 minutes after infection at 25°C, 4×10^{10} wild-type infected cells were labeled for 10 minutes with leucine-[14]-C, and 4×10^{10} ts 1.174 infected cells were labeled for 10 minutes with leucine-[3]H. Extracts were made from mixed cells and chromatographed on denatured calf thymus DNA-cellulose (0.7 ml packed volume containing ca. 1 mg of DNA). (*a*) Elution profile from a DNA column. Fractions were collected every 15 minutes, with a flow rate of 2 ml/hour. (*b*) Sodium dodecyl sulfate-containing polyacrylamide gel electrophoresis of the 2.0-*M* NaCl-containing fractions. The major band is T_4-32 protein. Reprinted from *Federation Proceedings* (20) with permission.

12

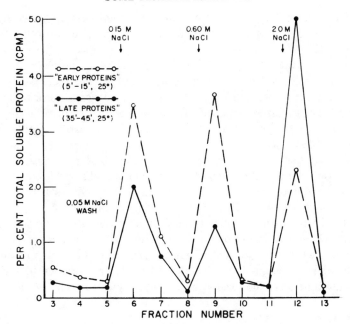

Figure 1.6. Chromatography of doubly labeled T_4-infected crude cell extracts on a mixed (1,1) native T_4 plus denatured calf thymus DNA–cellulose column. Pancreatic DNAse I treatment of the cell sonicate was utilized for DNA removal, and 3 ml of extract from 8×10^{10} infected cells were applied to 1-ml column containing about 1 mg of DNA. Calf thymus rather than T_4 denatured DNA–cellulose was used because it is difficult to prevent the T_4 DNA single strands from renaturing during the drying procedure used to adsorb DNA to the cellulose. Only about 0.2% of the total labeled protein applied to a control, DNA-free cellulose column appeared in column eluates. Reprinted from *Federation Proceedings* (20) with permission.

When these two 0.6-M NaCl eluants were cochromatographed on Sephadex G-200, one protein band contained only ^{35}S, but not ^{3}H, again implying that this protein was present only in the wild type T_4 phage but not in the gene 32-deficient phage. These results demonstrated again that this protein must be the product coded for by gene 32. Subsequently, the gene 32 product has been purified to homogeneity (24) by using both DNA-cellulose chromatography and ion-exchange chromatography. The protein has the unusual property of melting DNA, that is, separation of the two complimentary DNA strands (25), and is involved in the replication of T_4 phage DNA (26).

1.4.2. Plasminogen Purification

The purification of plasminogen with the use of affinity chromatography is representative of the many applications of bioselective adsorption that employ agarose as the matrix. It is, like DNA–cellulose chromatography, a widely used procedure, generally recognized as the most convenient and useful purification of plasminogen.

Plasminogen is the zymogen, or proenzyme, of the proteolytic enzyme plasmin, and it is converted to plasmin by proteotytic activation. Plasmin, itself, is a serum protease responsible for dissolving blood clots and thus is of considerable medical interest. If plasminogen could be activated to plasmin in persons with thrombosis, or undesired blood clots, the resulting plasmin would be able to dissolve the blood clot.

Like many serum proteins, the purification of plasminogen by classical protein purification procedures, such as gel-filtration and ion-exchange chromatography, have proved to be tedious and often somewhat irreproducible, with varying amounts of denatured plasminogen contaminating the product.

One of the best-known inhibitors of plasmin activation is ϵ-aminocaproic acid, a drug with widespread use in blood-clotting disorders. Lysine is very similar in structure to ϵ-aminocaproic acid. Deutsch and Mertz (27,28), utilizing this structural similarity, immobilized lysine on activated agarose and have demonstrated its usefulness as a bioselective adsorbent for plasminogen.

$$NH_2CH_2CH_2CH_2CH_2-\overset{\overset{\displaystyle O}{\|}}{C}-OH \qquad H_2NCH_2CH_2CH_2CH_2\underset{\underset{\displaystyle NH_2}{|}}{CH}-\overset{\overset{\displaystyle O}{\|}}{C}-OH$$

ϵ-aminocaproic acid   Lysine   Point of attachment to agarose

To effect the coupling of lysine to agarose, the agarose was first treated with cyanogen bromide at pH 9.0. The resulting activated agarose reacts rapidly with amine groups. When it was reacted at pH 8.9 with lysine, α-aminolysyl agarose resulted (Figure 1.7).

To a column (0.9 cm $\times$ 1.5 cm) of the lysine agarose was applied 10 ml of plasma diluted with 15 ml of water, followed by washing with 0.3-M phosphate, pH 7.4, until no further protein was detected in the eluant. Plasminogen was then eluted with 0.2-M ϵ-aminocaproic acid. Because of the high affinity of plasminogen for ϵ-aminocaproic acid, the plasminogen eluted as a sharp peak. The ϵ-aminocaproic acid could be subsequently removed by dialysis or by gel-permeation chromatography.

The yield of plasminogen prepared by means of bioselective adsorption on lysyl-agarose followed by bioselective elution with ϵ-aminocaproic acid was between 70 and 90% (29), in contrast to conventional techniques that yielded only

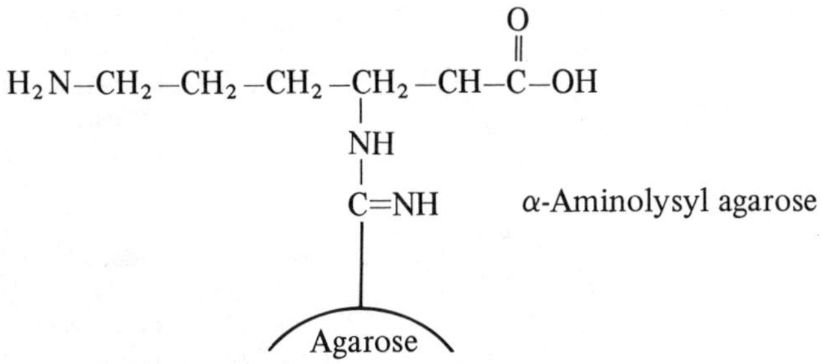

$$H_2N-CH_2-CH_2-CH_2-CH_2-\overset{\displaystyle |}{\underset{\displaystyle NH}{CH}}-\overset{\displaystyle O}{\overset{\displaystyle \|}{C}}-OH$$

NH
|
C=NH   α-Aminolysyl agarose
|

Agarose

Figure 1.7. α-Aminolysyl-agarose.

14 to 59%. Plasminogen preparations obtained via bioselective adsorption were at least as pure as those obtained by use of classical purification procedures, and both methods yield essentially pure plasminogen. Even though it is enzymatically pure, the purified enzyme is electrophoretically heterogeneous, apparently because of the existence of isoenzymic forms of plasminogen.

Interestingly, a second type of heterogeneity can also be shown by way of affinity chromatography. If plasminogen is eluted from lysyl–agarose by using a linear gradient of ε-aminocaproic acid, two distinct protein peaks emerge. Both peaks contain plasminogen as their sole component, both are capable of being activated to plasmin, and both can be resolved into isomeric forms on gel electrophoresis. It would appear from these observations that there are two distinct forms of plasminogen in plasma, each having different affinities for ω-amino acids. The physiological importance of these isomeric forms has not yet been thoroughly investigated.

1.4.3. FMN Purification on Apoflavodoxin–Agarose

In an example of the purification of a small biomolecule by way of bioselective adsorption on an immobilized macromolecule, Mayhew and Strating (30) immobilized apoflavodoxin and utilized it to purify FMN and several of its analogues (Figure 1.8). Basically, this is the reverse of the purification of apoflavoenzymes on flavin cellulose derivatives mentioned earlier and is just one example of the many instances where complex, expensive, or difficult-to-purify small biomolecules have been purified by using bioselective adsorbents.

Riboflavin-5-phosphate from various commercial preparations is only 70 to 80% pure. Half of this impurity can be removed by use of ion-exchange chromatography, but the remainder, 12 to 15% of the product, resists removal by way of conventional techniques. However, apoflavodoxin, which binds only flavins with an N-10 side chain and a C-5′ phosphate group, does not bind the impurity. Thus an investigator can adsorb FMN from a partially purified preparation on

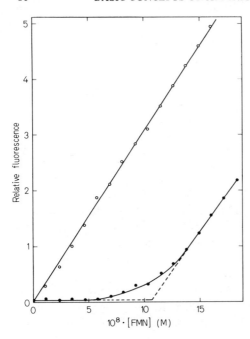

Figure 1.8. Determination of dissociation constant for dissociation of FMN from immobilized flavodoxin. Aliquots of FMN, 11.6 μM, were titrated into a suspension of immobilized flavodoxin, 0.214 mmole in 2 ml 25-mM sodium acetate buffer, pH 6.5, and 0.4-M sodium chloride, which was stirred with a magnetic stirrer. The temperature was 23°C. The fluorescence of the solution (● ●) was recorded after each addition when all fluorescence changes were complete (ca. 60 min). The solution was not stirred during fluorescence measurements. Values for the dissociation constant of immobilized flavodoxin were calculated from points 7 to 10, which lie off the linear regions of the titration curve. The fluorescence changes during a control experiment, in which FMN was added to Sepharose 4B that did not contain bound apoflavodoxin, are shown by (0 0). Reprinted with permission from Mayhew and Strating (30).

apoflavodoxin batchwise, dialyze the enzyme, and remove the bound FMN by treating the flavodoxin with either 5% trichloroacetic acid, 0.1-N acetic acid, or 0.1-N hydrochloric acid.

Such batchwise procedures however are tedious and consume considerable amounts of the apoflavodoxin. For that reason Mayhew and Strating (30) immobilized flavodoxin holoenzyme on cyanogen bromide-activated agarose.

FMN

The structure of flavin mononucleotide (FMN). Reprinted with permission from *Affinity Chromatography* by C. R. Lowe and P. D. G. Dean, Copyright 1974 from John Wiley. Reprinted with permission of John Wiley and Sons, Ltd.

TABLE 1.2. Purification of FMN and FMN analogues [a]

| Treatment | Residual Fluorescence of FMN Derivative, Percent Initial | | | | | |
| | FMN | | 3-Me– | 3-CH$_2$COO$^-$ | iso- | 8-OH– |
	Sigma	Boehringer				
None	20.5	25.5	30	31	15	25
DEAE–cellulose chromatography	13.0	16.0	23	21	7	–
Affinity chromatography	0.7	0.7	1	3	0.6	0.6

Source: Reprinted with permission from Mayhew and Strating (30).

[a] Flavins were purified by chromatography on DEAE–cellulose, or by affinity chromatography on a column of immobilized apoflavodoxin. Their purity was tested by titration with soluble apoflavodoxin to determine the extent of fluorescence quenching. The residual fluorescence is due to flavin that is not bound to the apoprotein and is the flavin fluorescence measured after adding 2 moles of apoflavodoxin per mole of flavin, except in the case of 3-CH$_2$ COO–FMN to which 4 moles of apoprotein was added per mole of flavin.

The immobilized enzyme was washed with 10-mM potassium phosphate, pH 7.0, containing 0.3-mM EDTA, and then, to prepare the apoflavodoxin, a column of the immobilized holoenzyme was prepared and eluted in dim light with 5% trichloroacetic acid containing 0.3-mM EDTA. The FMN that had originally been bound to the flavodoxin eluted as a sharp yellow band. After the elution of the FMN, the apoenzyme column was further washed with buffer as before and then stored in buffer containing 0.1% dithiothreitol plus 0.1% sodium azide.

For purification of FMN and its analogues, the crude flavin was applied to the column in 0.1-M potassium phosphate, pH 7.0, containing 0.3-mM EDTA. When the FMN binding sites were occupied, the eluate became yellow, and application of further crude flavin was stopped and the column allowed to stand for 1 hour. Excess FMN, flavin derivatives, and other compounds not binding to the apoflavodoxin were subsequently eluted with 10-mM potassium phosphate, pH 7.0, containing 0.3-mM EDTA, followed by elution of FMN (or those FMN analogues that are bound by apoflavodoxin) with 5% trichloracetic acid. The trichloracetic acid was extracted with ether, the pH of the flavin solution was adjusted to pH 5 to 7, and the purified FMN was stored at -20°C. The resulting FMN was homogeneous as determined by way of chromatographic and enzymatic methods. The results of the purification of FMN and several N-10 analogues with C-5' groups as shown in Table 1.2.

These examples are only a few of the hundreds of uses of affinity chromatography.

References

1. R. H. Allen and P. W. Majerus (1972), *J. Biol. Chem.* **246**, 7709.

2. R. H. Allen and P. W. Majerus (1972), *J. Biol. Chem.* **247**, 7695.

3. P. Cuatrecasas, M. Wilchek, and C. B. Anfinsen (1968), *Proc. Nat. Acad. Sci. (USA)* **61**, 636.

4. P. H. Reiner and A. Walsh (1971), *Chromatographia* **4**, 578.

5. J. Porath (1973), *Biochemie* **55**, 943.

6. W. H. Scouten (1974), *Am. Lab.* **6**, 23.

7. E. Starkenstein (1910), *Biochem. Z.* **24**, 210.

8. O. Holmbergh (1933), *Biochem. Z.* **258**, 134.

9. P. S. Thayer (1953), *J. Bacteriol.* **66**, 656.

10. L. F. Leloir and S. H. Goldenberg (1960), *J. Biol. Chem.* **235**, 919.

11. L. S. Lerman (1953), *Proc. Nat. Acad. Sci. (USA)* **39**, 232.

12. C. Arsenis and D. B. McCormick (1964), *J. Biol. Chem.* **239**, 3093.

13. C. Arsenis and D. B. McCormick (1966), *J. Biol. Chem.* **241**, 330.

14. R. Axen, J. Porath, and S. Ernback (1967), *Nature (Lond.)* **214**, 1491.

15. R. Axen and P. Vretblad (1971), *Acta Chem. Scand.* **25**, 2711.

16. B. M. Pogell (1962), *Biochem. Biphys. Res. Commun.* **7**, 225.

17. S. Pontremoli (1966), in *Methods in Enzymology* (W. B. Wood, ed.) **2**, 625.

18. C. R. Lowe and P. D. G. Dean (1974), *Affinity Chromatography*, Wiley, London.

19. W. B. Jakoby and M. Wilchek, in *Methods in Enzymology: Affinity Techniques*, **34**, Academic, New York, 1974.

20. B. M. Alberts (1920), *Fed. Proc., Fed. Am. Soc. Exp. Biol.* **29**, 1154.

21. B. M. Alberts and G. Herrick (1971), in *Methods in Enzymology* (L. Grossman and K. Moldave, eds.) **210**, 198.

22. R. M. Litman (1968), *J. Biol. Chem.* **243**, 6222.

23. B. M. Alberts and L. M. Frey (1970) *Nature (Lond.)* **227**, 1313.

24. B. Alberts, L. Frey, and H. Delius (1972) *J. Molec. Biol.* **68** 139.

25. B. M. Alberts, F. Amodio, M. Jenkins, E. D. Gutmann, and F. L. Ferris (1968), *Cold Spring Harbor Symposium on Quantitative Biology*, p. 289.

26. J. A. Huberman, A. Kornberg, and B. M. Alberts (1971), *J. Molec. Biol.* **62**, 39.

27. D. G. Deutsch and E. T. Mertz (1970), *Fed. Proc., Fed. Am. Soc. Exp. Biol.* **29**, 647

28. D. G. Deutsch and E. T. Mertz (1970), *Science* **70**, 1095.

29. W. J. Brockway and F. J. Castellino (1972), *Arch. Biochem. Biophys.* **151**, 194.

30. S. G. Mayhew and M. J. J. Strating (1975), *Eur. J. Biochem.* **59**, 539.

31. Lowe, C. R. (1979), An Introduction to Affinity Chromatography, Elsevier-North Holland, Amsterdam

32. Turková, J. (1978), Affinity Chromatography, Elsevier, Amsterdam.

MATRICES AND SPACER ARMS

Affinity chromatography is the result of the interaction of an immobilized biological ligand with an enzyme protein or other biomacromolecule that we desire to isolate and study. However, the affinity chromatography *system* includes much more than just the enzyme and the bioligand; it also includes the matrix, the "spacer arm" (or other means of attachment of bioligand to matrix), and the buffer or solvent system. Unfortunately, the enzyme to be purified usually has *some* interaction with *all* of the components of the system, and each must be chosen with care.

The matrix is clearly one of the most important components of an affinity chromatographic medium as it composes, for the most part, the largest volume of the adsorbent. Further, certain characteristics of the matrix are essential: (1) it must be insoluble in the solvents or buffers that may be employed; (2) it must be mechanically and chemically stable, with good flow properties; and (3) it must be easily coupled to the ligand or to a spacer arm onto which the ligand may be attached (see properties listed in Table 2.1). It is also advisable that it have a large surface area accessible to the enzyme or the other biomolecules that are being purified.

Matrices currently employed can be divided into two groups, the first-generation matrices that are mainly single-composition matrices (e.g., agarose, glass, or cellulose) and the second-generation matrices (i.e., dual-composition and/or chemically modified matrices). Among the dual-composition matrices are glycidoxy-coated glass, agarose-coated polyacrylamide (Ultragel), and poly-acrylic-coated iron particles (Enzacryl). Examples of chemically modified matrices are crosslinked agarose and hydroxyethyl polyacrylamide (Spheron). Most of the matrices used in affinity chromatography to date are the first gen-

TABLE 2.1. Ideal Matrix Properties

Easily deviatized
Both mechanically and chemically stable
Possess surfaces with easy ligand accessibility
Good flow characteristics
Stable to eluant, e.g., denaturing buffers

---

eration matrices, but the newer matrices, which attempt to combine the better aspects of two or more matrices, offer new horizons for improvement of affinity systems. The best matrix for any given procedure will depend on the nature of the enzyme to be purified and on any specific demands placed on the procedure, for example, rapid separation time, high yields, and high purity.

2.1.   AGAROSE

By far the most widely used (and perhaps least understood) matrix is beaded agarose, which has gained predominance chiefly because of its ready availability to the researcher. Most investigators have agarose in some form in their stockrooms, and the methods for derivatizing it are widely known. Furthermore, a large number of partially prepared bioselective adsorbents, based on agarose, are commercially available. And finally, because of the hydrophillicity and minimal nonspecific adsorption of nonderivatized agarose ("virgin" agarose), many researchers expect the same properties to persist in the final bioselective adsorbent. This is now known not to be true and illustrates the fact that it is the final bioselective adsorbent, not the starting matrix, that must be free of non-specific factors, for example, ionogenic groups. Even if you initially employ a good matrix, such as agarose, you can readily form a very poor adsorbent in the process of attaching the bioligand. To avoid such problems, you must carefully consider each step in the preparation of the affinity chromatographic medium.

Agarose, per se, is a linear polysaccharide consisting of alternating residues of D-galactose and 3-anhydrogalactose (1). It is a naturally occurring compound isolated in a heterogeneous form from certain types of sea kelp. Unrefined agarose contains many ionic residues, chiefly carboxylate and sulfate residues, for example. Commercial agarose beads contain up to 0.37% sulfur, thus indicating that the number of sulfate groups are considerable in this matrix (Figure 2.1).

Removal of ionic groups by way of sodium borohydride reduction is a reasonable pretreatment of the agar whenever any such groups could potentially inter-

Gel Type	Adsorption Capacity	Percent Sulfur
Commercial agarose 6%	0.080	0.118
ECD–agarose	0.008	0.021
Commercial Agar 6% (beads)	0.240	0.371
ECD–agar 6% (beads)	0.060	0.049
Reduced ECD–agar 6% (beads)	0.004	0.012

Figure 2.1.   Effect of agar gel type and sulfur content on adsorption capacity for cytochrome C. Reprinted with permission from J. Porath (1973), *Biochemie* **55**, 943.

fere with the separation. For example, when the affinity of the enzyme for the ligand is weak, even a minute residual ionic character to the matrix might produce a chromatographic packing whose chief mode of operation is ion-exchange chromatography, rather than bioaffinity chromatography.

To reduce the agarose beads, they are treated with either $NaBH_4$ in alkaline solution or $LiAlH_4$ in dioxane or diethyl ether (1). For example, 7 ml of agarose is suspended in 5 ml of 1-N NaOH containing 2 mg/ml of $NaBH_4$. The reduction is continued at room temperature overnight, and the beads are collected and washed.

Reduction by $LiAlH_4$ is much more complete, but $LiAlH_4$ must be very carefully handled since it is very pyrogenic.

Other problems involved in the use of agarose include lack of thermal stability, shrinkage or swelling due to changes in ionic strength or dielectric constant of the medium, an inability to be frozen or dried easily, and ready solubility in the presence of denaturing or chaotropic ions. It is also impossible to use agarose with many organic solvents because its structure changes drastically and irreversibly under such conditions.

The properties described previously are a result of the secondary structure of agarose. As has been stated, agarose consists of linear polysaccharide chains with no covalent cross-linking between them. Such linear saccharides are normally soluble in aqueous media. Thus, for example, when agarose gels are heated to about 40°C, they melt (or solubilize). The structural integrity, hardness, and porosity of the agarose-gel depends on the secondary structure caused by noncovalent bonds (hydrogen bonds) between various agarose chains. The importance of this secondary structure has only recently been studied in detail, largely because of the failure of conventional techniques to yield detailed pictures of the agarose network. With the development of the scanning electron microscope and with the refinement of classical electron microscopy, it has been established that agarose consists of pentagonal pores created by the bridging of triple helical agarose chains.

Figure 2.2 shows a diagram of the proposed structure of the agarose network. The pentagonal pores are the essential feature of agarose supports. These pores are large enough to be readily penetrated by protein with molecular weights of several million yet are strong enough to permit the shaping of agarose into definite spherical particles with good flow characteristics. However, the stability of the pores is dependent on hydrogen bond formation between the three strands of the triple helix of agarose chains. Anything that is capable of disrupting these bonds will disrupt the entire network, and a solution of soluble monomeric agarose will result. The thermal instability of agarose is caused by the disruption of these bonds. These hydrogen bonds are also disrupted by urea, guanidinium hydrochloride, chaotropic ions, and certain detergents, any one of which may be

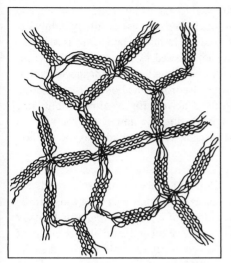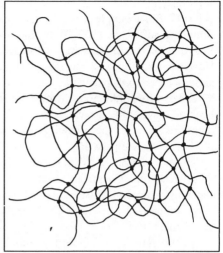

Figure 2.2. Schematic representation of agarose gel network (right), in comparison with a network such as Sephadex that is formed from random chains at similar polymer concentration. Note that the aggregates in agarose gels may actually contain 10 to 10^4 helices rather than the smaller numbers shown here. Reprinted with permission from Arnott, et al (1974), *J. Molec. Biol.* **90**, 269. Copyright by Academic Press Inc. (London) Ltd.

desired as an eluant in affinity chromatography. The strength of the bonds and, therefore, the porosity and size of the beads are altered by change in ionic strength.

This peculiarity of agarose is normally not too severely limiting in practice. However, any attempt to elute "residual protein" with strong denaturants increases the probability that they will dissolve the entire adsorbent. Therefore, to produce an agarose derivative that is free from such susceptibility to hydrogen bond disrupting agents, it is imperative to form *covalent* cross-links between the various strands of the agarose network. Several such procedures have been reported and a commerial agarose cross-linked with epichlorohydrin (ECD-agarose) is available. Since the porosity of cross-linked agarose with defined porosity, is less than that of the starting agarose, the investigator may well wish to start with the commercial cross-linked agarose rather than prepare it in the laboratory. In the case of the commercially cross-linked material, the manufacturer determines the porosiy of the final product and includes that information with the product. Gels made in the laboratory, on the other hand, vary considerably in porosity from perparation to preparation. Despite their restricted porosity, the resulting cross-linked agarose beads are mechanically much stronger, have better flow characteristics, and can be employed with a far wider range of buffers and solvents. For example, commercial non-cross-linked agarose is totally soluble in dimethyl sulfoxide (100°C) or in 4-M sodium iodide, whereas agarose cross-linked with epichlorohydrin is essentially unaffected by these solvent systems. However, cross-linking can introduce additional, albeit small, nonspecific interactions in agarose, particularly in the form of hydrophobic bonds. Furthermore, all excess cross-linking agent must be removed before use, or some protein may become covalently attached to the matrix during the initial application. To cross-link agarose, the procedure of Porath using epichlorohydrin is the method of choice (1) (see page 50).

Commercial agarose preparations are available from several sources (Sepharose, Pharmacia, Fine Chemicals, Agarose, and BioRad). These commercial beads are the most commonly used form of agarose, even though agarose beads can be prepared easily by mixing a hot aqueous solution of highly purified agar [usually at a concentration of 4 to 6% (w/w)] with mineral oil and a small percentage of detergent. The agar solution is rapidly stirred until it cools and hardens. A special mixing vessel is often employed to obtain more uniform bead size. However, separation of the various agarose bead sizes is necessary to obtain the desired maximum flow rates. The process of preparing agarose beads requires considerable skill and time that the ordinary laboratory finds unnecessary when adequate commercial preparations are readily available.

Poonian et al. (2) have gelled a solution of melted agarose and DNA to form a DNA-affinity matrix in which the large DNA molecules were physically trapped within the "cage" of agarose, and for this reason there was no loss of DNA from the matrix (see page 228). Their procedure involved making slabs of DNA–agarose and cutting them into minute pieces for use in chromatographic packing. However, such irregularly shaped beads give excessively slow flow rates. The author has found that beads of agarose–DNA can be easily formed (see page 227) and that these have flow characteristics identical, or nearly so, to that of commercial agarose. Other possible uses for this "entrapment" procedure include the trapping of large molecules, (trypsin, apoflavodoxin, etc.) for the use in purifying small molecules (e.g., soybean trypsin inhibitor or FMN, respectively).

Cyanogen bromide-activated agarose beads in which the vicinal hydroxyls of the agarose have been partially replaced by reactive imidocarbonate functions (see Figure 2.3) are the most commonly employed form of agarose. The reactive imidocarbonate is, in turn, reacted with amino functions of proteins or bioligands that are ultimately immobilized. The product of such a reaction is a positively charged isourea function, which, of course, leads to a positively charged bioselective adsorbent. In many instances the ion-exchange effect of such an adsorbent far exceeds the bioselective affinity properties. Fortunately, other, noniongenic methods for ligand attachment exist.

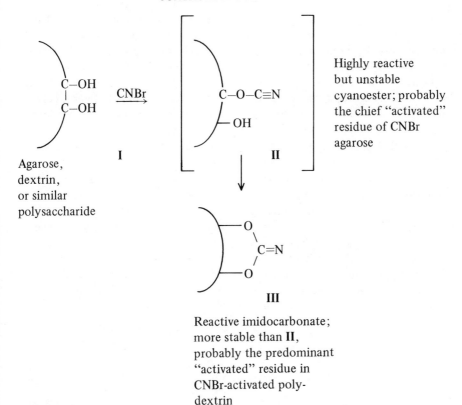

Figure 2.3. Imido-carbonate intermediates in cyanogen bromide activation of polysaccharides.

2.2. CONTROLLED PORE GLASS

Controlled pore glass is the most commonly employed inorganic matrix for the immobilization of biological molecules. Most of the applications of glass as a matrix have involved the immobilization of enzymes, chiefly for later use as bio-catalysts for various processes of industrial or medical interest. A relatively small number of papers have reported the use of glass as an affinity chromatography matrix. Several explanations exist for the paucity of papers, such as (1) underivatized glass is known to absorb enzymes nonspecifically (although derivatized glass does not); (2) porous glass is not normally present in the biochemist's stockroom; and (3) there exists some real problems in its use, although no more problems than in the use of other, more widely accepted, support materials. Further, a certain mythology concerning the use of glass as a bioaffinity support seems to reoccur regularly in the literature. Actually, the advantages of using controlled pore glass far outweigh its disadvantages. Glass is mechanically durable, can be fabricated into various shapes (including spherical or nearly spherical particles) is thermally stable, is chemically compatible with most organic solvents one might desire to use in making the bioselective adsorbent, and, contrary to some reports in the literature, yields chromatographic columns with excellent flow rates. Its major disadvantage is nonspecific adsorption, which, however, can be minimized with little difficulty.

Controlled pore glass is synthesized by heating certain borosilicate glasses to 500 to 800°C for prolonged periods of time (3,4). These glass mixtures (see Figure 2.4) separate on such heat treatment into borate and silicate rich phases. The borate phase can then be dissolved by treatment with acid, leaving a network of extremely small tunnels and pores (25 to 70 Å) throughout a glass "sponge" (which, in fact behaves much like a sponge toward various solvents). The pores are very precise in size (±10%) and can be enlarged by removing some of the alkali-soluble rich phase (4). The resulting glass can be obtained with pores from 45 to 4000 Å. When the acid and alkali leaching is carefully controlled, the pore size distribution of the controlled pore glass is extremely narrow, as shown in Figure 2.5, possibly the most uniform porosity of any

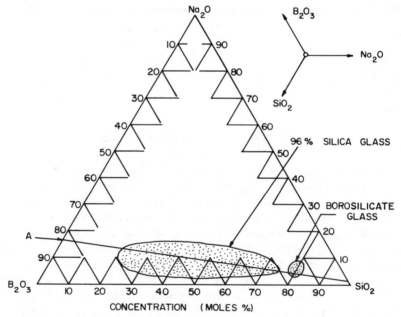

Figure 2.4. Phase diagram of sodium borosilicate system. Reprinted with permission from Weetal and Filbert, (6).

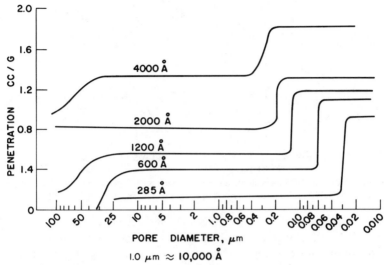

Figure 2.5. Pore size distribution of controlled-pore-diameter glass supports. All values were determined by the mercury intrusion method. Reprinted with permission from Weetal and Filbert (6).

material available. Therefore, it is employed in gel-permeation chromatography, and certain derivatives, treated to remove nonspecific, ionic adsorption sites, have proven to be useful for high-pressure gel-permeation chromatography of proteins and other biomolecules.

Virgin (i.e., untreated) controlled pore glass consists of 96% silica and about 4% borate plus traces of other inorganic oxides. The surface is composed chiefly of silanol functions, which have pK_a values around pH 9. The silanols thus bear a slight negative charge under conditions usually employed for bioselective adsorption. In addition to the Si–OH functions, the surface of the controlled pore glass contains tightly bonded water molecules and some silyl ether residues. The process of complete dehydration of the glass surface takes place at very high temperatures and is very slowly reversible. For example, heating of glass in a vacuum at 850°C, causes all the surface silanol groups to dehydrate and form silyl ether residues. Concomitantly, a hydrophobic surface results (5).

Further complicating the determination of the nature of the controlled pore glass surface structure is the presence of borate functions. The total boron composition of the glass as a whole is only about 4%, but the surface of the glass contains most of the boron, up to 30% borate (6). The numerous boron Lewis acid sites will adsorb nucleophiles such as protein amines and/or adsorb ammonium ions to yield positively charged centers that lead to nonspecific adsorption (see Figure 2.6).

Recent developments in derivatization of glass have essentially eliminated this

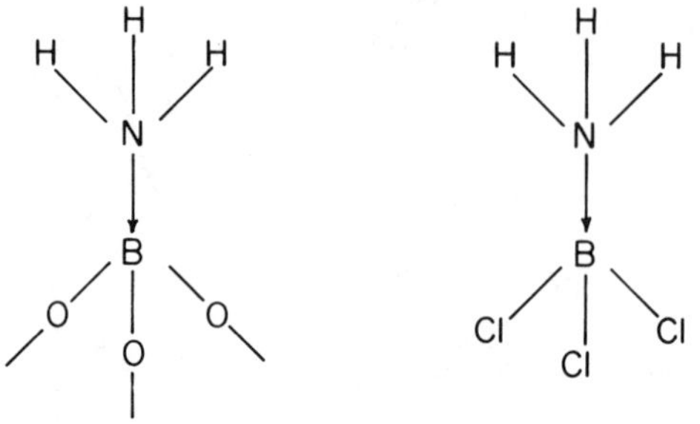

Figure 2.6. Concept of Lewis acid site on the surface of glass developed by analogy to inorganic Lewis acid halides. Reprinted with permission from Weetall and Filbert (6).

problem (11,12). Nonetheless, virgin glass readily adsorbs proteins, usually without altering their biological function; that is, an enzyme remains virtually 100% in the active form. Messing (7) first observed this phenomenon when attempting to utilize a porous glass test tube as a selective membrane. The pores in the tube were sufficiently small that the protein shouldn't pass through it, but buffer should. Messing tried to force a solution of bovine serum albumin through the tube, in the belief that the bovine albumin, since it was too large to penetrate the porous glass, would become concentrated in the tube. The result, however, was the disappearance of the enzyme from the solution on *either* side of the glass membrane. Further investigation indicated that the serum albumin was adsorbed within the porous glass membrane itself. Subsequent studies revealed that many enzymes could be adsorbed on porous glass while still retaining their activity.

To prevent ionic adsorption of enzymes on glass surfaces, the glass is reacted with organic coatings that cover the silanol groups, thus minimizing, if not eliminating, the ion-exchange character of glass. Derivatization of controlled pore glass is most often performed by refluxing the beads in either aqueous (10%, pH 3.45) (8) or toluene (10%) solutions of γ-aminopropyltriethoxysilane (9) (see Figure 2.7).

The resulting amino function can be further derivatized to form the final bioselective adsorbant desired. However, to avoid nonspecific adsorption, several steps must be taken. First, the beads must be treated very carefully from this point on since any breakage due to rough handling will result in the exposure of fresh SiOH functions, with resulting nonspecific adsorptions. Second, the triethoxysilane chain grows as a polymer around the glass surface, with some strands overlapping each other. This creates a multilayered surface and, most importantly, leaves some silanol groups exposed. To completely block these, the amino alkyl glass can be further treated with trimethyl chlorosilane. Many

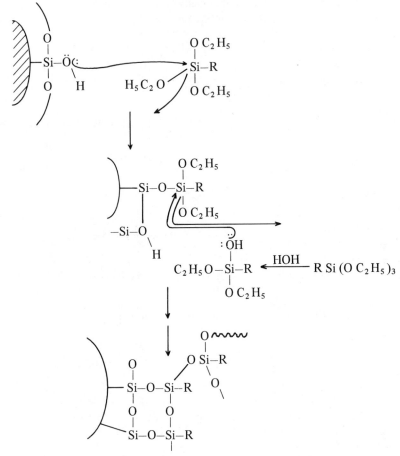

Figure 2.7. Mechanism of silanization of porous glass.

alkoxysilanes, other than γ-aminopropyl silanes, are commercially available (see Table 2.2). They are employed in a manner similar to that for the γ-aminopropyltriethoxysilane.

Mercaptopropyltriethoxysilane yields a very hydrophobic glass, but one that can be further reacted with epibromohydrin or bisoxirines to yield very stable hydrophilic glass derivatives with no evidence of ion exchange or similar nonspecific adsorption. Further, both bisoxirane and epibromohydrin yield derivatized beads with reactive epoxide functions that are easily modified to yield bioselective adsorbents (10).

Other derivatives of controlled pore glass may be made by using triethoxypropyl glycidoxysilane as the alkysilane (11,12). This forms the basis for a

**TABLE 2.2.   Some Commercially Available Alkoxysilanes
with Reactive Functions**

γ-Aminopropyltriethoxy silane	$H_2N-(CH_2)_3-Si(OC_2H_5)_3$
N-β-(Aminoethyl)-γ-aminopropyltrimethoxy silane	$H_2N-(CH_2)_2-NH-(CH_2)_3-Si(OCH_3)_3$
β-(Chloromethylphenyl)-ethyltrimethoxy silane	$Cl-CH_2-\langle\bigcirc\rangle-CH_2-CH_2-Si(OCH_3)_3$
γ-Chloropropyltriethoxy silane	$Cl-(CH_2)_3-Si(OC_2H_5)_3$
γ-Cyanopropyltrimethoxy silane	$NC-(CH_2)_3-Si(OCH_3)_3$
β-(3,4-Epoxycyclohexyl)-ethyltrimethoxy silane	$O\overset{\triangle}{\underset{}{\langle S \rangle}}-(CH_2)_2-Si(OCH_3)_3$
γ-Glycidoxypropyltrimethoxy silane	$\underset{\diagdown\diagup}{\overset{}{CH_2CH-CH_2}}-O-(CH_2)_3\,Si(OCH_3)_3$ $\qquad\quad O$
γ-Mercaptopropyltrimethoxy silane	$HS-(CH_2)_3-Si(OCH_3)_3$
γ-Methacryloxypropyltrimethoxy silane	$\overset{\overset{\displaystyle O}{\parallel}}{H_2C-C-C-O}-(CH_2)_3-Si(OCH_3)_3$ $\qquad\underset{\displaystyle CH_3}{\mid}$

Source- Reprinted with permission from M. Lynn, in *Immobilized Antigens, Antibodies, and Peptides,* H. H. Weetall, Ed. Marcel Dekker, 1975. Courtesy of Marcel Dekker, New York.

second-generation, two-component, affinity chromatography support available commercially as glycidoxy-controlled pore glass (Pierce Chemical Company, Rockford, Ill.). Glycidoxy glass is a product of reaction of the oxirine function of the glycidoxysilane glass with a bisoxirane (using BF_3 as a catalyst) to effect cross-linking between the individual silane functions. Thus the silane matrix is completely covered with a pliable cross-linked polymer. When the silica is removed, for example, with hydrofluoric acid, the remaining organic polymer remains intact, although, of course, it possesses no mechanical strength of itself.

Glycidoxy controlled pore glass combines with the best aspects of glass as a matrix, such as mechanical and thermal stability, and wide solvent compatability with the best aspects of agarose, such as a hydrophilic surface, and nonionic character. Further, glycidoxy glass has the additional advantage of possessing a reactive epoxy function, and, like agarose, it is available in a wide range of commercially prepared forms.

Like all controlled pore glass products, glycidoxy glass is reasonably priced for laboratory purposes but prohibitively costly for use on an industrial scale. For industrial scale, or even pilot plant scale, a fused-silica particle is available. This material is prepared by fusing together fine, uniform-sized silica particles by heating them to a temperature just below their melting point (13). The resulting product is a porous particle of very high purity. The surface of such fused silica is essentially free of borate, unlike controlled pore glass. The pore size distribution, however, is not as shraply controlled as that of controlled pore glass. Nonetheless, for the purpose of affinity chromatography, pore size distribution is not of great significance, provided that most of the pores are significantly larger than the protein to be purified.

Little, if any, use of fused silica bodies has been made in affinity chromatography, but the methods of derivatization by way of alkyl silanes is identical to that used for glass. Thus glycidoxy, alkyl amine, or mercaptopropyl fused silica should be similar, and probably superior (because of its lower cost and more uniform composition) to controlled pore glass in all uses except gel-permeation chromatography.

Glass and silica are generally thought of as insoluble, stable materials. Actually, in the underivatized state, glass is very slightly soluble at alkaline pH (>8.0). Because of their large surface area, porous glass and silica materials are, in practice, appreciably soluble, actually going into complete solution under alkaline conditions (see Figure 2.8). There are two solutions to this difficulty. The first possibility is to derivatize the glass with a cross-linked organic coating, as in the utilization of glycidoxy-, mercaptopropyl-coated or polysaccharide-coated (dextran, etc.) glass. Such coatings appear to minimize the problem of solubility. Alternatively, inorganic coating of other metal oxides can be employed. Zirconium clad glass, with a ZrO_2 surface, for example, is commercially available, as are fused titanium and zirconium oxide particles. Each of these can be derivatized with alkysilanes in the same fashion as glass, but the resulting product is much more inert in alkaline solutions. Controlled pore titania and zicronia have also been prepared, but they would probably have little advantage over fused titania or zirocnia particles and would be far more expensive.

Once the type of inorganic matrix to be used has been determined, the proper pore diameter and mesh size must be carefully selected. One recurrent myth is the reportedly slow flow characteristics of inorganic matrices in affinity chromatographic applications. For example, Cuatrecasas (14) noted in one report that estradiol glass beads were not useful for purification of estradiol receptor protein because of slow flow rates and the ease with which it became clogged with particulate matter. This observation was probably due to the small bead

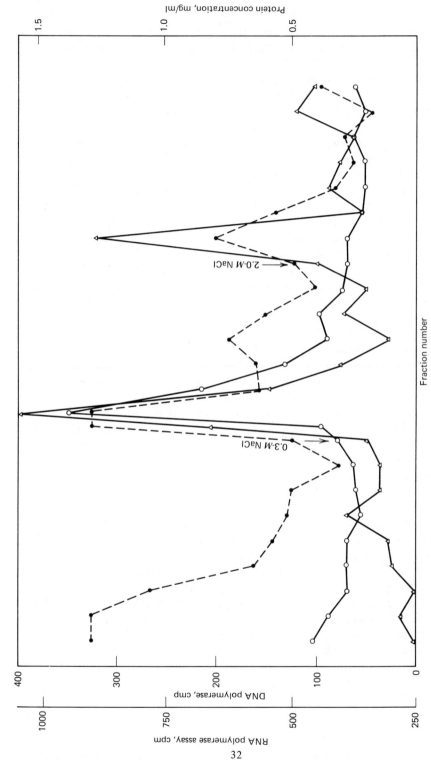

Figure 2.8. Purification of *E. coli* DNA polymerase and RNA polymerase on DNA glass. DNA polymerase and RNA polymerase were purified from 30 g of *E. coli* cells. Reprinted with permission from W. H. Scouten, *Am. Lab.* **6**, 28 (1974). Copyright © International Scientific Communications, Inc.

size and very large pore size of the glass that was employed. Glass or other carriers with very large pores (>2500 Å) are very friable and tend to break under pressure. Furthermore, small glass particles, particularly if they are not relatively homogeneous in size, can become clogged with particulate matter from crude cell extracts. When glass of 40 to 80 mesh with pore diameters of 550 Å are employed, even crude cell extracts containing considerable amounts of particulate matter can be handled with ease. Flow rates of up to 40 ml/min can thus be achieved. Moreover, glass beads can even be prepared for use in high-pressure liquid chromatography. For example, the author has purified DNA-binding proteins on DNA covalently coupled to very precisely sized, spherical porous glass (Spheron, Waters Associates) beads by employing pressures of 250 to 350 psi (15).

2.3. POLYACRYLAMIDE

Polyacrylamide beads are synthetic polymers of acrylamide crosslinked with bisacrylamide to form a reasonably rigid, hydrophilic matrix. At first glance they would appear to be superior to either agarose or porous inorganic matrices, since they possess excellent chemical stability (due chiefly to their polyethylene backbone) and contain very few ionic residues. Further, they are commercially available as spherical beads with defined mesh size and pore diameters. The backbone carbon–carbon bonds are stable over a wide pH range (1 to 10). Yet on slow hydrolysis of the amide function they yield a controlled number of carboxyl functions that could be used as attachment sites (16–18). Other derivatives of the amide function are also easily made (see page 62). In short, polyacrylamide can be easily derivatized and will yield a far wider range of degrees of substitution than will any other matrix. Furthermore, the ligands attached will be reasonably uniformly distributed throughout the gel.

Despite all these favorable properties, polyacrylamide, per se, is a rather poor matrix for most affinity chromatography separations. The first difficulty apparent in the application of polyacrylamide is its lack of mechanical stability. The most porous of the commercially available polyacrylamide gels, with an exclusion limit of about 300,000 for protein, is far less dimensionally stable than agarose; and agarose, in turn, is vastly inferior in this respect to controlled pore glass. Moreover, polyacrylamide adheres to glass surfaces so strongly that all glassware used in connection with polyacrylamide must be coated with silicone. If possible, plastic laboratory ware should be employed. These two properties make the use and handling of polyacrylamide difficult.

Of even greater importance in considering polyacrylamide as a support is the porosity of the final, derivatized product. As stated, the most porous underivatized commercial polyacrylamide gel, Biogel P-300, has an exclusion limit of about 300,000 molecular weight. This would be large enough for most bio-selective adsorption techniques. But once derivatized, the pores shrink so drastically that the largest protein that can be purified on Biogel P-300 is about 30,000 molecular weight rather than 300,000. For example, Steers et al. (19) have compared the purification of β-galactosidase from *E. coli* sonicates by using *p*-aminophenyl β-D-galactopryanoside bound to both agarose and acrylamide. Polyacrylamide, as expected, could be substituted to a much larger degree than agarose, with 20 μmoles of ligand per milliliter of polyacrylamide and 2 to 8 μmoles of ligand per milliliter of agarose. However, β-galactosidase emerged from the derivatized polyacrylamide column with the breakthrough protein peak, and none was retained or even retarded by the column. Conversely, when the agarose columns, properly derivatized, were employed, the bioselective adsorbent retained all the enzyme applied. The enzyme could be removed by way of bioselective elution with the inhibitor, *p*-aminophenyl-β-D-galactopyranose, although it was most readily eluted with a pH-10.5 borate buffer.

Although β-galactosidase (molecular weight 520,000) could not be purified by using polyacrylamide as the matrix, a number of smaller enzymes (e.g., staphylococcal nuclease (molecular weight 17,000) and dihydrofolate reductase (molecular weight 29,000) have been purified by using it. Polyacrylamide would seem to be especially useful for purifying smaller proteins, particularly those that have a weak affinity for the immobilized bioligand. The higher degree of substitution of polyacrylamide would make it the best matrix to be employed in such a case.

Polyacrylamide gels have been improved in the last few years. Modifications to polyacrylamide beads have largely eliminated the problems previously described here for untreated polyacrylamide gel. The modification of polyacrylamide takes two forms: (1) a change in the composition of the monomeric units used in the polymerization process and (2) the use of a two-part matrix composed of agarose-coated polyacrylamide beads.

Among the modified polyacrylamides of the first type are Enzacryl, a copolymer of *N*-acryloylmorpholine and *N,N'*-methylene bisacrylamide (20,21) and Spheron (22–24), a copolymer of hydroxyethyl methacrylate and ethylene bismethacrylate. A comparison of the structures of acrylamide-, acryloylmorpholine-, and hydroxyethyl methacrylate-based polymers reveals that the latter two are much more hydrophilic than the first. Both have unusual mechanical stability and can be prepared with very large pore sizes. Enzacryl is commercially available as rough particles with molecular exclusion limits for protein of up to 2 million, whereas Spheron is available as spherical beads with exclusion limits up to 5 million. Each is also commercially available in reactive forms containing a wide variety of available functional groups including amino, hydrazine, thiol, and activated carboxyl forms.

Enzacryl (polyacrylomorpholine) was the first of these modified acrylamide carriers to be marketed commercially (Figure 2.9). It is available with a wide range of functional groups preattached to the acrylamide matrix. Furthermore, various inorganic matrices, which provide greater mechanical strength, have been coated with Enzacryl. Among the most interesting of these is a magnetic iron oxide, coated with Enzacryl. These beads can be used in stirred batch reactors or

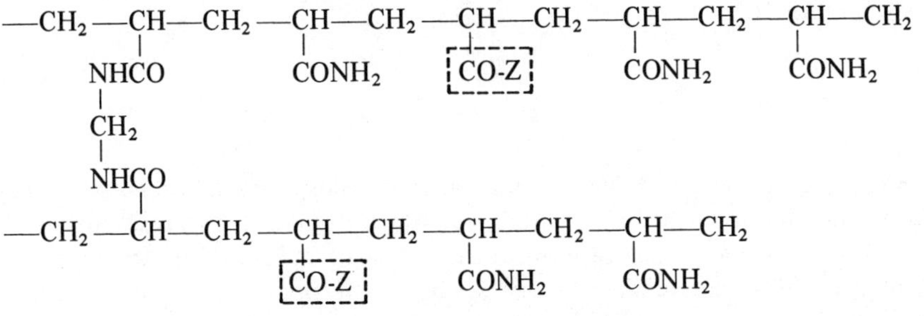

$$-CH_2-CH-CH_2-CH-CH_2-CH-CH_2-CH-CH_2-CH-CH_2$$

$$\underset{\begin{array}{c}NHCO \\ | \\ CH_2 \\ | \\ NHCO\end{array}}{|} \quad CONH_2 \quad \boxed{CO\text{-}Z} \quad CONH_2 \quad CONH_2$$

$$-CH_2-CH-CH_2-CH-CH_2-CH-CH_2-CH-CH_2$$

$$\boxed{CO\text{-}Z} \quad CONH_2 \quad CONH_2$$

a. Enzacryl AA $Z = -NHC_6H_4NH_2$
b. Enzacryl AH $Z = -NHNH_2$
c. Enzacryl Polythiol $Z = -NHCH(COOH)CH_2SH$
d. Encacryl Polythiolactone $Z = -NHCH-CO$
                                $\quad\quad\quad | \quad |$
                                $\quad\quad CH_2-S$

$$-CH_2-CH-CH_2-CH-CH_2-CH-CH_2-$$

$$\underset{\begin{array}{c}NH\text{-}CO \\ | \\ CH_2 \\ | \\ NH\text{-}CO\end{array}}{|} \quad \underset{\begin{array}{c}CONH \\ | \\ CH_2 \\ | \\ CH(OCH_3)_2\end{array}}{|} \quad \underset{\begin{array}{c}CONH \\ | \\ CH_2 \\ | \\ CH(OCH_3)_2\end{array}}{|}$$

$$-CH_2-CH-CH_2-CH-CH_2-CH-CH_2-$$

$$\underset{\begin{array}{c}CONH \\ | \\ CH_2 \\ | \\ CH(OCH_3)_2\end{array}}{|} \quad \underset{\begin{array}{c}CONH \\ | \\ CH_2 \\ | \\ CH(OCH_3)_2\end{array}}{|}$$

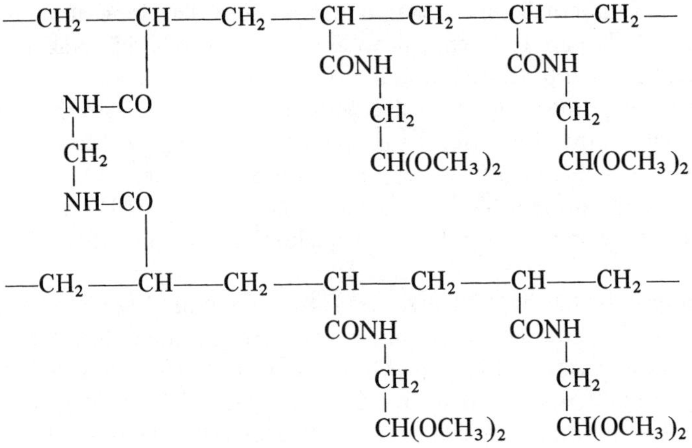

Enzacryl Polyacetal

Figure 2.9. Reprinted with permission of Koch-Light Laboratories Limited, England.

fluidized beads, where the magnetic properties of the beads allow them to be easily recovered after the purification is completed.

Although Enzacryl has been widely used for enzyme immobilization (20, 25-26), there are few examples of its application to affinity chromatography.

Spheron, on the other hand, is a macroporous resin derivative of hydroxyethyl methacrylate, which has been used as the matrix in many affinity chromatographic investigations (25-27). It is available with hydroxyl, carboxyl, sulfonyl, and aminoaryl functions and with porosities up to a molecular exclusion limit of 10^8.

2.4. CELLULOSE

Cellulose has been widely used as a matrix for biospecific adsorbents, although with the development of methods to attach ligands to agarose the use of cellulose as a matrix has receded. Today it is the matrix of choice only if the bioligand is a macromolecule *per se* or if the application, for example in industrial-scale use, demands a very inexpensive carrier.

In many ways cellulose would be an ideal matrix. It can be prepared such that it bears no ionogenic functions, has little nonspecific adsorption, is mechanically and thermally stable, and is compatible with most organic solvents. Unfortunately, cellulose possesses very small pores that are not uniformly distributed throughout the matrix. Thus although beaded porous cellulose may be able to couple 500 times as many small molecules (e.g., succinic dihydrazide) as can be bound to agarose, the capacity for the same cellulose to bind macromolecules (e.g., serum albumin) is actually lower than that of agarose. This is because most of the pores of cellulose are large enough to allow the succinic dihydrazide to enter, but not large enough for the albumin.

Any single batch of porous cellulose has a wide distribution of pore sizes and available functional groups, thus resulting in regions of both high and low ligand concentration and is, consequently, a very nonuniform adsorbent. Only when the bioligand is a large protein, or other macromolecule such as DNA, is the ligand distributed uniformly, or nearly so, throughout the support (See Figure 2.10).

Among the bioligands that have been successfully immobilized as bioselective adsorbents with cellulose as the matrix are soybean trypsin inhibitor, used to purify trypsin (36), and acrosin (37), and chitin, used to purify lysozyme (38), and poly(deoxythymidine)$_{12-18}$, used to purify RNA-dependent DNA polymerase (39). By far the most important application of cellulose as an affinity chromatography matrix is in the purification of DNA-binding proteins as described in Chapter 1. Even in this application cellulose is not the most efficient

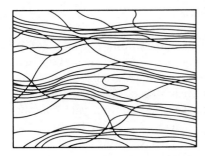

Figure 2.10. Nonuniformity of cellulose matrices. Diagrammatic representation of the organization of molecules within cellulose structure showing areas of order and disorder. Reproduced from C. S. Knight, *Adv. Chromatogr.* (1967), **4**, 61, by courtesy of Marcel Dekker, Inc.

matrix. Deoxyribonucleic acid–cellulose is 50-fold less efficient per unit volume than is DNA agarose (40) or DNA glass (41) in binding DNA polymerase or similar DNA-binding proteins.

There are two other basic problems that diminish the usefulness of cellulose as a matrix. First, because of its fibrous nature, it is easily clogged by particulate material and easily compressed by the application of even moderate pressure. Thus, it is of limited use in rapid, high-pressure application or with very crude cell extracts. Further, commercial cellulose is often contaminated with a wide variety of impurities. These impurities can be removed in many ways (see page 209). Alberts and Herrick (42) have detailed several methods for purification of cellulose and also suggest that glycerol, incorporated in the eluting buffers, will further limit nonspecific adsorption on cellulose.

2.5. OTHER MATRICES

Of the more than 120 different affinity chromatography purifications prepared before 1975, at least 95% were performed on agarose, glass, or cellulose, and almost 90% of these were performed by using agarose. Most of these purifications were done before the ion–exchange and hydrophobic chromatographic properties of agarose derivatives were widely understood. Even so, very few affinity chromatographic purifications have been performed by using other matrices, despite the fact that a large number of potential affinity matrices exist. Among the matrices that have been used are the cross-linked dextrins (43,44), insolubilized proteins (45,46), and chitin (47). By analogy with immobilized enzymes, nylon (48), metal oxide (49), polystyrene (50), and collagen (51) could be used. Obviously, no one matrix will be suitable for all purifications, but the hydrophillic gels, such as agarose, glycidoxy glass, and acrylamide appear likely to be the first choice for most applications for some time to come.

2.6. SPACER ARMS

Once the matrix to be used has been determined, the investigator must decide on the method of attaching the bioligand to the matrix. In some cases the bioligand is attached directly to the matrix; however, in most instances the matrix sterically hinders the enzyme from bnding to such a ligand. To allow the ligand to reach into and bind the active site of the enzyme, the bioligand must be attached to a flexible "arm" or "spacer" (see Figure 2.11). The length of the spacer arm and the method of attaching it to both matrix and the bioligand must be chosen carefully. A spacer arm that is too long may contain potential internal hydrophobic interaction sites and, thus, may well fold back on itself, making the *effective* length of the spacer much shorter than it really is (see Figure 2.12).

Furthermore, a spacer arm may impart ion-exchange or hydrophobic properties to the bioselective adsorbent and thus result in a nonspecific adsorption that is so great as to mask any bioselective adsorbent properties the chromatographic packing may have.

Despite these potential difficulties, the usefulness of spacer arms has been well documented. Recently Wilchek and Miron (52) introduced a method for preparing macromolecular spacer arms that are attached to the matrix at multiple attachment sites. Such spacer arms not only eliminate steric hindrances to a ligands accessibility, but also minimize leakage of the ligand from the matrix. Polylysine (53), polyglutamic hydrazide (54), and polyacrylic hydrazides (55) have all been used with considerable success.

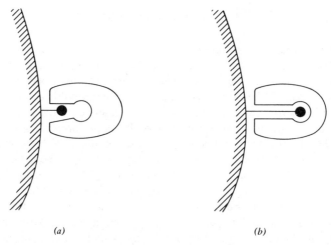

(a)   (b)

Figure 2.11. Effect of spacer arms on enzyme binding to immobilized ligands: (a) direct attachment-enxyme-active site is sterically prevented from reaching the substrate; (b) attachment through a spacer arm—enzyme active site can freely interact with the substrate.

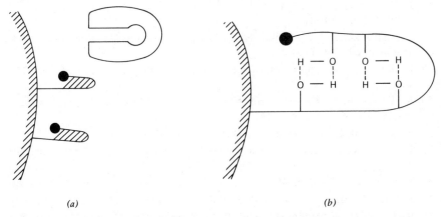

(a)   *(b)*

Figure 2.12. Ligands buried by folding of spacer arms on themselves—a shorter spacer arm would be more effective: (*a*) hydrophobic folding of spacer arm; (*b*) hydrogen bonding of spacer arm.

References

1. J. Porath and R. Axen (1976), in *Methods in Enzymology* (K. Mosbach, ed.) **44**, 19.

2. M. S. Poonian, A. J. Schlabach, and A. Weissbach (1971), *Biochemistry* **10**, 420.

3. A. Weissbach and M. S. Poonian (1974), in *Methods in Enzymology* (W. B. Jakoby and M. Wilchek, eds.) **34**, 463.

4. W. Haller (1965), *J. Chem. Phys.* **42**, 686.

5. H. P. Hood and M. E. Nordberg (1940), U.S. Patent 2,221,709.

6. H. H. Weetall and A. M. Filbert (1974), in *Methods in Enzymology* (W. B. Jakoby and M. Wilchek, eds.) **34**, 59.

7. R. A. Messing (1969), *J. Am. Chem. Soc.* **91**, 2370.

8. H. H. Weetall and N. B. Havewala (1972), *Biotech. Bioeng. Symp.* **3**, 241.

9. H. H. Weetall (1969), *Science* **766**, 615.

10. W. H. Scouten (1977), in *Encyclopedia of Polymer Science and Technology* (N. Bikales, ed.) Supplement **2**, 19.

11. F. E. Regnier and R. Noel (1976), *J. Chromatogr. Sci.* **14**, 316.

12. S.-H. Chang, R. Noel and F. E. Regnier (1976), *Anal. Chem.* **48**, 1839.

13. D. L. Eaton (1974), in *Immobilized Biochemicals and Affinity Chromatography* (R. B. Dunlap, ed.), Plenum, New York, p. 241.

14. P. Cuatrecasas (1970), *J. Biol. Chem.* **245**, 3059.

15. V. Levi, T. Frielle and W. H. Scouten (1978), submitted for publication.

16. J. Coupek, J. Labsky, S. Kalal, J. Turkova, and O. Valentova (1977), *Biochem. Biophys. Acta* **481**, 289.

17. J. K. Inman (1974), in *Methods in Enzymology* (W. B. Jakoby and M. Wilchek, eds.) **34**, 30.

18. G. Manecke and J. Schlunsen (1976), in *Methods in Enzymology* (K. Mosbach, ed.) **44**, 107.

19. E. Steers, Jr., P. Cuatrecasas, and H. B. Pollard (1971), *J. Biol. Chem.* **246**, 196.

20. R. Epton, B. L. Hibbert, and T. H. Thomas (1976), in *Methods in Enzymology* (K. Mosbach, ed.) **44**, 84.

21. R. Epton, B. L. Hibbert, and G. Marr (1975), *Polymer* **16**, 314.

22. J. Hradil, M. Krivakova, P. Stary, and J. Coupek (1973), *J. Chromatogr.* **79**, 99.

23. J. Turkova (1976), in *Methods in Enzymology* (K. Mosbach, ed.) **44**, 66.

24. J. Coupek, M. Krivakova, and S. Pokorny (1973), *J. Polym. Sci.* **42**, 185.

25. R. Epton, J. V. McLauren, and T. H. Thomas (1973), *Biochim. Biophys. Acta* **328**, 418.

26. S. A. Barker, J. F. Kennedy, and R. Epton (1974), U.S. Patent 3,794,563.

27. J. Turkova and A. Seifertova (1978), *J. Chromatogr.* **148**, 293.

28. J. Turkova, S. Vavreinova, M. Krivakova, and J. Coupek (1975), *Biochim. Biophys. Acta* **386**, 503.

29. J.Turkova, O. Hubalkova, M. Krivakova, and J. Coupek (1973), *Biochim. Biophys Acta* **322**, 1.

30. O. Valentova, J. Turkova, R. Lapka, J. Zima, and J. Coupek (1975), *Biochim. Biophys. Acta* **403**, 192.

31. J. Vaneckova, J. Barthova, T. Barth, L. Krejci, and I. Rychlik (1975), *Collect. Czechoslov. Chem. Commun.* **40**, 1461.

32. Z. Slovak, S. Slovakova, and M. Smzr (1975), *Anal. Chim. Acta* **75**, 127.

33. L. M. Sirakov, J. Bartova, T. Barth, S. P. Ditzov, K. Jost, and I. Rychlik (1975), *Collect. Czechoslov. Chem. Commun.* **40**, 775.

34. M. Cech, M. Jelinkova, and J. Coupek (1976), *J. Chromatogr.* **135**, 435.

35. J. Borak and M. Smrz (1977), *J. Chromatogr.* **133**, 127.

36. G. D. Jameson and D. T. Elmore (1971), *Biochem. J.* **124**, 66p.

37. W. C. Schleuni, H. Schiessler, and H. Fritz (1973) *Hoppe-Seyler's Z. Physiol. Chem.* **354**, 550.

38. T. Imoto and K. Yagishita (1973), *Agric. Bio. Chem.* **37**, 1191.

39. B. I. Gerwin and J. B. Milstein (1972), *Proc. Nat. Acad. Sci. (USA)* **69**, 2599.

40. H. Schaller, C. Nusslein, F. F. Bonhoeffer, C. Kurz, and I. Neitzschmann (1972), *Eur. J. Biochem.* **26**, 474.

41. L. Jervis and N. M. Pettit (1974), *J. Chromatogr.* **97**, 33.

42. B. Alberts and G. Herrick (1971), in *Methods in Enzymology* (K. Moldave and L. Grossman, eds.) **21**, 198.

43. P. Roschlau and B. Hess (1972), *Hoppe-Seyler's Z. Physiol. Chem.* **353**, 441.

44. D. Rickwood (1972), *Biochim. Biophys. Acta* **269**, 47.

45. G. B. Brown (1976), in *Methods in Enzymology* (K. Mosbach, ed.) **44**, 263.

46. A. F. S. A. Habeeb and R. Hiramato (1968), *Biochim. Biophys. Acta* **126**, 16.

47. R. Bloch and M. M. Burger (1974), *Biochem. Res. Commun.* **58**, 13.

48. W. E. Hornby and L. Goldstein (1976), in *Methods in Enzymology* (K. Mosbach, ed.) **44**, 118.

49. H. A. Weetall and L. S. Hersh (1970), *Biochim. Biophys. Acta* **206**, 54.

50. W. Ledingham and W. Hornby (1969), *FEBS Lett.* **5**, 118.

51. W. R. Vieth and K. Venkatasubramanian (1976), in *Methods in Enzymology* (K. Mosbach, ed.) **44**, 243.

52. M. Wilchek and T. Miron (1974), in *Methods in Enzymology* (W. B. Jakoby and M. Wilchek, eds.) **34**, 72.

53. M. Wilchek (1973), *FEBS Lett.* **33**, 70.

54. J. Kovacs, V. Bruckner, and K. Kovacs (1953), *J. Chem. Soc.* **1953**, 145.

55. V. W. Kern, T. Hucke, R. Hollander, and R. Schneider (1957), *Makromol. Chem.* **22**, 31.

SYNTHETIC METHODS

There are normally two stages in preparing a bioselective adsorbent; (1) the attachment of a spacer arm to the matrix and (2) the attachment of the ligand to the spacer arm. Step 1 depends to a large extent on the nature of the matrix, whereas the second step is dependent on the specific ligand to be used. These two stages of synthesis allow us to divide synthetic methods into (1) those methods directly involving the matrix and (2) those methods directly involving the ligand. A few of the most widely used methods are outlined in the following paragraphs with references given to those less often employed. Hopefully, this brief discussion will provide the investigator with the means to devise his or her own methods for preparing a heretofore unknown bioselective adsorbent.

3.1. ATTACHMENT OF SPACERS TO MATRICES

3.1.1. CNBr-Activated Agarose

Cyanogen bromide reacts with the vicinal diols of agarose to produce an "activated" agarose that can be subsequently coupled to spacer molecules containing primary amines. This activation process is the most widely employed of all attachment methods. (See Figure 3.1.) Unfortunately, because of the formation of charged isourea functions (1-6), this method also invariably yields a bioselective adsorbent with anion-exchange properties.

The isourea formed in this fashion is positively charged at physiological pH, [$pK_a \sim 9.5$.] If the alkyl group attached to the isourea function is a long hydrophobic group, the dielectic constant around the isourea groups will be very low, thus causing a large dispersal of the charge around nearly the entire gel surface. The ion-exchange properties of alkyl amine agarose are likely to be much more pronounced (at least when proteins are being employed) than is that of a comparable commercial ion exchanger that possesses an idential amount of substitutions but with charges that are concentrated on the individual ion-exchange centers. (See Figure 3.2.)

Even with these limitations, cyanogen bromide-activated agarose has been widely and successfully employed in the synthesis of many bioselective adsorbents. Because of the inherent dangers of cyanogen bromide, for example, the

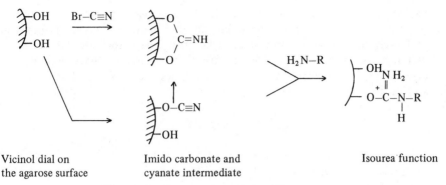

| Vicinol dial on the agarose surface | Imido carbonate and cyanate intermediate | Isourea function |

Figure 3.1. Cyanogen bromide coupling method.

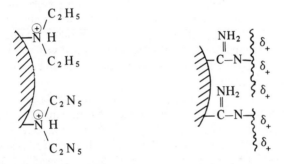

DEAE–cellulose—the positive charge is concentrated at the individual ion-exchange sites

N-alkyl sepharose—the charge is dispersed over much of the gel surface

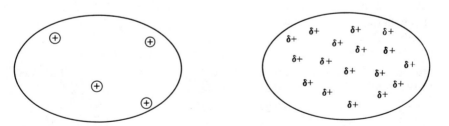

Figure 3.2. Comparison of DEAE–cellulose with point charges and aminoalkyl-agarose with positive charges distributed over the surface.

release of HCN on acidification, the use of commercially preactivated cyanogen bromide agarose has gained wide acceptance. For many uses, however, it may be preferable to activate agarose in the laboratory just before use. Of the many methods for cyanogen bromide activation of agarose that have been devised (7–11), the one outlined in the list that follows is possibly the least laborious and most reproducible (12).

Reagents

2-M Potassium phosphate buffer, pH 11.5 to 12.0
Aqueous cyanogen bromide, 100 mg/ml
0.25-M Sodium bicarbonate, pH 9.0
Agarose (6% or 4%) (either untreated or cross-linked)

In this procedure 10 g of agarose is washed with 25 ml of phosphate buffer and then filtered to remove the interstitial water. The agarose is added to a tube containing the amounts of buffer listed in Table 3.1. Aqueous cyanogen bromide is added slowly in the amounts listed, and the mixture is shaken gently for 10 minutes at 5°C. When the reaction is finished, the product is collected on a fine-sintered glass funnel and washed with ice-cold distilled water until the filtrate is essentially neutral. The activated agarose should be immediately coupled to the ligand or spacer arm desired.

To couple the matrix to ligands or to spacers containing free primary amines, 20 ml of the CNBr-activated agarose (or the same amount of commercial CNBr-preactivated agarose swollen according to the manufacturer's directions) is washed with 50 ml of 0.25-M NaHCO$_3$. The washed agarose is suspended in 20 ml of 0.25-M NaHCO$_3$, and 50 to 500 mg of the amine is added. The mixture is

TABLE 3.1.　CNBr-Activated Agarose

Amount of Derivatization Desired		Amount of Buffer Needed (ml)	Amount of Aqueous CNBr to be Used, ml
Highly derivatized	4% gel	10	4
Highly derivatized (~300 mmole/g)[a]	6% gel	15	6
Moderately derivatized (~140 mmole/g)[a]	4% gel	0.8	0.8
	6% gel	1.2	1.2
Weakly derivatized (~100 m mole/g)[a]	4% gel	0.6	0.2
	6% gel	0.7	0.3

[a]Cross-linked agarose (e.g., Sepharose CL) will yield only 25 to 30% as much derivatization as will non-cross-linked agarose.

gently shaken or stirred at 0 to 25°C for 12 to 24 hours. To ensure that all of the activated agarose hydroxyls are derivatized, the researcher may wish to block any unreacted agarose–imidocarbonate by shaking the beads in 1-M glycine, pH 8.0, for 6 to 12 hours at 0°C.

Diamines such as 1,4-butanediamine or hexamethylexediamine are very commonly exployed as spacer arms. However, because such derivatives produce matrices with hydrophobic and ion-exchange properties, the use of adipic dihydrazide (3, 13–23) or polyhydrazides (24) is favorable. The hydrazide has a pK of about 2.5 and thus is uncharged at neutral pH, and the adipoyl group appears to be the longest hydrocarbon spacer that can be employed without creating considerable difficulty with hydrophobic interactions.

3.1.2. Adipoyl Dihydrazide Agarose (23)

Diethyladipate
Hydrazine hydrate (98%)
0.1-M Na$_2$CO$_3$, pH 9.5
2-M NaCl
CNBr-activated agarose

Adipoyl dihydrazide is prepared by refluxing 100 ml of diethyldipate with 200 ml of hydrazine hydrate and 200 ml of absolute ethanol for 3 hours (13). The product, adipoyl dihydrazide, precipitates and is collected after cooling to room temperature. It can be further purified, if desired, by recrystallization from ethanol–water [melting point (169 to 171°C)].

To synthesize adipoyl dihydrazide agarose, the CNBr-activated agarose beads, prepared as described previously, are mixed with 2 volumes of a cold saturated solution of adipoyl dihydrazide in 0.1-M Na$_2$CO$_3$ (pH 9.5). The beads are stirred or shaken gently overnight and then washed thoroughly, first with distilled water, then with 2-M NaCl, and finally with distilled water.

To prevent "leakage" or hydrolysis of ligand from the matrix, the alkyl diamine or dihydrazide are often replaced with polyamines, for example, polylysine (18), or with polyhydrazides, for example, polyglutamic hydrazide or polyacrylic hydrazide (24). These are attached to the agarose in the same fashion as the monomeric diamines or dihydrazides, with the exception that they become attached to CNBr-activated agarose at many points. Again, the polyhydrazides are preferable to polyamines since the polyhydrazide–agarose doesn't contain ion-exchange centers. Polyhydrazide–agarose is also more attached to bioligands than are the polyamine derivatives.

3.1.3. Periodate–Hydrazide Agarose

The large number of vicinal *cis*-hydroxyl groups of agarose and many other polysaccharides are susceptible to oxidation by periodate to yield aldehydes

that can be converted to secondary amines by reductive amination or to hydrazides by reaction with dihydrazides. Such derivatives are very stable and easily
prepared. The hydrazide–agarose preparations possess no undesirable hydrophobic or ionic interactions (13,23,54).

Reagents

1-M NaIO$_4$
5-M NaBH$_4$ (freshly prepared)
1-M NaCl
Adipoyl dihydrazide or succinoyl dihydrazide
Alkyl diamine (e.g., hexamethylenediamine)
1.5% Trinitrobenzene sulfonic acid (TNBS or picrylsulfonic acid)
Saturated sodium borate
Agarose beads (4 to 6%)

The oxidation of agarose hydroxyls is performed by adding 100 ml of agarose
(washed with water and freed of interstitial water via suction filtration) to 80 ml
of H$_2$O. To this is added 20 ml of 1-M sodium periodate. The suspension is
gently shaken for 2 hours at room temperature, preferably in the dark. The
oxidized agarose is then washed with water and further derivatized via diamines
or dihydrazides as described in the following paragraphs.

To prepare dihydrazide–agarose, oxidized agarose from the procedure just
described is suspended in 100 mL of 0.1-M alkyl dihydrazide and the pH is
adjusted to pH 5.0 with 1-M HCl (Figure 3.3). Adipoyl dihydrazide (see page
45), succinic dihydrazide (26), or any similar alkyl dihydrazide may be employed. The mixture is shaken for 2 hours and then the pH adjusted to 9.0 with
sodium carbonate. The mixture is cooled to 0 to 4°C and 10 ml of freshly
prepared 5-M sodium borohydride is added in small aliquots over a 10- to 12-
hour period.

The resulting beads must be thoroughly washed with 1-M sodium chloride.
One procedure (27) used is to elute the beads on a sintered glass funnel for 3 to
5 hours without suction, followed by incubation of the beads in fresh 1-M NaCl
overnight, and finally eluting the beads again slowly for 3 to 5 hours with 1-M
NaCl.

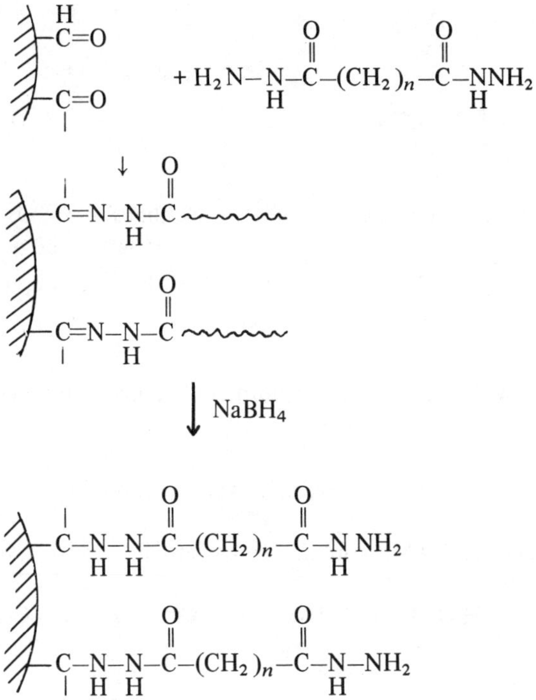

$$-C=O \quad + H_2N-N-C-(CH_2)_n-C-NNH_2$$

$$\downarrow$$

$$-C=N-N-C\sim\sim\sim$$

$$-C=N-N-C\sim\sim\sim$$

$$\downarrow NaBH_4$$

$$-C-N-N-C-(CH_2)_n-C-N\,NH_2$$

$$-C-N-N-C-(CH_2)_n-C-N-NH_2$$

Figure 3.3. Coupling adipic dihydrazide to periodate oxidized agarose.

To test for the presence of dihydrazide in the filtrate, 1 to 2 ml of the filtrate is added with stirring to 5 drops of a 3% (w/v) solution of 2,4,6-trinitrobenzene-sulfonic acid (picryl sulfonic acid) in a small test tube (30). Primary hydrazides yield a rust color, which changes on standing for 30 minutes to a magenta-colored solution. The beads can also be tested for the presence of the hydrazide function by suspending a few beads in water and treating them as described previously. If derivatization is successful, the beads, but not the solution, should turn to a rust–magenta color within 10 to 30 minutes.

Preparation of N-alkyl amine agarose is performed in a similar fashion. The oxidized agarose (100 ml), prepared as described previously, is added to 100 ml of an aqueous 2-M solution of the alkyl diamine, pH 5.0. After 6 to 10 hours at room temperature the pH is raised to 9 and the beads are reduced at 0 to 4°C by adding 10 ml of 5-M sodium borohydride over 12 hours. The beads are then washed as described for the preparation of hydrazide agarose. Excess amine in the eluate forms a bright yellow color in the TNBS test.

Analogous to the cyanogen bromide activation, the use of hydrazide agarose is preferred over alkyl amine agarose in most applications since the amine function ($pK_a \simeq 10.4$) is positively charged at neutral pH, whereas the hydrazide moeity

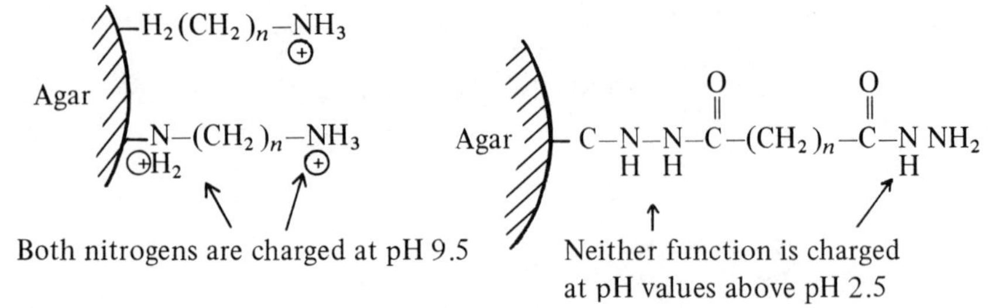

$$\text{Agar} \left\{ \begin{array}{l} -H_2(CH_2)_n - \overset{\oplus}{N}H_3 \\[2ex] -\underset{\oplus H_2}{N} - (CH_2)_n - \overset{\oplus}{N}H_3 \end{array} \right.$$

Both nitrogens are charged at pH 9.5

$$\text{Agar} \left\{ \begin{array}{l} -\overset{O}{\overset{\|}{C}} - \underset{H}{N} - \underset{H}{N} - \overset{O}{\overset{\|}{C}} - (CH_2)_n - \overset{O}{\overset{\|}{C}} - \underset{H}{N}\,NH_2 \end{array} \right.$$

Neither function is charged
at pH values above pH 2.5

Figure 3.4. Comparison of alkyldiamines and alkyl dihydrazides coupled to CNBr activated agarose.

(pK_a = 2.45) is uncharged within the normal pH range of bioselective adsorption chromatography. (See Figure 3.4.)

3.1.4. Bisoxirane–Agarose (28–30)

Bisoxiranes, such as 1,4-butadioldiglycidoxyether:

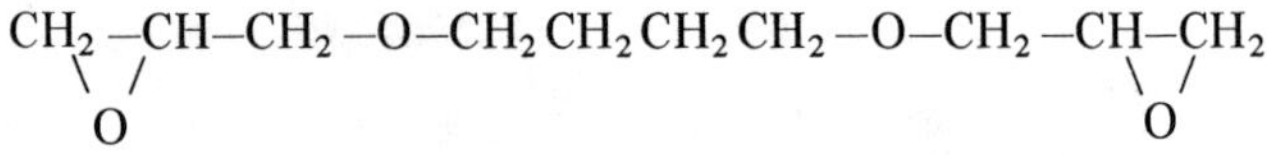

$$CH_2 - CH - CH_2 - O - CH_2\,CH_2\,CH_2\,CH_2 - O - CH_2 - CH - CH_2$$
$$\underset{O}{\diagdown \diagup} \qquad\qquad\qquad\qquad\qquad \underset{O}{\diagdown \diagup}$$

react readily at alkaline pH to yield derivatives containing a long-chain hydrophilic function with a reactive oxirane (epoxide), which, in turn, can be reacted with nucleophilic ligands (amines, phenols, etc.) to form bioselective adsorbents with a minimum number of hydrophobic and ionic groups. The single, most unfortunate, drawback to the technique is the extreme difficulty encountered in attempting to remove excess unreacted oxirane functions from the reacted agarose beads. Any residual unreacted oxirane groups may react with and immobilize any protein subsequently applied to the adsorbent.

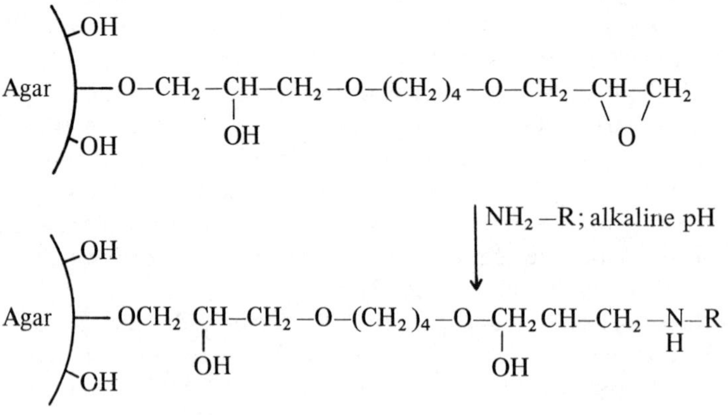

$$\text{Agar} \left\{ \begin{array}{l} OH \\[2ex] -O-CH_2-\underset{\underset{OH}{|}}{CH}-CH_2-O-(CH_2)_4-O-CH_2-\underset{O}{\underset{\diagdown \diagup}{CH}}-CH_2 \\[2ex] OH \end{array} \right.$$

$$\Big\downarrow NH_2-R;\ \text{alkaline pH}$$

$$\text{Agar} \left\{ \begin{array}{l} OH \\[2ex] -OCH_2\ \underset{\underset{OH}{|}}{CH}-CH_2-O-(CH_2)_4-O-\underset{\underset{OH}{|}}{CH_2CH}-CH_2-\underset{H}{N}-R \\[2ex] OH \end{array} \right.$$

Generally very porous agarose should be used as the starting material, for example, 4 to 6% agarose beads. The preparation of the bisoxirane–agarose is very straightforward. The procedure does result in cross-linking of hydroxyls within the agarose, thus decreasing their porosity slightly while simultaneously considerably improving the mechanical stability and flow characteristics.

In this procedure 10 ml of beaded agarose is washed with distilled water to free it from various compounds used as preservatives, and then the excess water is removed on a sintered glass filter. The washed gel is suspended in 5 ml of 1-M NaOH containing 2 mg/ml of $NaBH_4$. Then 5 ml of 1,4-butandioldiglycidoxy-ether is added, and the mixture is gently shaken or stirred for 5 to 10 hours at room temperature. The activated gel is washed *thoroughly* with distilled water and stored at 5°C. Loss of oxirane groups will occur slowly during storage; thus, the activated gel should be used as soon as possible. It has been reported that little or no loss of reactive functions occurs under these conditions for at least 1 week (17), although alkaline conditions may accelerate the decomposition of the oxirane. Amines, hydroxyls, and other nucleophilles can be readily coupled to the oxirane agarose. For example, glycyl leucine is coupled by suspending 3 ml of oxirane agarose in 2 ml of 0.2-M Na_2CO_3, pH 11, containing 300 mg of glycyl leucine (18). The mixture is kept at 4°C for 15 hours and then washed with 1-M NaCl. Aromatic amines may be similarly reacted. At higher temperatures (e.g., 40°C) and at pH 11 to 13, hydroxylic compounds can also be coupled.

Because of the relative stability of the oxirane group, removal of excess epoxide may be troublesome. Porath has suggested (28) that storage for 1 week in Na_2CO_3 buffer, pH 11 to 12, or prolonged treatment with concentrated hydroxylamine at neutral pH is recommended to remove excess reactive functions. However, much briefer treatment of oxirane–agarose (4 to 6 hours at pH 7.5) with mercaptans will result in complete removal of the excess reactive epoxide (2-mercaptoethanol, ethanedithiol, and thiourea have all been used). The presence of cysteine, thiosulfate, and thioglycolic acid should be avoided during the preparation since they will introduce ionic functions into the final product.

The potential usefulness of the bisoxirine spacer arm should not be underestimated. Recently it has been shown that sulfanilamide coupled to agarose by way of such a spacer arm has high capacity for carbonic anhydrase (26), whereas bioselective adsorbents prepared by use of cyanogen bromide coupling to agarose or dextran polymers have very low capacity for this enzyme.

3.1.5. Phloroglucinol–Agarose

An agarose derivative with increased binding capacity and increased stability can be prepared by coupling phloroglucinol to agarose by way of an epichlorohydrin bridge (28,29). (See Figure 3.5.)

Porath and Sundberg (28) coupled soybean–trypsin inhibitor to agarose by

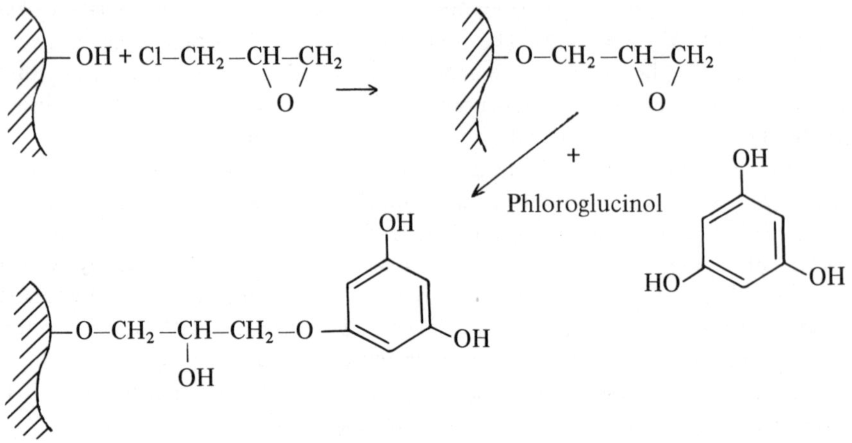

Figure 3.5. Coupling ligands to agarose through a phloroglucinol bridge.

way of a phloroglucinol intermediate in a three-step procedure. First, agarose was incubated with both epichlorohydrin and phloroglucinol in 1-N NaOH and the product washed and treated with 5% divinyl sulfone for 30 minutes at pH 11. This product was again washed and incubated with soybean–trypsin inhibitor at pH 9.0. The resulting immobilized soybean inhibitor exhibited a 2.5-fold higher capacity than did cyanogen bromide- or divinylsulfone-activated agarose. Coombe and George (30) used the same procedure to prepare immobilized methacillin–agarose for the purification of penicillin. Methacillin–agarose, prepared by this method, was compared with methacillin immobilized by way of divinylsulfone–agarose and cyanogen bromide–agarose procedures. The stability and the capacity of gels produced by using phloroglucinol intermediates were as good as or better than similar adsorbents prepared by any other procedure.

Caron et al. (29) coupled phloroglucinol to agarose by a similar procedure. Agarose (100 ml) was incubated with 15 g of phloroglucinol and 10 ml of epichlorohydrin in 1-M NaOH for 2.5 hours at 50°C, and the resulting phloro-glucinol–agarose was activated with CNBr and coupled to antihuman α-feto-protein antibodies to prepare an immunosorbent for the purification of α-fetoprotein. The amount of immunosorbent bound was about two fold above that prepared by direct cyanogen bromide activation. Further, the adsorbent was more stable than that prepared by CNBr activation.

Cross-linking with Epichlorohydrin

Agarose is an excellent matrix for preparation of bioselective adsorbents, but with two reservations: (1) it possesses a small number of sulfate ion-exchange centers; and (2) it is held together as a porous network by way of hydrogen bonds and thus is destroyed by the presence of many hydrogen bond breaking agents and conditions. For example, denaturants such as dimethylsulfoxide (DMSO), urea, iodide and thiocyanate, as well as temperatures above 40°C (see page 22), all destroy the gel-forming ability of untreated agarose. Cross-linking with epichlorohydrin in the presence of sodium borohydride both removes the sulfate ion-exchange centers and cross-links the polysaccharide chains, such that hydrogen bonds are not the chief basis for the gel structure. The result is a highly stable matrix, even in the presence of denaturants, which is slightly less permeable than the original agarose (32–34). Similar products can also be prepared by using 2,3-dibromopropanol (34).

The cross-linking is performed by suspending 100 ml of agarose beads in 1 l of 1-M NaOH solution containing 10 ml of epichlorohydrin and 0.5 g of NaBH$_4$. The mixture is heated to 60°C for 2 hours with stirring. This can be conveniently done by attaching the flask containing the mixture to a rotary flash evaporator and immersing the rotating flask in a 60°C water bath. (*Caution:* do *not* apply a vacuum.) This method of stirring is to be preferred over stirring with a magnetic stirrer, since magnetic stir bars may damage the beads under some circumstances.

After reaction, the beads are washed in hot water until the wash is neutral. The beads may then be used as they are, but some carboxylate groups, created during the cross-linking, may be present. To remove these, (31) the beads are autoclaved for 1 hour at 120°C in 50 ml of 2-M NaOH containing 0.25 g of NaBH$_4$, followed by washing with 150 ml of 1-M NaOH containing 0.5% NaBH$_4$. The gel is then washed with cold water and finally is suspended in cold water and the suspension brought to pH 4.0 with acetic acid. The beads should be stored at 4°C (not frozen), and if storage for more than a few weeks is probable, NaN$_3$ (final concentration, 0.02%) should be added to prevent microbial growth.

As shown in Table 3.2, agarose that has been epichlorohydrin cross-linked and desulfated as described previously (ECD–agar) is now very stable in such normally gel-destroying denaturants as DMSO, I⁻ and SCN⁻.

TABLE 3.2. Amount of Carbohydrate Dissolved Under Conditions Given in the Text (g/100 g Dry Substance)

	DMSO 100°C	4-M I⁻	3-M SCN⁻
ECD-Bacto agar 2%	0.5	0.2	< 0.03
Agarose gel 2%	100	100	~ 90

Source: Reprinted with permission from Porath (31).

3.1.6. Divinylsulfone (8)

Divinylsulfone may be used to couple nucleophilic ligands to agarose. The vinyl groups are more reactive than epoxy functions; therefore, the coupling proceeds rapidly and completely. However, the stability of the products presents a problem. The product with hydroxyl functions is unstable above pH 9, and the product with amine functions is unstable above pH 8. (See Figure 3.6.)

Reagents

6% Agarose beads
1-M Sodium carbonate buffer, pH 11.0
Divinylsulfone

Agarose (10 ml) is washed and freed of interstitial water by way of suction filtration. The agarose is suspended in 10 ml of 1-M sodium bicarbonate, pH 11.0, and 2 ml of divinylsulfone is added to the solution. The reaction is continued with stirring for approximately 1 hour at room temperature, after which the beads are washed with water.

The freshly prepared divinylsulfone agarose is coupled directly to the ligand. An amino- or hydroxyl-containing ligand (250 mg/ml) is dissolved in 1-M sodium carbonate. This solution is added to 10 ml of activated divinylsulfone agarose and stirred gently. The time needed for complete reaction depends on the temperature employed (e.g., 2 hours at 50°C or 24 hours at 4°C); the reacted beads are then washed with 10 ml of 1-M NaCO$_3$, pH 11.0, then with 500 ml each of 0.2 glycine, pH 3.0, containing 1-M NaCl, and finally with 500 ml of 1-M NaCl.

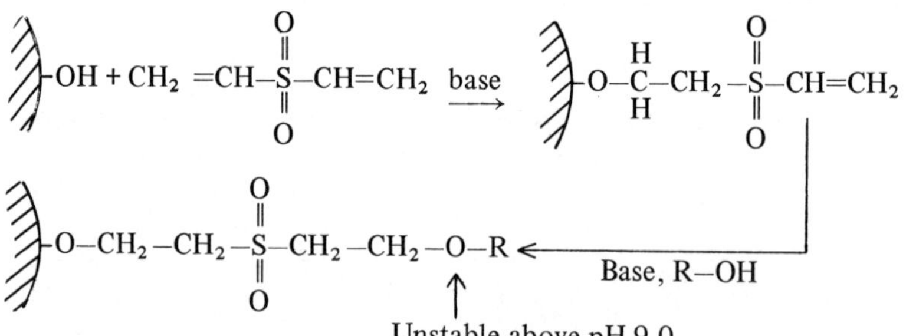

Figure 3.6. Immobilization of ligands to agarose by divinylsulfone activation

3.1.7. Other Methods for Activation of Agarose

It has been known for a long time that proteins are highly reactive toward plant polyphenols. Indeed, the removal of polyphenols is an essential step in the purification of most plant proteins, and many enzymes are irreversibly inactivated in their presence. The basic reaction is the addition of any nucleophilic residue on the enzyme by way of a Michael-type reaction to the quinone form of the polyphenol. (See Figure 3.7.)

Brandt et al. (35) have applied this reaction to the immobilization of nucleophilic ligands to agarose. The method is particularly valuable in the attachment of proteins to agarose or to similar materials, in that it can be applied over a wide range of pH values. Moreover, it has found application in other protein-protein coupling procedures including coupling enzymes to antibodies to form enzyme–antibody complexes that are employed in enzyme-linked immunnoassay (36,37).

The procedure is a two step process, the first involving the activation of the agarose matrix by reaction with p-benzoquinone followed by coupling the activated matrix with the desired protein or nucleophilic ligand. To activate the agarose, agarose is mixed with an equal volume of a pH-8 buffer to which is added p-benzoquinone dissolved in ethanol. The final concentrations should be 20% in ethanol and 50 mM in benzoquinone. Activation proceeds rapidly, and although usually 1 hour of incubation is employed, the reaction is over 90% complete within 20 minutes at room temperature.

After activation, the gel is washed thoroughly by suction filtration with several volumes each of 20% ethanol, water, 1-M NaCl, and again water. The washed gel is suspended with an equal volume of an appropriate buffer, usually pH 7 to 9, although adequate coupling can be obtained at pH 3 to 10; thus the choice of buffer is dependent chiefly on the stability and kind of ligand to be immobilized. The activated gel–ligand mixture is incubated with stirring at 4°C for 24 hours. (Mixing in slowly rotating test tubes is preferred over the use of magnetic stirrers.) After incubation the ligand–agarose matrix is collected by way of

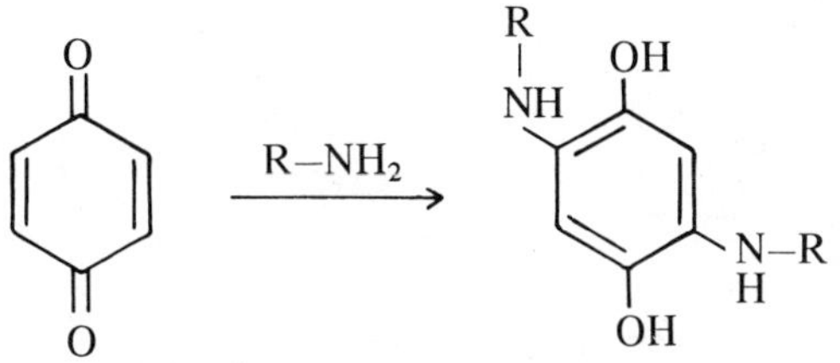

Figure 3.7. Crosslinking of proteins with a quinone (p-benzoquinone).

suction filtration and washed thoroughly with the coupling buffer. (See Figures 3.8 and 3.9.)

Triazines, such as 2 amino-4,6 dichlors-s-triazine, have been used to couple ligands to agarose (38–42), although this reagent has been used more widely with cellulose than with agarose. Gribnau (43,44) has conducted a thorough investigation of the use of a wide variety of substituted fluorotriazines and fluoropyrimidines for activation of agarose. Chromophores containing triazines have been widely employed as reactive dyes that import "fast" colors to various cellulosic fabrics, and for that reason their chemistry has been well studied. The most successful activating reagent employed by Gribnau was 2,4,6-trifluoro-5-chloropyrimidine (termed "FCP"). This compound is widely employed to synthesize reactive dyes for wool and cellulosic fibers and can be purchased cheaply and in large quantities. Coupling seems to be dependent on the nucleophilicity of the ligand, and protein residues appear to follow the following order of decreasing reactivity: cysteine thiol, N-terminal α amino, histidine imidazol, lysyl amine, serine hydroxyl, tyrosyl hydroxyl, arginine guanidino, and threonine hydroxyl. The rate of reaction is pH dependent, and the tryptophan

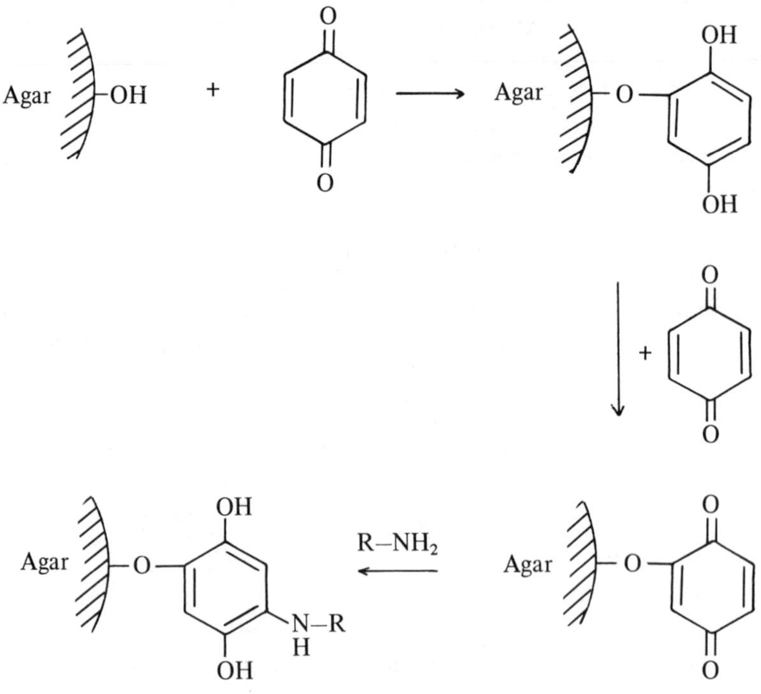

Figure 3.8. Putative mechanism of activation using quinones.

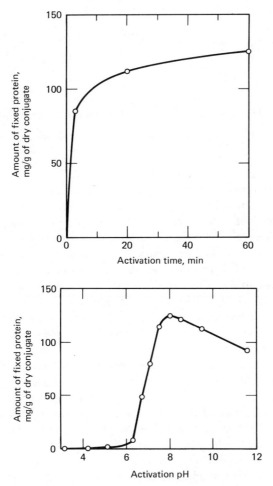

Figure 3.9. (*a*) Amount of bovine serum albumin fused to Sepharose 4B gel as a function of the activation time by use of *p*-benzoquinone activation. (*b*) Amount of bovine serum albumin fused to Sepharose 4B gel as a function of pH during the activation reaction with *p*-benzoquinone activation. Reprinted with permission from Brandt et al. (35).

indole reacts only at very high pH. Cystine and methionine are totally un-reactive. (See Figure 3.10.)

To activate agarose with FCP, 25 g of agarose (wet weight) was suspended in 50 ml of 3-*M* NaOH and rotated slowly for 15 minutes at room temperature. The alkali was removed by suction filtration, and the beads were resuspended (without washing) in 40 ml of a 1:1 mixture of xylene:dioxane. The suspension was agitated in the apparatus shown in Figure 3.10 until the temperature

2-Amino-4.6-dichloro-s-triazine   2,4,6 Trifluoro-5-chloropyrimidine (FCP)

Figure 3.10. Heterocyclic halide activating reagents used with agarose.

reached 0°C, after which 3 ml of FCP in 4 ml of xylene–dioxane was added dropwise over a 1-hour period, taking care that the temperature remained below 0.5°C. After 1 hour, 4.2 ml of glacial acetic acid was added and the beads were collected and washed successively with 200 ml each of xylene and dioxane and 1 l of water. Some cross-linking of the agarose occurs during this procedure, as evidenced by the fact that the material did not melt or dissolve in water at temperatures over 40°C. The activated agarose was stored at 4°C.

To couple ligands to FCP–agarose, several approaches can be employed. As one example, 25 ml of Kallikrein–trypsin inhibitor in 30 ml of 0.1-M sodium phosphate, pH 7.0, was mixed with 2.5 g of FCP–agarose and the mixture stirred for 20 hours at room teperature. The trypsin inhibitor–agarose was then collected by filtration and washed with the same buffer, with and without 0.5% NaCl. Low-molecular-weight inhibitors that are alkali stable are more readily immobilized at pH 10.6. For example, p-aminobenzamidine was coupled through a glycyl glycine spacer to prepare an adsorbent for trypsin and chymotrypsin. First, 11 g (wet weight) of FCP–agarose was reacted with 740 mg of glycyl glycine in 0.5-M sodium carbonate, pH 10.6, with shaking for 27 hours at room temperature. The product was collected by filtration, washed with 250 ml of water, and suspended in 5 ml of water containing 1.25 g of p-aminobenzamidine. The pH was adjusted to pH 5.0 and 1.15 g of water-soluble carbodiimide [1-(3-dimethylaminopropyl)-3-ethylcarbodiimide] was added. The mixture was held for 2 hours at room temperature, after which the beads were collected and washed sequentially with water, with 100 ml of 0.5-M NaCl, and again with 250 ml of water. the resulting matrix (108 μmoles ligand/g dry matrix) had a capacity of 10 mg of trypsin/g gel (wet weight). (See Figure 3.11.)

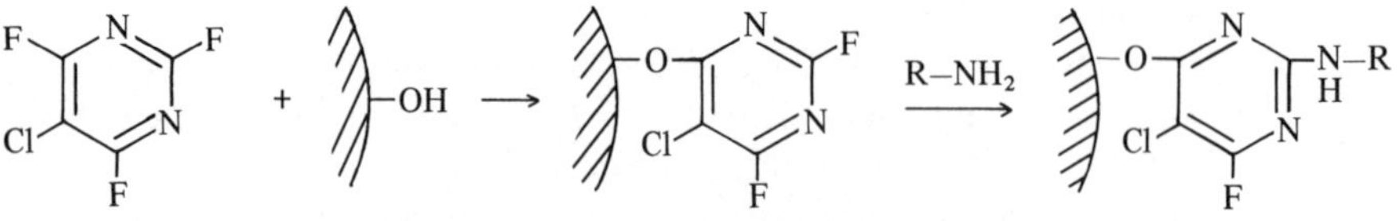

Figure 3.11. Immobilization of nucleophilic ligands by activation with FCP.

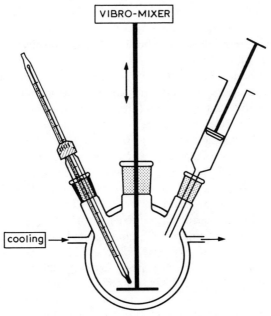

Figure 3.12. Experimental setup for the FCP activation of polysaccharides. Reprinted with permission from Gribnau (43).

Gribnau describes (43) a wide variety of hormones, enzymes, and bioselective adsorbents immobilized by use of this and other halogen-containing triazines and pyrimidines, each of which was at least as useful as those prepared via the CNBr technique and in two respects much superior, namely: (1) the absence of iogenic groups in the product and (2) a much higher stability toward nucleophiles, denaturants, and elevated temperature. (See Figure 3.12.)

3.1.8. Alkylamine Derivatives of Porous Glass and Fused Silica (45–48)

Proper preparation of the glass bead surface is one of the most crucial aspects in the application of controlled pore glass. The beads must be scrupulously clean, free from fragile projections, and possess appropriate pore diameter. To prepare the beads for derivatization, 10 g of beads are sonicated in 50 ml of 1-N HCl for 15 to 30 minutes at the highest sonifier output; thus the beads are cleaned and fragile projections and irregularities are removed from the surface. The beads are then added to 200 ml of 1-N HCl in a 250-ml graduated cylinder and allowed to settle. The settling process is repeated until the supernatant is clear.

These sonicated beads are then placed in a two-necked 500 ml flask fitted with a small overhead stirrer in the center hole and a reflux condenser in the side neck. The stirrer is positioned well above the beads. (*Never use a magnetic stirrer with glass beads*—it will grind the beads to a fine powder.) Then 100 ml of 1-N HNO$_3$ is added and the beads are refluxed for 30 minutes. The stirring rate is adjusted to the minimum needed to prevent excessive bumping of the liquid.

The beads are collected by suction filtration and dried at 90 to 110°C for 1 to 3 hours. The beads (10 g) are then refluxed for 16 hours in 10% γ-aminopropyltriethoxysilane in dry toluene (45), with overhead stirring. The beads are filtered on a Buchner funnel and washed with 100 ml of 95% alcohol. An alternative procedure is to reflux the beads in an aqueous medium (10% γ-aminopropyltriethoxysilane at pH 3.45) in place of dry toluene (46). All aspects of the process are otherwise identical.

After silanization the beads are refluxed for 15 minutes in 95% alcohol to complete the removal of unreacted γ-aminopropyltriethoxysilane. After reflux they are washed with 500 ml of 95% ethanol, followed by 500 ml of distilled water. To test for the presence of primary amine functions (either on the glass beads or in the filtrate), a few beads are placed in 0.5% ninhydrin in butanol and heated gently. If primary amines are present on the glass, a deep blue color develops on the beads while the solution remains yellow for at least 1 hour. A similar test using a few drops of the wash filtrate should be colorless. If a blue color develops in the filtrate or if the ninhydrin solution around the beads turn blue in less than 30 minutes, the beads are again refluxed in 95% ethanol.

3.1.9. Glycidoxy–Glass Beads (49–50)

Hydrophilic glycidoxy coated glass beads can be prepared from clean controlled pore glass beads by reflux in γ-glycidoxypropyltrimethoxysilane (Dow Corning–Z-6040 Silane). The resulting glycidoxy coated beads are then cross-linked by using 1,4-butandiol and boron trifluoride gas. Porous glass beads coated in this fashion are available commercially (Pierce Chemical Company, Rockford, Ill.), and the process has been patented (49). The glycidoxy glass beads prepared by this process bear vicinal hydroxyl groups that are readily derivatized via cyanogen bromide as previously described for agarose.

An alternative method to achieve a stable hydrophilic coating on glass is accomplished by refluxing porous glass beads in a solution of γ-glycidoxypropyltrimethoxysilane containing *p*-nitrophenoxide anion. This process is followed by reduction of the nitro group to a reactive arylamine (51,52). Further cross-linking can be accomplished by the addition of diols and/or bisoxiranes. The use of phlorglucinol plus γ-glycidoxypropyltrimethoxysilane also results in a highly cross-linked surface with high capacity for further derivatization.

$$
\text{Si–OH} \quad \rightarrow \quad
\begin{array}{l}
\text{–Si–O–Si–CH}_2\text{CH}_2\text{CH}_2\text{–O–CH–CH}_2\text{–CH}_2\text{–O–}\langle\bigcirc\rangle\text{–NH}_2 \\
\qquad\qquad\quad\;\; \text{O} \qquad\qquad\quad\; \text{O} \\
\text{–Si–O–Si–CH}_2\text{CH}_2\text{CH}_2\text{–O–CH–CH}_2\text{–CH}_2 \\
\qquad\;\;\diagup\text{O} \qquad\qquad\qquad\qquad\qquad \text{O–}\langle\bigcirc\rangle\text{–NH}_2
\end{array}
$$

Reagents

Glycidoxypropyltrimethoxysilane, 5% in toluene
p-Nitrophenol, 1% in ethanol
Sodium p-nitrophenoxide, 1% in ethanol
Sodium dithionite, 1% aqueous solution
Ethanol

One gram of controlled pore glass or fused silica beads are refluxed in 10 ml of 5% γ-glycidoxypropyltrimethoxysilane in toluene containing p-nitrophenol and sodium p-nitrophenoxide (10 ml of a 1% ethanol solution of each) for 16 hours (52). The beads are collected by filtration and washed with distilled water (60 ml). They are then refluxed for 1 hour in 1% aqueous sodium dithionite, followed by washing with distilled water (60 ml) and drying at 110°C overnight.

The aryl amine beads prepared in this fashion can be coupled to an appropriate ligand via diazotization. They may also be converted to a carboxylic acid-containing derivative by refluxing them with succinic anhydride or succinyl dichloride.

A third method of preparing a hydrophilic glass with a highly cross-linked organic surface involves the silanization of the beads with γ-mercaptopropyltrimethoxysilane (53). The resulting mercapto glass beads can be further derivatized by refluxing them in an alkaline solution of epibromohydrin or a bisoxirane. The oxirane glass that results may be coupled directly to a nucleophilic ligand.

$$
\begin{array}{llll}
\text{Si–OH} & & \text{OCH}_3 & \\
\text{Si–OH} & \text{H}_3\text{CO–Si–CH}_2\text{CH}_2\text{CH}_2\text{SH} & & \text{Si–O–Si–CH}_2\text{CH}_2\text{CH}_2\text{SH} \\
\quad\text{O} & \qquad\quad\text{OCH}_3 & & \quad\text{O} \quad\;\; \text{O} \\
\text{Si–OH} & \xrightarrow{\hspace{2cm}} & & \text{Si–O–Si–CH}_2\text{CH}_2\text{CH}_2\text{SH} \\
\quad 1 & \qquad\quad 2 & & \qquad\quad\;\; \text{O} \;\; 3
\end{array}
$$

3 plus $CH_2-CH-CH_2O-(CH_2)_4-O-CH_2-CH-CH$

$$\downarrow$$

$$Si-O-Si-(CH_2)_3-S-CH_2-CHCH_2O-(CH_2)_4-O-CH_2-CH-CH_2$$

Other methods for coating porous glass or fused silica matrices can be easily devised by using the commercially available alkoxysilanes listed in Table 2.2.

3.1.10. Periodate-Activated Cellulose (54-55)

Because cellulose is a naturally occurring polymer, it suffers, like agarose, from the presence of small but troublesome amounts of ionic impurities. Several methods have been developed for washing cellulose and for reducing carboxylate groups (the chief ionic constituent of cellulose) with sodium borohydride to remove the ionic groups. Alberts and Herrick (56) employ a particularly high purity cellulose, Munktell 410 (Bio-Rad, Richmond, Calif.) which is washed several times with boiling ethanol, and then treated briefly with successive washes of $0.1\text{-}M$ NaOH, $1\text{-}M$ EDTA, and $0.01\text{-}M$ HCl. Excess HCl is removed by washing the cellulose with distilled water until the pH of the filtrate is approximately 7. Other authors (57) have reduced the purified cellulose by treating it with 2 g/l of $NaBH_4$ in $0.5\text{-}M$ NaOH at 60°C for 1 hour. Reduction with $NaBH_4$ was regarded as unnecessary for the preparation of DNA cellulose (56), but might be advisable when the ligand to be employed is not anionic. (When DNA is the ligand, a few extra anionic functions are not likely to be troublesome).

Periodate cleavage of cellulose, like the same treatment for agarose, allows the convenient coupling of spacer arms containing amine or hydrazide functions. The hydrazide, although perhaps less stable than the amine, provides a neutral spacer molecule of exceptional versatility.

Reagents

Purified cellulose, (fibrous, microcrystalline, or beaded)
Sodium metaperiodate; freshly prepared $0.5\text{-}M$ aqueous solution
Adipoyl hydrazide (or similar dihydrazide)

Sodium borohydride
0.01-M Acetic acid

In this procedure 10 g of cellulose are shaken in 1 l of 0.5-M sodium periodate at room temperature for 2 hours; then 30 ml of ethylene glycol are added and shaking is continued for an additional hour. The cellulose is then collected by filtration and washed with 1 l of water. The washed gel is mixed with an equal volume of water. To this slurry approximately 30 moles of dihydrazide, dissolved in a minimum of water, are added and the pH of the resulting mixture adjusted to pH 5 with HCl. After 1 hour of further agitation the pH is readjusted to pH 9.0 with sodium carbonate. Then 1 g of sodium borohydride is added with stirring and the slurry is allowed to react by shaking the mixture in an *open* container. (*Caution:* hydrogen gas is evolved.) The cellulose is collected by filtration and washed with 200 ml of 0.1-M acetic acid, followed by washing with distilled water until the filtrate is neutral.

3.1.11. Bromoacetyl–Cellulose (58–60)

Of the many affinity chromatographic purifications performed using cellulose as the matrix, the vast majority have utilized a macromolecule (e.g., DNA, protease inhibitor, or antibody) as the ligand. When macromolecular bioligands are to be immobilized, bromoacetyl cellulose works easily and effectively. Moreover, the synthesis of bromoacetyl cellulose is not complicated, but unfortunately it does yield derivatives with anionic charges.

Reagents

Purified cellulose, prewashed with acetone and anhydrous dioxane and stored desiccated over P_2O_5
Bromoacetic acid
Bromoacetyl bromide
0.1-M Sodium bicarbonate

In this procedure 10 g of dry cellulose are added to 100 g of bromoacetic acid, which is dissolved in 30 ml of dioxane. The resulting slurry is stirred for 20 hours at room temperaturel. Bromoacetyl bromide (75 ml) is added and stirring is continued for an additional 6 to 8 hours. The cellulose is then poured into a large volume (5 to 6 l) of ice water, stirred quickly, and collected by suction filtration. The collected cellulose is washed with 0.1-M NaHCO$_3$ until the filtrate is alkaline followed by additional washing with several volumes of water. The resulting bromoacetyl cellulose can be further derivatized by reacting it with sulfhydryl groups at pH 6.0 or with amine functions at pH 8.0. To attach proteins, the optimal pH for the attachment process must be experimentally ascertained since a complex array of complicating factors, such as protein conformation, denaturation, and inactivation are also pH dependent.

3.1.12. Derivatives of Polyacrylamide

Derivatives of polyacrylamide can be prepared by including a monomer containing a reactive function (e.g., a primary amine or oxirane group) in the polymerization reaction. Alternatively, polyacrylamide-based bioselective adsorbents can be prepared by derivatization of normal polyacrylamide polymers. A wide variety of polyacrylamide derivatives prepared by the first method are available commercially (Koch-Light, Ltd. and Aldrich Chemical Company), thus rendering synthesis in the laboratory unnecessary. Derivatives made from plain commercial polylacrylamide and described by Inman and Dintzis (61,62) are available in spherical beads. Despite the limitations to the uses of polyacrylamide, which are chiefly due to its limited porosity, the biochemist who wishes to employ polyacrylamide beads in preparing a bioselective adsorbent should be aware of the simplicity of its derivatization.

The most versatile derivative of polyacrylamide is polyacrylhydrazide, formed by heating of polyacrylamide gel in hydrazine. The resulting hydrazide is easily converted to an activated carboxyl function immediately prior to coupling to the desired ligand.

$$\underset{\overset{\parallel}{C}-NH_2}{O} \quad \xrightarrow[90^\circ C]{NH_2NH_2} \quad \underset{\overset{\parallel}{C}-NHNH_2 + NH_3 \uparrow}{O} \quad \left(\begin{array}{l} \text{To other derivatives} \\ \text{by various} \\ \text{coupling procedures} \end{array} \right)$$

Reagents

Hydrazine, 6 M
Polyacrylamide beads (Bio Gel P-3000)
Plastic labware or silanized glassware
0.2-M Sodium chloride
3% Aqueous (w/v) 2,4,5-trinitrobenzenesulfonic acid (picryl sulfonic acid)
Saturated aqueous sodium borate solution

Hydrazidoacrylamide is prepared by adding the dry acrylamide beads to a 6-M aqueous solution of hydrazine made by adding hydrazine hydrate (20.4 M) or 97% hydrazine (30.4 M) very slowly to ice-cold water. (*Caution:* The heat of hydration is very large; therefore, hydrazine must be added *to* water!)

The hydrazine–acrylamide solution is held at a constant temperature until the reaction is stopped, after the desired incubation period, by collecting the beads by suction filtration and washing them with 0.2-M NaCl until the filtrate no longer contains hydrazine as determined using the 2,4,5 trinitrobenzoate test described previously. To the length of time needed, you can use the relationship $D = Dcth$, where D is the density (in millimoles per gram) of the resulting hydrazide, c is the hydrazine (millimolar) concentration, t is the reaction time (hours), and h is a temperature coefficient (at 250°C; $h = 0.013$; 50°C $h = 0.123$). A typical hydrazidoacrylamide preparation that yields an approximate hydrazide concentration of 1 mmole/g of acrylamide will require 15 hours at 25°C using 6-M hydrazine. Under these circumstances potential side reactions are minimal. For this reason these are good conditions to be used in initial preparations of hydrazidoacylamide, although higher hydrazido densities can be achieved by using elevated temperatures.

Incorporation of bioligands or chemically reactive groups in monomers used for forming synthetic polymers is a very promising technique for preparing bioselective adsorbents. Horejsi and Kocourek (63–65) have copolymerized allyl glycosides with acrylamides to prepare an affinity chromatography matrix for the purification of lectins. Schnaar and Lee (66) have prepared an activated polyacrylamide using N-hydroxysuccinimide and N-hydroxyphthalamide acrylate esters as monomers. The resulting activated polyacrylamide was reacted with 6-aminohexyl-2-acetamido-2-deoxy-β-D-glucopyranoside to form an affinity chromatography matrix for the purification of wheat germ lectin. Other activated acryl compounds have been prepared by Mosbach et al. (67–70), Epton et al. (71–76), and Mannecke et al. (77–82). These have been used chiefly in the immobilization of enzymes, and a few such activated acrylates are commercially available (Enzacryl, Koch-Light and Affi-Gel, Bio-Rad, Richmond, Calif.). Most of these preactivated acrylamide (and polystyrene) materials could be potentially employed for the immobilization of any nucleophilic bioligand, and the product could be employed in bioselective adsorption. The advantage is the inherently high ligand loadings that can be obtained. Furthermore, the investigator can control such factors as porosity and mechanical strength of the beads by synthesizing the carrier in the laboratory. Several "representative" examples of the synthesis of ligand-containing or preactivated matrices are given in the following paragraphs.

Glycosides Immobilized in Polyacrylamide by Way of Glycoside Precursors

To prepare o-glycoside polyacrylamide, 10 g of methylene bisarylamide (other birsacrylamides could be substituted, such as cleavable bisacrylamides) are dissolved, together with 4 g of allyl glycoside. Ammonium persulfate (0.1 g) is added and the mixture is heated on a boiling water bath. Within about 5 minutes the solution polymerizes, and the resulting gel is allowed to cool. After cooling, the gel is homogenized in a glass homogenizer and the fine particles are removed from the suspension by repeated suspension and decantation of the beads in water. The de-fined gel is then washed with water in a chromatography column until the eluate yields a negative Molesch test and can thus be considered free of nonpolymerized allyl glycoside (63–66).

These gel particles can be used directly in affinity chromatography, or they can be dried by first replacing the gel fluid with ethanol followed by final drying in a stream of warm air. Excellent purification of several lectins have been reported with the use of bioselective adsorbents made by this method. It would also seem plausible that beaded glycoside–acrylamide beads could be prepared by adapting this procedure to the two-phase polycarylamide bead forming method of Mosbach et al. (67).

Formation of Activated Forms of Acrylate Monomers and of Acrylate–Ligand Monomers (66)

Activated acrylic acid derivatives are formed by the same simple chemical procedures used to activate any carboxylic acid. The formation of N-hydroxysuccinimidyl acrylate is only one of many such procedures. Among the many possible are p-nitrophenyl esters (83), monacylated dihydrazides, for example, the monoacrylate of tartaric dihydrazide (71), thiolactones (71), acrylyl hydrazide (71), and the amide of fluoroaniline or 2,4-dinitrofluoroaniline (76).

To prepare N-hydroxylsuccinimidyl acrylate, 4.64 g (40 mM) of N-hydroxysuccinate was refluxed in 18 ml (220 mM) of acrylol chloride for 3 hours in an anhydrous system. The product was evaporated to yield a thick syrup to which 100 ml of cold water was added to hydrolyze excess acid chloride. The mixture was stirred for 30 minutes at $4°C$ and then extracted with 100 ml of chloroform. The chloroform layer was separated and washed successively with 100-ml portions of water until the pH of the wash was about pH 5.0. The aqueous washes were combined and extracted once with 50 ml of chloroform. The chloroform layer was added to the previous chloroform layer, and the combined extract was dried with anhydrous sodium sulfate. The dried chloroform solution was again evaporated to a syrup and crystals obtained by storing the syrup at $-20°C$ overnight. The crystals were triturated with ether and collected by suction filtration. The procedure yielded 2.3 g (34%) of N-hydroxysuccinimidyl acrylate, mp 60.5 to $62.0°C$. This product, or N-hydroxypthalimidyl acrylate, could also be prepared from acrylic acid using dicyclohexyl carbodiimide as the condensing agent.

N-Hydroxysuccinimyl acrylate (NHS-acrylate) was used as a commonomer with acrylamide and methylene bisacrylamide to form a gel that was subsequently homogenized and washed as described previously for allyl–glycoside polymers. N-Hydroxysuccinimyl-acrylate (1.15 g in 20 ml of ethanol) was added to 37.5 ml of sodium phosphate buffer, pH 6.0, containing 15 g of acrylamide and 0.75

g of methylenebisacrylamide. After mixing, 31.5 ml of water, 1.0 ml of 4% tetraethylmethylenediamine, and 10 ml of 5.6% ammonium persulfate were added with vigorous stirring. The mixture was quickly poured into a 100-ml graduated cylinder and allowed to polymerize (20 to 30 minutes). After polymerization, the gel was cut into 1- to 2-cm pieces and homogenized in 4 volumes of cold water in a Virtis homogenizer (medium speed, 15 minutes). The particles were collected by filtration and dehydrated by stirring in 10 volumes of dry dimethylformamide for 2 hours at 4°C; The particles were again collected and the dehydration procedure repeated. The final particles were stored in fresh, *dry* dimethylforamide at 4°C. Hydrolysis of the active ester was not seen after storage under these conditions for up to 6 months.

Formation of Polyacrylates in Bead Form

Whenever gels are formed by polymerization in beads, followed by mechanically forming gel particles, the product is invariably heterogeneous in size, shape, and mechanical durability. Chromatography through columns of such particles is slow relative to chromatography through columns of similarly sized spherical beads. Many matrices are commercially available in bead form, but special polymers, such as those described previously, which to date have usually been prepared as gel particles, could readily be cast in bead form as described by Mosbach et al. (67).

To do so the polymerization is carried out in a well-stirred two-phase system consisting of a monomer containing (1) an aqueous phase, (2) a hydrophobic phase of the same density, and (3) a small percentage of detergent. A nitrogen atmosphere is usually employed to prevent oxygen inhibition of the polymerization in such a rapidly stirred system. Either apparatus shown in Figure 3.13 would be adequate for this purpose.

For a 1 1 polymerization vessel, 400 ml of a mixture of toluene and chloroform (290/110, v/v) is added and the mixture cooled to 1 to 5°C. Detergent (1 to 3 ml of Sorbitan sesquinoleate or a similar detergent) is added, and stirring under nitrogen is begun (ca. 260 rpm). At the same time an aqueous monomer phase, consisting of 58.5 ml of Tris–HCl (0.1 M, pH 7.0) containing 5.7 g of the desired preactivated monomer and 0.3 ml of the cross-linking acrylamide, is prepared. (The latter is usually N,N'-methylenebisacrylamide, but any one of several cleavable cross-linking agents could be used if an investigator desired a matrix that could be dissolved under defined conditions, e.g., periodate treatment or alkaline pH). To the monomer misture 0.5 ml of TEMED (N,N,N',N'-tetramethylenediamine) and 1.0 ml of ammonium persulfate (600 mg/ml) are added with stirring, and when the mixture begins to polymerize, as shown by an increased viscosity, it is added to the stirred hydrophobic phase. The stirring is continued until the beads are polymerized (usually 30 minutes), after which the beads are collected and washed successively with toluene, water, and then

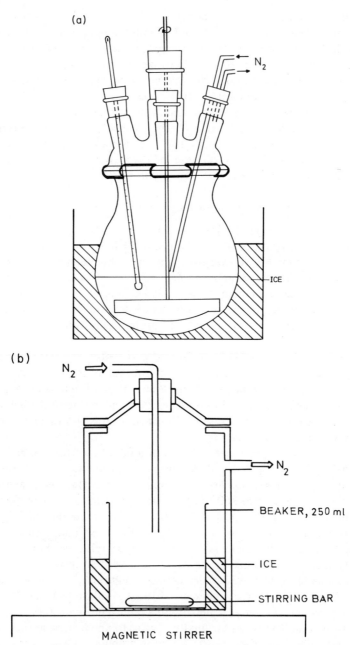

Figure 3.13. Apparatus used for the preparation of bead-formed acrylic copolymers. Reprinted with permission from Mosbach et al. (67).

buffer. The beads may be freeze dried or stored as a suspension, depending on the stability of the monomer.

This procedure can be varied in any type of monomer or cross-linker, in bead size (by changing the stirring rate and/or the detergent used), and in porosity (by changing the amount of monomer used). If a hydrophobic monomer is used, alcohol must be added to the monomer mix (up to 50%), and, if necessary, the monomer may be first dissolved in 100% methanol and then added to at least an equal volume of buffer. The alcohol serves to keep the monomer in the aqueous phase.

3.2. ATTACHMENT OF LIGANDS TO SPACER ARMS AND TO GEL MATRICES

The attachment of bioligands to spacer arms (which are themselves usually preattached to an inert matrix) involves a large variety of simple organic reactions. No one attachment method that is generally applicable to all, or even most, ligand–spacer arm systems has ever been described. Nonetheless, most ligand attachments are categorized by less than a dozen basic reactions. The methods detailed in this section are those that are most likely to be used to attach ligands bearing a reactive functional group, such as a carboxyl, hydroxyl, or amino, to a spacer arm and/or a gel matrix. The usefulness and specificity of any bound ligand must be demonstrated empirically since the bound ligand must retain its original affinity for the biomolecule that one wishes to purify. Of the several reactive sites that may exist on the bioligand, one or more may be known to be uninvolved (or at least not essential) in ligand binding. If this is the case, the investigator should initially concentrate on utilizing these nonessential reactive functions as sites for attaching the ligand to the matrix and/or spacer arms. By doing so, those sites involved in the ligand–enzyme interaction are left undisturbed. In any case, it must be established that the ligand matrix affinity for the enzyme is specific and that the binding is at the active site, allosteric center, and so on before the purification process can be legitimately termed "bioselective adsorption" or "affinity chromatography."

Selected examples of specific methodology for attaching bioligands to spacer arms and/or matrices are given in detail in the following paragraphs. Many procedural details involving the type of solvent, pH, and so on may be adjusted within reasonable limits to compensate for any unique requirements of the ligand–matrix system that is to be synthesized.

3.2.1. Attachment by Amide Formation

Basically, amide formation involves two phases of development: (1) formation of an activated or high-energy form of the carboxylic acid and (2) addition of the activated carboxylic function to the amine (84–88). Potentially, the reverse of this procedure could be utilized; that is, the amine could be activated and then reacted with the carboxylate (89). In practice, however, the activation of the amine is far more difficult and is rarely employed.

Succinylation of Aminoalkyl Agarose (90)

Sometimes it is necessary to attach a ligand with a free amine function to a spacer arm matrix that also contains a free amine group. In such cases the first step in the coupling procedure is to convert the spacer arm amine function into a carboxyl group by reacting the amine with an activated form of a dicarboxylic acid. Aqueous buffered solutions of succinic anhydride are commonly employed for this purpose when agarose is used as the matrix (90).

Inorganic matrices are more readily converted to the carboxylate by using dicarboxylic acid dichlorides, for example, succinyl chloride.

Succinylation of amino groups by way of succinic anhydride in aqueous systems does not always go to completion. To test whether the reaction is completed, a few reacted beads should be added to 0.1% ninhydrin in ethanol and heated to 70°C for 30 minutes. A light blue color implies the existence of residual primary amine functions.

Reagents

Succinic anhydride
Aminoalkyl agarose
5-M Sodium hydroxide
1-M Sodium phosphate, pH 6.0

Aminoalkyl agarose is suspended in an equal volume of water, and succinic anhydride is added at least 10 fold excess over the amino content of the agarose derivative. The pH is maintained at pH 6.0 by the addition of the 5-M NaOH solution. The pH is monitored constantly until it remains stable for at least 30 minutes without further addition of NaOH. Ice is added periodically to keep the temperature between 25 and 30°C. The mixture is then incubated for 5 hours at 4°C, followed by filtration and washing with distilled water. The beads are then tested with ninhydrin to determine whether the reaction was complete. If a light blue–purple color develops, the succinylation procedure is repeated by using a larger excess of succinic anhydride and the reacted suspension is held for 5 hours at room temperature. Titration with 5-M NaOH may be eliminated by suspending the aminoalkyl agarose in 1.0-M sodium phosphate, pH 6.0, in place of water. In this case the agarose–buffer–succinic anhydride mixture is kept at room temperature for 12 hours and thereafter treated the same as beads prepared by the NaOH titration procedure given previously.

It has also been recommended (91) that the final succinylated agarose be incubated with 0.1-N NaOH for 30 minutes at 24°C to remove any "labile carboxyl groups" that may have been formed. The labile carboxyl groups may lead to slow release of the substituted ligand during the bioselective adsorption process. However, the beads should be assayed again with ninhydrin after the NaOH treatment since some primary amines could have become exposed during the alkali treatment, thereby imparting some cationic nonspecific adsorbent properties to the final bioselective adsorbent.

3.2.2. Succinylation of Amino Alkyl Porous Glass and Other Amine Inorganic Matrices (45)

Succinyl chloride can impart a much more complete and rapid derivatization of amine functions on glass surfaces than can be effected with succinic anhydride. This procedure can also be utilized to succinylate other matrices that are compatible with anhydrous solvents and reflux temperatures. This procedure also could be applied to agarose if a suspension of aminoalkyl agarose in anhydrous dioxane were used.

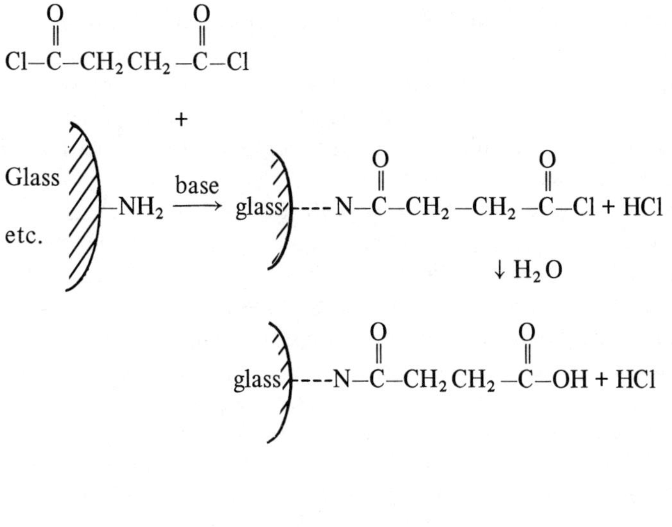

$$\underset{\|}{\overset{O}{Cl-C}}-CH_2CH_2-\underset{\|}{\overset{O}{C}}-Cl$$

+

Glass etc. $\rangle\!-NH_2 \xrightarrow{\text{base}} glass\rangle---N-\underset{\|}{\overset{O}{C}}-CH_2-CH_2-\underset{\|}{\overset{O}{C}}-Cl + HCl$

$\downarrow H_2O$

$glass\rangle---N-\underset{\|}{\overset{O}{C}}-CH_2CH_2-\underset{\|}{\overset{O}{C}}-OH + HCl$

Reagents

10% Succinyl chloride in dry chloroform
10% Triethyl amine in dry chloroform
Aminoalkyl porous glass

The aminoalkyl porous glass is added to a two-necked flask, fitted with an overhead stirrer in the center and a reflux condenser on the side arm. Next, the beads are covered with at least twice their volume of 10% succinyl chloride in dry chloroform. (The volume used should reach the stirrer blades, which are positioned far enough above the bottom of the flask as to prevent any danger of the beads being ground between the stirrer and the wall of the flask.) Then an equal volume of 10% triethylamine in dry chloroform is added slowly to the suspension with continuous stirring. The beads are refluxed with slow stirring for 30 minutes and collected by filtration on a sintered glass funnel. They are washed with chloroform, and a few are then tested with ninhydrin (see page 58), to determine whether all primary amine functions have been reacted. If some residual primary amines still remain (as shown by a light blue or purple color in the ninhydrin test), then the process just described is repeated.

The alkylation procedure must be performed very carefully to avoid any "bumping" of the solvent during the reflux. An unstirred bead–chloroform mixture will splash violently, whereas, conversely, too rapid stirring may damage the matrix.

To convert the resulting acid chloride into a matrix-bound carboxylic acid, the beads are again refluxed in water or in dilute neutral buffers. Alternatively, the acid chloride reactive group may be employed directly as the species that is to be reacted with the ligand. If this is the case, the beads should be refluxed for 15 minutes with 10% thionyl chloride, in dry chloroform followed by thorough rinsing with chloroform and then used immediately. Storage in a vacuum desiccator over phosphorous pentoxide for short periods is possible, but the complete exclusion of moisture from the derivatized beads is essential to prevent hydrolysis of the acid chloride and the consequential formation of anionic carboxylate functions.

This procedure is not limited to use with porous glass materials, but also can be employed with any aminoalkyl group attached to a matrix, organic or inorganic, that is compatible with anhydrous, nonprotic solvents. The precise reaction conditions, such as solvent and temperature, can be adjusted to the solvent and thermal stability of the matrix employed. However, if reflux conditions cannot be employed, the matrix–solvent–succinyl chloride mixture should be degassed by using a gentle vacuum to ensure that the solvent–succinyl chloride mixture reaches the interior of the porous matrix.

3.2.3. Activation of Matrix-Bound Carboxylate Groups

The first method for attaching a nucleophilic functional group of a ligand to a carboxylate or carboxylate derivative that is bound to a matrix is by the formation of a high-energy "activated" carboxyl derivative. Such derivatives provide the thermodynamic driving force for the subsequent coupling reaction. Formation of an activated carboxyl may be accomplished in the same solution in which the coupling reaction is carried out, for example, as is done with diimidate coupling (page 78), or it may be performed in two separate steps, as with acyl azide formation (page 79). In the latter case, activation of the carboxyl group is carried out first, followed by subsequent reaction with the nucleophilic ligand that is to be bound. There are four major groups of activated carbonyls that are commonly employed: acyl azides, acid chlorides, activated esters (e.g., N-hydroxysuccinimide or p-nitrophenolate esters), and carbodimide reagents.

Acid Chloride Formation (45,46)

Acid chlorides are exceedingly reactive compounds, such that even traces of moisture hydrolyzes them to HCl and the corresponding carboxylic acid. Thus it is essential that the greatest care be exercised to keep them totally dry. Moreover, acid chlorides are seldom useful with matrices that contain a hydroxyl function. Even so, their very high degree of reactivity makes them useful intermediates wherever they can be employed.

$$\text{Matrix} \quad \left.\right\} -\overset{\overset{\text{O}}{\parallel}}{\text{C}}-\text{OH} \quad \xrightarrow{\text{SOCl}_2} \quad \text{Matrix} \quad \left.\right\} -\overset{\overset{\text{O}}{\parallel}}{\text{C}}-\text{Cl} + \text{SO}_2 + \text{HCl}$$

Reagents

10% Thionyl chloride in dry chloroform
Alkylcarboxy porous glass

Carboxylate containing porous glass is refluxed with at least a 10-fold volume of 10% thionyl chloride in dry chloroform. Reflux is continued for 4 hours, followed by collection of the beads on a sintered glass funnel. They are next washed with 3 to 5 volumes of dry chloroform and stored in a vacuum desiccator over P_2O_5 (exclusion of moisture is essential).

Sulfonyl groups bound to various matrices may also be converted to acid chlorides, but (unlike carboxyl groups) dimethylformamide is added as a catalyst to the thionyl chloride. Likewise, Carboxylic acid chlorides may be prepared without refluxing if dimethylformamide is added to the thionyl chloride. The beads containing either a free sulfonyl group or a free carboxyl group are washed quickly with 1% aqueous HCl and dried at 110°C overnight. They are then cooled to room temperature and 10 volumes of 10% thionyl chloride in dry chloroform are added. Then 2 to 3 drops of dimethyl formamide (ca. 0.05 ml) is added per gram of matrix and the solution gently degassed by applying vacuum from a water pump aspirator several times for 5 to 10 minutes each.

The mixture is then allowed to sit overnight at room temperature; alternatively, to shorten the time involved, it can be refluxed for 4 hours if the matrix will permit. The beads are collected, washed, and stored as described before.

The acid chlorides are best reacted with nucleophilic ligands or matrices in nonaqueous, nonprotic solvents. Triethylamine or pyridine are added as organic bases and the acid chloride-containing material is reacted with the ligand in such a solvent overnight at room temperature. Then a large excess (100 fold) of dry ethanolamine is added to the mixture and further reacted for 6 to 12 hours at room temperature. This latter step is performed to convert all the acid chloride groups that did not react with the ligand to amides by their reaction with ethanolamine. Small amounts of moisture present in the process will convert some of the acid chloride to carboxylate functions, imparting an anionic character to the material. These carboxylate functions are not very significant if the ligand being employed is anionic, but otherwise they may produce an ion-exchange character in the affinity chromatographic material (92).

If proteins (or similar macromolecules) are to be immobilized by using this type of support material and if some anionic character in the resin is not incompatible with the experiment, aqueous solutions of the protein may be used. To do so, add 50 to 100 mg of protein in 0.5-M phosphate buffer, pH 8.0, and degass the beads for 5 to 10 minutes as described before, taking care that foaming is minimized. Check the pH occasionally, readjusting it to pH 8.0 with dilute sodium hydroxide if needed. After 1 to 2 hours of reaction, collect the beads by filtration on a sintered glass filter and rinse them with the same buffer as employed before. The final wash can be performed by utilizing the same buffer that will later be used in the affinity chromatographic procedure.

Acyl Azide Activation (93-97)

The formation of the acyl azide by way of an acryl hydrazide intermediate, is probably one of the best methods for attaching a nucleophile, especially an amine, to an inert matrix. The precursor acyl hydrazide is stable and is easily formed (in some cases it is the normal product of attaching a spacer arm to a matrix) and is readily converted to the azide in quantitative yields. The azide itself is relatively stable, as compared, for example, to acyl chlorides, and it can be stored for prolonged periods if kept from moisture. However, acyl azides of agarose, since they are normally formed in aqueous media, should be prepared immediately before use.

The initial step in acyl azide formation is the preparation of a hydrazido derivative of agarose, acrylamide, controlled pore glass, and similar substances. The hydrazido group is then converted to the azide by treating it with nitrous acid. The acyl azide reacts rapidly and selectively with amines at pH value near neutrality. First, hydrazido–agarose (97) beads are prepared by using cyanogen bromide (see page 44) or periodate oxidation (see page 46).

$$\text{Matrix} \longrightarrow \text{matrix}\!-\!\!\underset{\underset{H}{|}}{\overset{O}{\overset{\|}{C}}}\!-\!\underset{H}{\overset{}{N}}\!-\!NH \xrightarrow{HNO_2} \underset{}{\overset{O}{\overset{\|}{C}}}\!-\!N_3$$

The hydrazido–agarose is converted into acyl azide–agarose by suspending the derivatized beads in an equal volume of ice-cold 1-N HCl. To this an equal volume of ice-cold 0.5-M NaNO$_3$ is added dropwise with gentle stirring. The mixture is kept at 0°C for 15 to 20 minutes, collected by suction filtration and washed with ice-cold 0.1-N HCl. The filtrate is emptied from the suction flask to prevent foaming, and the beads are rapidly washed with ice-cold 0.1-M sulfamic acid and then with ice-cold water. The washed beads are suspended in 10 volumes of 0.2-M NaHCO$_3$, pH 9.5, containing the ligand that is to be bound. The concentration of the ligand is dependent upon its cost, its nature, and its availability. Proteins are usually used at a concentration of 5 to 20 mg/ml, whereas smaller amine containing bioligands are used at concentrations of 10 to 100 mM. If in the use of proteins, coupling through the α-amino group rather than ε-lysyl amines, is desired, the pH of the coupling solution may be lower (e.g., 0.5 potassium phosphate, pH 6.0). Acyl azide derivatives of polyacrylamide can be prepared and employed in essentially the same fashion (see page 63).

The hydrazido–acrylamide can be converted into the highly reactive acyl azide by using the same procedure as described with the use of hydrazido-agarose. Alternatively, either hydrazido–agarose or hydrazido–acrylamide may be reacted with an aldehyde or ketone at moderately low pH (ca. pH 5) to form a hydrazone that is then reduced with alkaline sodium borohydride (ca. pH 9.0) (98) or with sodium cyanoborohydride at pH 6.0 (99).

$$\text{Matrix}\!-\!\!\underset{\underset{H}{|}}{\overset{O}{\overset{\|}{C}}}\!-\!N\!-\!NH_2 \xrightarrow[pH\ 5.0]{R-\overset{O}{\overset{\|}{C}}-H} \quad \text{Matrix}\!-\!\!\underset{\underset{H}{|}}{\overset{O}{\overset{\|}{C}}}\!-\!N\!-\!N\!=\!\underset{\overset{|}{H}}{C}\!-\!R$$

$$\Big\downarrow\ NaBH_4,\ pH\ 9.0$$

$$\text{Matrix}\!-\!\!\underset{H\ \ H}{\overset{O}{\overset{\|}{C}}}\!-\!N\!-\!N\!-\!CH_2\!-\!R$$

The use of tartaric acid dihydrazide instead of a simple acyl dihydrazide offers some possible additional advantages in that the vicinol diol of the tartarate is cleavable with sodium periodate (100). Thus if the affinity for the enzyme for the immobilized bioligand is too great to permit elution easily, the entire enzyme-bioligand complex may be cleaved from the matrix by elution with buffer containing sodium metaperiodate. This could be of particular value in enzyme purification with immuno adsorbents. since antibody-enzyme complexes often require denaturing agents during the elution with subsequent loss of substantial amounts of enzyme (see page 273). In addition, immobilized protein modification reagents have been based on the periodate cleavage of tartaric acid dihydrazido-agarose (see page 320).

$$
\text{Matrix} \quad \left.\begin{array}{l}\\\\\\\end{array}\right\} -\underset{H}{N}-\underset{H}{N}-\overset{\overset{O}{\|}}{C}-CH-CH-\overset{\overset{O}{\|}}{C}-\underset{H}{N}-\underset{H}{N}-R
$$

$$
\overset{\hspace{3cm}}{\underset{OH\ OH}{}}
$$

$$\downarrow \ IO_4^-$$

$$
\text{Matrix} \quad \left.\begin{array}{l}\\\\\\\end{array}\right\} -\underset{H}{N}-\underset{H}{N}-\overset{\overset{O}{\|}}{C}-\overset{\overset{O}{\|}}{C}-H \ + \ H-\overset{\overset{O}{\|}}{C}-\overset{\overset{O}{\|}}{C}-\underset{H}{N}-\underset{H}{N}-R
$$

+ Other products

Another coupling method based on the hydrazide coupling procedure employs polymetharyl hydrazide in place of succinic dihydrazide or adipoyl dihydrazide. The procedure is otherwise identical. To form polymethacrylate hydrazide, 5 g of polymethylacrylic acid is mixed with 70 ml of 98% hydrazine and the mixture is heated with constant stirring to $100°C$ on a water bath. The reaction is held at $100°C$ for 3 hours, after which the solution is cooled and filtered through cheese cloth and the filtrate is added to 500 ml of ice-cold methanol acidified with 1 ml of glacial acetic acid. The precipitate is collected by suction filtration, washed with 500 ml of the same ice-cold methanol-glacial acetic acid mixture and dissolved in water. Any insoluble material is removed by suction filtration, and the clarified solution is used immediately or lyophilized and stored dry in a desiccator. The polymethacrylhydrazide can substitute for the acyl dihydrazide in any of the procedures described previously, but with the advantage that the resulting hydrazido-agarose will have a higher capacity because of the many hydrazide groups in the polymer. It will also be more stable since the polymeric hydrazide will be attached to the agarose at many points (23).

Activation of Immobilized Carboxylic Acid Groups
by Formation of Active Esters

Of the several activated esters that have been employed in affinity chromatography, possibly the most useful are the hydroxysuccinimide esters (101). The structure of the hydroxysuccinimide ester of acetic acid is as follows.

The N-hydroxysuccinimide esters of agarose, acrylamide, and controlled pore glass are all commercially available, easily prepared in the laboratory, and stable almost indefinitely when stored under anhydrous conditions. Yet when treated with primary or secondary amine containing compounds, they react to form amides over a pH range of 5 to 9 within 5 to 30 minutes.

Hydroxysuccinimide agarose is prepared by washing carboxylate-containing agarose [e.g., succinylated aminoalkyl agarose (102)] with anhydrous dioxane. The washing is carried out by suction filtration and should be extensive (100 or more volumes) so that the beads become anhydrous or nearly so. The agarose is then resuspended in 3 volumes of anhydrous dioxane, and solid N-hydroxysuccinimide is added with stirring to yield a solution with a final concentration of 0.1 M in N-hydroxysuccinimide. Then N,N'-dicyclohexylcarbodiimide or similar carbodiimide (see Table 3.3) is added with stirring (final concentration 0.1 M) and the mixture is stirred at room temperature for 70 minutes. The beads are collected by suction filtration, washed with 8 volumes of dioxane, and further washed with at least 3 volumes of methanol. The latter removes the precipated dicyclohexylurea that accumulates during the reaction and that must be completely removed before use or storage. Next the beads are washed with

TABLE 3.3. Use of Various Condensing Reagents in the Preparation of N-Hydroxysuccinimide Ester of Agarose.[a]

Coupling Reagent	L-Alanine Coupled, μmoles/ml of Packed Agarose
N,N'-Dicyclohexylcarbodiimide	1.5
1-Cyclohexyl-3-(2-morpholinoethyl)-carbodiimide metho-p-toluenesulfonate	4.2
N-Ethoxycarbonyl-2-ethoxyl-1,2-dihydroquinoline	3.5
1-Ethyl-3-(3-dimethylaminopropyl)-carbodiimide	4.0

[a]Succinylated agarose was incubated for 70 minutes at room temperature in dioxane with 0.1-M N-hydroxysuccinimide and 0.1 M of the indicated condensing reagent as described in detail in the text. The active agarose ester was reacted for 8 hours (4°C) in 0.1-M NaHCO$_3$ buffer, pH 8.5, containing 50 mM L-[^{3}H]alanine (8 Ci/mole). The coupled derivatives were washed with bicarbonate buffer and with sodium hydroxide as described earlier before determination of radioactive content.

Source: Reprinted with permission from Cautrecasas and Parikh (86).

3 volumes of dioxane and either used immediately or further washed with another 3 volumes of dioxane that has been dried by passing it over a neutral alumina column. For storage, the beads are suspended in 1 or 2 volumes of alumina-dried dioxane and placed in a brown or aluminum foil coated bottle, with molecular sieves (type 4A or 5A) added and capped tightly. In 4 months less than 10% of the N-hydroxysuccinimide ester is hydrolyzed if these precautions are taken.

To couple amine-containing ligands to the hydroxysuccinimide ester–agarose, the beads are collected by suction filtration, washed quickly with ice water, and added to 5 volumes of ligand in an appropriate buffer. Any pH from 5 to 9 may be employed if it is compatible with the ligand. Buffers containing a nucleophilic group (e.g., Tris, ethanolamine, glycine) cannot be employed since they would react with the activated agarose. Acetate, citrate, phosphate, borate, and bicarbonate buffers are all effective and do not interfere with the reaction. The reaction is continued for 10 minutes to 6 hours, depending on the pH, the ligand used, and other variables (see Figure 3.1) and is terminated by adding either ethanolamine or glycine to a final concentration of 1 M. These will react with any excess activated ester that might otherwise react with any amine subsequently applied to the matrix. After an additional 2 hours at room temperature, the beads are washed with the buffer that is to be used during the affinity chromatographic procedure. (See Figure 3.14.)

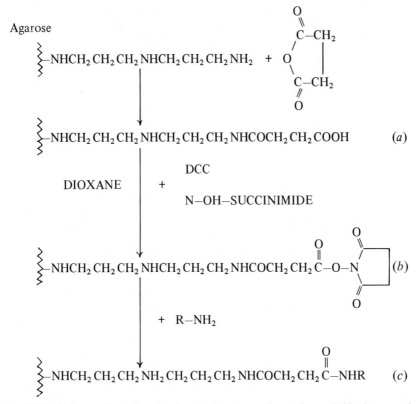

Figure 3.14. Scheme of reactions involved in the preparation and use of N-hydroxysuccinimide esters of agarose. Reproduced with permission from Parik and Cuatrecasus, *Biochemistry II*, 2291.

The same procedure can be equally well applied to carboxycellulose, carboxyacrylamide, and carboxylate derivatives of controlled pore glass.

The corollary to this reaction, namely, the activation of carboxylate-containing ligands and their subsequent addition to amine containing matrices have also proved to be a very useful procedure (103–106). For example, the hydroxysuccinimide ester of lipoic acid has been coupled to aminoalkylglass (103), aminoalkyl agarose (104), aminoalkyl acylamide (104), and other aminoalkylated matrices. Lipoic acid has also been activated by preparation of the acid chloride, followed by the reaction with aminopropyl glass (107), to form an N-propyllipoamide that has found use in affinity chromatography (103,107), as an immobilized reducing agent (108), and as a heavy-metal chelate resin (109).

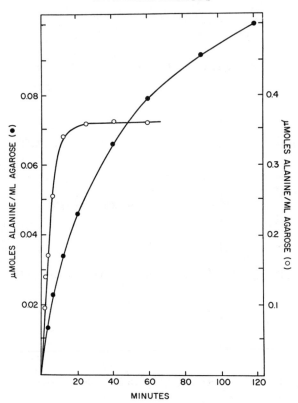

Figure 3.15. Rate of reaction of N-hydroxysuccinimide ester of agarose with 1-(^{4}H)-alanine at 4 in 0.1-M NaHCO$_2$ buffer, pH 8.6 (0), and in 0.1-M sodium acetate buffer, pH 6.3 (•). The reaction of N-α-acetyl-t-lysine in 0.1-M sodium acetate buffer, pH 6.3, yields 3 mmoles at 20 min, 10 mmoles at 60 min, and 22 mmoles at 120 minutes. Reproduced with permission from I. Parikh and P. Cuatrecasas, (1972), *Biochemistry, 11*, 2291.

Coupling Nucleophilic Ligands to Carboxylate-Containing Matrices by Using Carbodiimides

Carbodiimide-mediated coupling of amines to carboxylagarose (110–112) and carboxyacrylamide (113–114) was among the earliest of the coupling techniques employed in the synthesis of affinity chromatography supports and remains one of the most widely used of the many techniques that are presently available. It, like the hydroxysuccinimide method, can also be employed to couple carboxylate ligands to matrices that bear free primary or secondary amine functions.

Water-soluble carbodiimides (115–118) may be employed in aqueous media or dicyclohexylcarbodiimide may be used in organic solvents such as dioxane, dimethyl sulfoxide or 80% aqueous puridine (119). For example, 2 to 5 ml of carboxylate-containing material (e.g., carboxyalkylglass, carboxyalkylagarose) is added to 40 mg of either 1-cyclohexyl-3-(2 morpholinoethyl) carbodiimide or 1-ethyl-3-(3-dimethylamine propyl) carbodiimide dissolved in 10 ml of water and the pH is adjusted to 4.0. The ligand to be coupled (10 to 50 mg of protein or 1 μmole to 10 mmoles of amine-containing ligand) is added and the mixture is stirred at room temperature for at least 1 hour (some researchers employ periods as long as 20 hours, especially if the ligand concentration employed is very low). The reaction is terminated by filtering the beads and washing them with 10 volumes of water and 1 to 2 l of 40% aqueous dimethylformamide. More rapid washing can be achieved if the material is compatible with acetone, 100% dimethylformamide, or with reflux conditions. The most desirable wash procedure, for example, for glass beads is by several gentle reflux washings for 5 minutes each or by overnight extraction in a Soxhlet extracter with an appropriate solvent. This is particularly valid when carbodiimides that are not water soluble are employed.

The major difficulties in the application of the carbodiimide techniques are (1) the removal of the o-acyl urea that is formed during the reaction and (2) the presence of side products that may impart undesirable properties to the resulting bioselective adsorbent. These are formed during the course of the reaction as shown in Figure 3.16. The first step in the condensation is the addition of the

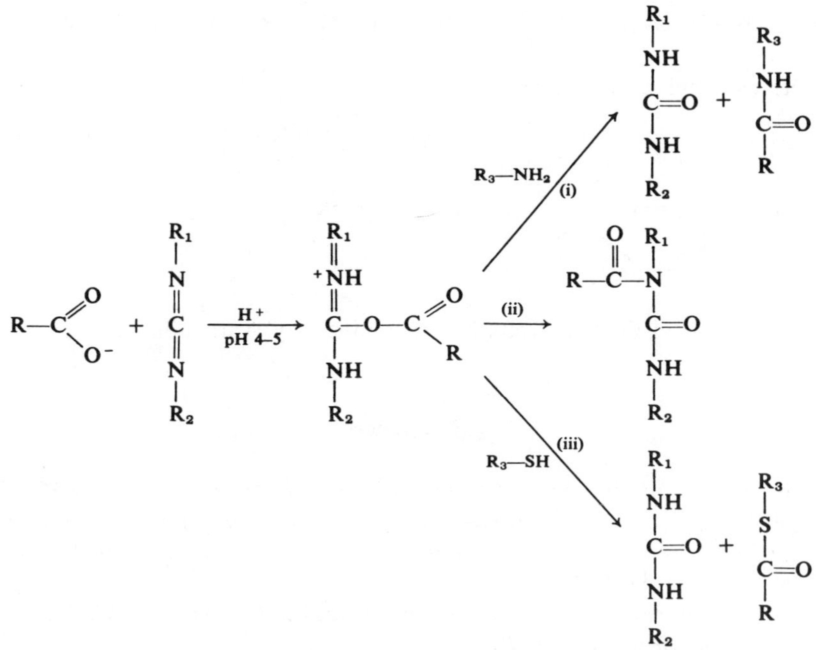

Figure 3.16. The carbodiimide-promoted reaction: (i) and (ii) mechanism, (iii) formation of thiol ester bonds. Reprinted with permission from Affinity Chromatography, by C. R. Lowe and P. D. G. Dean, Copyright © John Wiley and Sons, Ltd. (1974).

carboxylate to either C–N bond of the diimide, yielding the highly reactive and unstable o-acylurea intermediate. This, in turn, reacts chiefly with the amine by the first route shown to yield the desired amide plus the N,N'-dialkyl urea. If dicylohexyl carbodiimide is used, the resulting urea is very water insoluble, although it can be washed from the beads by using ethanol, acetone, or other organic solvents or solvent–water mixtures. If a water-soluble carbodiimide is employed, the resulting urea is much more readily removed by using aqueous solvents or water–organic solvent mixtures.

The major side reaction, the intramolecular rearrangement of the O-acyl urea for formation of a stable N-acyl urea, can be minimized if a large excess of amine-containing ligand is used. If that cannot be done, some of the resulting N-acyl urea will be bound to the matrix. If dicylohexylcarbodiimide is used the result of this side reaction would be the impartation of some hydrophobic character to the adsorbent, whereas if 1-cyclohexyl-3-(2 morpholinoethyl) carbodiimide is employed, the resulting matrix will contain a variable degree of anion-exchange ability. Thus whenever the bioligand concentration is low, other coupling methods, for example, the hydroxysuccinimide ester procedure, should be employed in preference to the carbodiimide condensation.

Dicyclohexyl carbodiimide

$$CH_3 CH_2 - N{=}C{=}N - CH_2 CH_2 CH_2 - N \begin{smallmatrix} CH_3 \\ \\ CH_3 \end{smallmatrix}$$

1-Ethyl-3-(3-dimethylaminopropyl)-carbodiimide

1-Cyclohexyl-3-(2-morpholinoe ethyl) carbodiimide

References

1. R. Jost, T. Miron, and M. Wilchek (1974), *Biochim. Biophys. Acta* **362**, 75.
2. M. Wilchek (1974), in *Immobilized Biochemicals and Affinity Chromatography, Advances in Experimental Medicine and Biology,* Vol. 42 (R. B. Dunlap, ed.), Plenum, New York, p. 15.
3. M. Wilchek and T. Miron (1974), in *Methods in Enzymology* (W. B. Jakoby and M. Wilchek, eds.) **34**, 72.
4. B. Svensson (1973), *FEBS Lett.* **29**, 167.

5. R. J. Yon and R. J. Simmonds (1974), *Biochem J.* **151**, 281.

6. G. I. Tesser, H.-U. Fisch, and R. Schwyzer (1974), *Helv. Chim. Acta* **57**, 1218.

7. R. Axen, J. Porath, and S. Ernbach (1967), *Nature (Lond.)* **214**, 1302.

8. J. Porath and R. Axen (1976), in *Methods in Enzymology* (K. Mosbach, ed.) **44**, 19.

9. P. Cuatrecasas (1970), *J. Biol. Chem.* **245**, 3059.

10. S. C. March, I. Parikh, and P. Cuatrecasas (1974), *Anal. Biochem.* **60**, 149.

11. D. E. Stage and M. Mannick (1974), *Biochim. Biophys. Acta* **343**, 382.

12. J. Porath, K. Aspberg, H. Drevin, and R. Axen (1973), *J. Chromatogr.* **86**, 53.

13. R. Lamed, Y. Leven, and M. Wilchek (1973), *Biochim. Biophys. Acta* **304**, 231.

14. E. Junowicz and S. Charm (1976), *Biochim. Biphys. Acta* **428**, 157.

15. C. M. Joyce and J. R. Knowles (1974), *Biochem. Biophys. Res. Commun.* **60**, 1278.

16. R. Lamed and A. Oplatka (1974), *Biochemistry* **13**, 3137.

17. R. Lamed, A. Oplatka, and E. Reisler (1976), *Biochim. Biophys. Acta* **427**, 688.

18. T. J. Larson, T. Hirabayashi, and W. Dowhan (1976), *Biochemistry* **15**, 974.

19. V. K. Moudgil and D. O. Toft (1975), *Proc. Nat. Acad. Sci. (USA)* **72**, 901.

20. A. Oplatka, A. Muhlrad, and R. Lamed (1976), *J. Biol. Chem.* **251**, 3972.

21. H. R. Burrell and J. Horowitz (1975), *FEBS Lett.* **49**, 306.

22. I. P. Trayer, H. R. Trayer, D. A. P. Small, and R. C. Bottomley (1974), *Biochem. J.* **139**, 609.

23. M. Wilchek and R. Lamed (1974), in *Methods in Enzymology* (W. B. Jakoby and M. Wilchek, eds.), **34**, 475.

24. E. S. Severin, S. N. Kochetkov, M. V. Nesterova, and N. N. Gulyaev (1974), *FEBS Lett.* **49**, 61.

25. V. Sica, E. Nola, I. Parikh, G. A. Puca, and P. Cuatrecasas (1973), *Nature New Biol.* **244**, 36.

26. O. D. Nelidova and L. L. Kisselev (1968), *Molec. Biol.* **2**, 60.

27. P. Dunnill and M. D. Lilly (1974), *Biotechnol. Bioeng.* **16**, 987.

28. J. Porath and L. Sundberg (1972), *Nature New Biol.* **238**, 261.

29. M. Caron, F. Fabia, A. Faure, and P. Corhillot (1973), *J. Chromatogr.* **87**, 239.

30. R. G. Coombe and A. M. George (1976), *Aust. J. Biol. Sci.* **29**, 305.

31. J. Porath (1973), *Biochemie* **55**, 943.

32. J. Porath, J.-C. Janson, and T. Läas (1971), *J. Chromatogr.* **60**, 167.

33. T. Kristiansen (1974), *Biochim. Biophys. Acta.* **362**, 567.

34. T. Läas (1975), *J. Chromatogr.* **111**, 373.

35. J. Brandt, L. O. Andersson, and J. Porath (1975), *Biochim. Biophys. Acta* **386**, 196.

36. T. Ternynck and S. Avrameas (1977), *Immunochemistry* **14**, 767.

37. S. Avrameas, T. Ternynck, and J.-L. Guesdon (1978), *Scand. J. Immunol.* **8**, Supplement **7**, 7.

38. K. Watanabe, K. Kimura, H. Marumo, and H. Samejima (1977), *Agric. Biol. Chem.* **41**, 547.

39. G. Kay and M. D. Killy (1970), *Biochim. Biophys. Acta* **198**, 276.

40. J. R. Wykes, P. Dunnill, and M. D. Lilly (1971), *Biochim. Biophys. Acta* **250**, 522.

41. R. J. H. Wilson, G. Kay, and M. D. Lilly (1968), *Biochem J.* **109**, 137.

42. R. J. H. Wilson, G. Kay, and M. D. Lilly (1968), *Biochem. J.* **108**, 845.

43. T. C. J. Gribnau (1977), "Coupling of Effector Molecule to Solid Supports," Ph.D. thesis, University of Nijmegen, Nijmegen, The Netherlands.

44. T. C. J. Gribnau (1979), *Solid State Biochem.* (in press).

45. H. H. Weetall and A. M. Filbert (1974), in *Methods in Enzymology* (W. B. Jakoby and M. Wilchek, eds.) **34**, 59.

46. H. H. Weetall (1976), in *Methods in Enzymology* (K. Mosbach, ed.) **44**, 134.

47. H. H. Weetall and R. A. Messing (1972), in *The Chemistry of Biosurfaces* (M. L. Hair, ed.) Marcel Dekker, New York, Chapter 12.

48. W. H. Scouten (1974), in *Methods in Enzymology* (W. B. Jakoby and M. Wilchek, eds.) **34**, 288.

49. F. E. Regnier (Purdue Reserach Foundation) U.S. Patent 3,983, 299.

50. F. E. Regnier and R. Noel (1976), *J. Chromatogr. Sci.* **14**, 316.

51. C. Lewis and W. Scouten (1976), *J. Chem. Ed.* **53**, 395.

52. C. Lewis and W. Scouten (1976), *Biochim. Biophys. Acta* **444**, 326.

53. M. Lynn (1975), in *Enzymology: Immobilized Enzymes, Antigens, Antibodies and Peptides* (H. H. Weetall, ed.) **1**, 1.

54. I. Parikh, S. March, and P. Cuatrecasas (1974), in *Methods in Enzymology* (W. B. Jakoby and M. Wilchek, eds.) **34**, 77.

55. J. F. Kennedy (1974), *Adv. Carbohyd. Chem. Biochem.* **29**, 306.

56. B. M. Alberts and G. Herrick (1971), in *Methods in Enzymology* (L. Grossman and K. Moldave, eds.) **21D**, 198.

57. J. Porath and S. Hjerten (1962), in *Methods of Biochemical Analysis*, Vol. 9 (D. Glick, ed.) Wiley-Interscience, New York, p. 193.

58. A. T. Jagendorf, A. Patchornik, and M. Sela (1963), *Biochim. Biophys. Acta* **78**, 516.

59. J. B. Robbins and R. Schneerson (1974), in *Methods in Enzymology* (W. B. Jakoby and M. Wilchek, eds.) **34**, 703.

60. J. B. Robbins, J. Haimovich, and M. Sela (1967), *Immunochemistry* **4**, 11.

61. J. K. Inman and H. M. Dintzis (1969), *Biochemistry* **8**, 4075.

62. J. K. Inman (1974), in *Methods in Enzymology* (W. B. Jakoby and M. Wilchek, eds.) **34**, 30.

63. V. Horesji, M. Ticha, and J. Kocourek (1977), *Biochim. Biophys. Acta* **499**, 290.

64. V. Horesji and J. Kocourek (1973), *Biochim. Biophys. Acta* **297**, 346.

65. V. Horesji and J. Kocourek (1974), in *Methods in Enzymology* (W. B. Jakoby and M. Wilchek, eds.) **34**, 361.

66. R. L. Schnaar and Y. C. Lee (1975), *Biochemistry* **14**, 1535.

67. R. Mosbach, A.-C Koch-Schmidt, and K. Mosbach (1976) in *Methods in Enzymology* (K. Mosbach, ed.) **44**, 53.

68. H. Nilsson, R. Mosbach, and K. Mosbach (1972), *Biochim. Biophys. Acta* **268**, 253.

69. A.-C. Johansson and K. Mosbach (1974), *Biochim. Biophys. Acta* **370**, 339.

70. K. Mosbach (1970), *Acta Chem. Scand.* **24**, 2084.

71. R. K. Epton, B. L. Hilbert, and T. H. Thomas (1976), in *Methods in Enzymology* (K. Mosbach, ed.) **44**, 84.

72. R. Epton, J. V. McLaren, and T. H. Thomas (1973), *Biochim. Biophys. Acta* **328**, 418.

73. R. Epton, J. V. McLaren, and T. H. Thomas (1971), *Biochem. J.* **123**, 21p.

74. R. Epton, J. V. McLaren, and T. H. Thomas (1974), *Polymer* **15**, 564.

75. R. Epton, J. V. McLaren, and T. H. Thomas (1972), *Carbohyd. Res.* **22**, 301.

76. R. Epton, B. L. Hibbert, and G. Marr (1973), *Polymer* **16**, 314.

77. G. Mannecke and J. Schlünsen (1976), in *Methods in Enzymology* (K. Mosbach, ed.) **44**, 107.

78. G. Mannecke and S. Singer (1960), *Makromol. Chem.* **39**, 13.

79. G. Mannecke and G. Günzel (1972), *Makromol. Chem.* **51**, 199.

80. G. Mannecke and H.-J. Förster (1966), *Makromol. Chem.* **91**, 136.

81. G. Mannecke, G. Gunzel, and H.-J. Förster (1970), *J. Polym. Sci., Polym. Symp.* **30**, 608.

82. G. Mannecke (1974), *Chimia* **28**, 467.

83. N. Dattagupta and H. Buenemann (1973), *J. Polym. Sci.* **11B**, 189.

84. C. R. Lowe and P. D. G. Dean (1974), *Affinity Chromatography*, Wiley, London, p. 222.

85. D. G. Hoare and D. E. Koshland (1966), *J. Am. Chem. Soc.* **88**, 2057.

86. P. Cuatrecasas and I. Parikh (1972), *Biochemistry* **11**, 2291.

87. P. V. Sundaram (1974), *Biochem. Biophys. Res. Commun.* **61**, 717.

88. R. Benesch and R. E. Benesch (1956), *J. Am. Chem. Soc.* **78**, 1597.

89. G. W. Anderson, J. Blodinger and A. D. Welcher (1962), *J. Am. Chem. Soc.* **74**, 5309

90. E. Steers, P. Cuatrecasas, and B. Pollard (1971), *J. Biol. Chem.* **246**, 196.

91. P. Cuatrecasas and I. Parikh (1972), *Biochemistry* **11**, 2291.

92. W. H. Scouten, V. Levi, and T. Frielle, unpublished observations.

93. E. E. Rickli and P. A. Cuendet (1971), *Biochim. Biophys. Acta* **250**, 447.

94. H. L. Weith, J. L. Wiebers, and P. T. Gilham (1970), *Biochemistry* **9**, 4396.

95. L. A. Cohen (1974), in *Methods in Enzymology* (W. B. Jakoby and M. Wilchek, eds.) **34**, 102.

96. D. L. Robberson and N. Davidson (1972), *Biochemistry* **11**, 533.

97. C. J. Sanderson and D. V. Wilson (1971), *Immunology* **20**, 1061.

98. M. B. Wilson and P. K. Nakane (1976), *J. Immunol. Meth.* **12**, 171.

99. C. F. Lane (1975) *Synthesis* **(1975)**, 135.

100. A. S. Perlin (1969), in *Oxidation* (R. L. Augustine, ed.), Vol. 1, Marcell Dekker, New York, p. 189.

101. M. Robert-Gero and J. P. Waller (1972), *Eur. J. Biochem.* **31**, 315.

102. M. Robert-Gero and J.-P Walker (1974), in *Methods in Enzymology* (W. B. Jakoby and M. Wilchek, eds.) **34**, 506.

103. W. H. Scouten, F. Torok, and W. Gitomer (1973), *Biochim. Biophys. Acta* **309**, 521.

104. M. Gorecki and A. Patchornik (1973), *Biochim. Biophys. Acta* **303**, 36.

105. P. Corvol, C. Devaux, and J. Menard (1973), *FEBS Lett.* **34**, 189.

106. M. Sokolovsky (1974), in *Methods in Enzymology* (W. B. Jakoby and M. Wilchek, eds.) **34**, 411.

107. W. H. Scouten (1974), in *Methods in Enzymology* (W. B. Jakoby and M. Wilchek, eds.) **34**, 288.

108. W. H. Scouten and G. L. Firestone (1976), *Biochim. Biophys. Acta* **453**, 277.

109. W. H. Scouten and J. Michelson, unpublished results.

110. P.-O Larsson and K. Mosbach (1971), *Biotechnol. Bioeng.* **13**, 393.

111. C. R. Lowe, M. J. Harvey, D. B. Craven, and P. D. G. Dean (1973) *Biochem. J.* **133**, 499.

112. C. A. Arsenis and D. B. McCormick (1966), *J. Biol. Chem.* **241**, 330.

113. J. S. Erickson and C. K. Mathews (1971), *Biochem. Biophys. Res. Commun.* **43**, 1164.

114. E. Brown and R. Joyeau (1978), *J. Chromatogr.* **150**, 111.

115. S. L. Marcus and E. Balbinder (1972), *Anal. Biochem.* **48**, 448.

116. G. Schmer (1972), *Hoppe-Seyler's Z. Physiol. Chem.* **353**, 810.

117. B. T. Kaufman and J. V. Pierce (1971), *Biochem. Biophys. Res. Commun.* **44**, 608.

118. D. N. Salter, J. E. Ford, K. J. Scott, and P. Andrews (1972), *FEBS Lett.* **20**, 301.

119. S. Gutteridge and D. A. Robb (1973), *Biochem. Soc. Transact.* **1**, 519.

THEORY AND PROBLEMS IN AFFINITY CHROMATOGRAPHY

Several mathematical and theoretical models of affinity chromatography have been proposed (1–12), although no completely satisfactory quantitative model has been developed. The development of an adequate model has been hindered because of the large number of possible interactions involving ligand–enzyme binding, for example, ionic, hydrophobic, and hydrogen bonding. Further complications can be ascribed to the existence of numerous potential nonbio-affinity interactions of proteins with the matrix and spacer arms. Because of the complexity of this problem, we will first present a simple, qualitative, conceptual model, which will be followed by a discussion of the various attempts to formulate a quantitative approach to the theory of affinity chromatography.

4.1. A QUALITATIVE APPROACH TO BIOSELECTIVE ADSORPTION

The active site of an enzyme can be defined either as those amino acid residues directly in contact with the substrate or those amino acid residues indirectly involved in substrate binding through an intervening water molecule or amino acid side chain. There are many such residues that may be in contact with any single substrate. These may bind the substrate through various combinations of hydrophobic, ionic, or hydrogen bonding or charge-transfer interactions. The specificity of an enzyme to bind a particular substrate is based chiefly on the steric placement of each amino acid at the active site such that the substrate and the enzyme are properly aligned (e.g., cationic sites on the substrate are adjacent to anionic sites of the enzymes). Some substrates or inhibitors may "fit" into this active site, with some substrates fitting better than others. The affinity of an enzyme for the substrate is usually expressed in terms of the dissociation constant, K_D:

$$E + S \rightleftharpoons ES; \qquad K_D = \frac{[ES]}{[E]\,[S]}$$

$$E + I \rightleftharpoons EI; \qquad K_{D_i} = \frac{[EI]}{[E]\,[S]}$$

Often the kinetically derived Michaelis constant K_M is nearly equal to the dissociation constant and may be used to approximate it. In some instances, however, K_M may bear little resemblance to the dissociation constant, and therefore the K_M should be employed with caution when describing enzyme substrate interactions. Where it is appropriate, K_M and K_D can be shown to be inversely related to the enzyme affinity for the substrate (or inhibitor) (13).

In an ideal situation, where there are no enzyme-substrate interactions other than the formation of the desired enzyme–substrate complex, a very simple Langmuir adsorption isotherm will describe the enzyme–matrix interaction (see Figure 4.1). Nonetheless, almost without exception, proteins will also be attracted to the adsorbent in a nonspecific fashion. That is, the adsorbent may have interactions with many other proteins applied to the column because of ionic, hydrogen bond, hydrophobic, or other bonding sites found on the surface of the protein. Such interactions, however, are probably not at the enzyme active site and have no relationship to normal substrate binding.

For maximum purification to be achieved, nonspecific adsorption of proteins to the bioselective adsorbent must be as low as possible. If this is not achieved, many proteins other than the enzyme being purified will be adsorbed to the matrix. Figure 4.2 illustrates a schematic model of both an ideal bioselective adsorption process and of a poor bioselective adsorbent that possesses numerous nonspecific protein interactions.

This schematic model of bioselective adsorption is far too simplistic to be of general use since only a few adsorption parameters, for example, K_i, V_m, K_m, are included. These parameters are normally determined for substrate or inhibitor in free solution, and the values in free solution may not be close approxi-

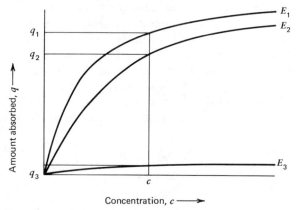

Figure 4.1. Absorption isotherms for three enzymes (E_1, E_2, and E_3) interacting with a single immobilized ligand. Reproduced with permission from *Affinity Chromatography* by C. R. Lowe and P. D. G. Dean, Copyright © John Wiley and Sons, Ltd. (1974).

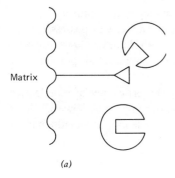

(a)

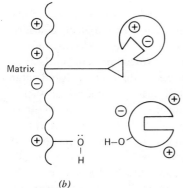

(b)

Figure 4.2. Comparison of an ideal bioselective adsorbent with a poor bioselective adsorbent. (*a*) An ideal bioselective adsorbent. The only interaction possible is the enzyme–substrate complex formation. (*b*) A nonideal bioselective adsorbent. Many interactions other than enxyme–substrate complex formation are possible (e.g., ionic, hydrophilic, hydrogen bonding). The $\Delta F°$ for enzyme-substrate complex formation must appreciably exceed the $\Delta F°$ for all nonspecific interactions if bioselective adsorption is to be useful.

mations to the enzyme–inhibitor affinities when an immobilized inhibitor is employed rather than the inhibitor in free solution. The inhibitor, bound to the matrix, may be sterically prevented from binding to the enzyme and the actual K_i much higher than the K_i predicted from solution experiments. In a very few cases the converse is true; that is, the matrix may actually increase the affinity of the enzyme for the immobilized ligand so that the immobilized inhibitor has a greater affinity for the enzyme than does the unbound inhibitor.

Concentration effects may also cause immobilized ligands to differ from their soluble counterparts. The effective concentration of the immobilized inhibitor at the surface of the matrix is much higher than would be predicted by assuming that the ligand was uniformly distributed throughout the volume occupied by the entire matrix. It is often difficult to precisely determine the actual concentration of ligand at the surface. When certain matrixes are employed, the surface distribution of the ligand is not at all uniform. Some regions of cellulose (and similar materials) will be much higher derivatized than are other regions.

Of course, every matrix will inherently exhibit some lack of uniformity, but most synthetic polymers, such as glass and polyacrylamide, will possess reasonably uniformly derivatized surfaces. Agarose also is rather uniformly derivatized, whereas starch, cellulose, chitin, and cross-linked collagen all possess very heterogeneous surfaces.

For these and other reasons, the capacity and the efficiency of bioselective adsorption is still best determined experimentally. In some cases, however, good predictions of chromatographic behavior have been made by using mathematical approaches.

4.2. QUANTITATIVE APPROACHES TO THE THEORY OF BIOSELECTIVE ADSORPTION

Chromatographic theory developed for ion-exchange, partition, or similar "usual" chromatographic procedures are not applicable to bioselective adsorption. As previously noted, the adsorption isotherms are nonlinear and the adsorption process per se is complex. Moreover, protein–protein interactions may play just as important a role in the process as do protein–ligand interactions. Usual chromatographic theory does not readily account for such a complex process. Therefore, several new approaches have been incorporated into classical chromatographic theory to account for bioselective adsorption. Graves and Wu (1) have employed a model based on batch adsorption and, using a perfectly mixed single-stage model, have developed a theory for bioselective adsorption. Wankat (2), on the other hand, has used a column model based on discrete-staged transfer and equilibrium steps. Denizot and Delange (3) have developed a statistical theory based on the method of moments developed by Giddings and Eyring (14) and initially employed for classical chromatographic techniques. The statistical method has the ability to yield, in addition to predicting chromatographic behavior, rate constants for association and dissociation of enzyme-inhibitor complexes. These rate constants are in the range usually determined by using stopped-flow techniques.

Each of these models can predict the effect of changes in variables on the bioselective adsorption process, provided the bioselective adsorption is sufficiently simple (i.e., minimal protein–protein interaction, only one enzyme present that recognizes the immobilized ligand, etc.). None of the techniques, however, is capable of handling the very complex relationships very often found in bioselective adsorption. When several enzymes in a crude enzyme preparation are competing for the same ligand, as, for example, with a group-specific ligand, we have the very common example of one of the situations that as yet are not adequately treated by the present theoretical approaches. However, each of these theories could, potentially, be modified to include several of these parameters. As a result, the mathematic expressions for retention time, adsorption isotherm, and so on would become much more complex, requiring computer analysis, and the result would also be a closer approximation to an accurate description of commonly employed bioselective techniques.

4.2.1. The Equilibrium Model (1)

The simplest models of affinity chromatography is the single mixed-stage perfect equilibrium mode of Graves and Wu (1). Although each model possesses the inherent errors discussed earlier, the equilibrium model possesses the advantage that each of the properties described, and the assumptions made, are readily seen in the final equation. Moreover, the result is normally just as useful as either of the other more complex models in designing experiments and explaining results.

Graves and Wu (1) begin by asking how, in the simplest case, an enzyme is distributed between the solution phase and the solid phase, where it is bound to the immobilized ligand. Under such conditions, at equilibrium and under constant conditions of pH, temperature, and ionic strength, the adsorption can be described simply by

$$E + L \underset{k_2}{\overset{k_1}{\rightleftharpoons}} EL$$

where (E) is the enzyme concentration at equilibrium, (L) is the equilibrium concentration of the ligand, and (EL) is the concentration of enzyme–ligand complex. This yields an equilibrium constant,

$$K_i = \frac{(E)\,(L)}{(EL)}$$

We can assume that there is a volume of ligand matrix that contains a volume v of solution entrapped within the pores and that the ligand is present at a concentration of L_0 in moles per volume v. Initial conditions dictate that no enzyme be present in the system and then a volume V of enzyme be added whose concentration is E_0 in moles per liter. The system is allowed to reach equilibrium.

Under these circumstances, Graves and Wu have shown that the concentration of bound enzyme, at equilibrium, is

$$\text{Enzyme bound} = (EL) = \frac{(L_0)}{2}\,(B)\quad(1 - \sqrt{1 - A})$$

where

$$B = \frac{K_i(V + v)}{L_0 v} + \frac{E_0 V}{L_0 v} + 1$$

and

$$A = \frac{4E_0}{B^2 L_0 v}$$

If $E_0 0$, the initial enzyme concentration, is low relative to L_0, then A becomes much less than unity. In that case $\sqrt{1 - A}$ will become nearly $1 - (A/2)$ and

$$(EL) \cong \frac{L_0 E_0 V}{K_i(V + v) + L_0 v + E_0 V}$$

This approximation is normally quite reasonable because the concentrations of enzyme in solutions are usually about one thousandth of the concentration of the immobilized ligand.

The fraction of enzyme bound to the ligand under these conditions is probably the most useful parameter that can be measured. This was shown to be

$$\frac{\text{Bound enzyme}}{\text{Total enzyme}} = \frac{(EL) v}{E_0 V} \cong \frac{L_0 v}{K_i(V + v) + L_0 v + E_0 V}$$

It is further assumed that $E_0 V \lll L_0 v$ and that $V \lll v$, conditions that in practice are almost universally true for bioselective adsorption, then

$$\frac{\text{Bound enzyme}}{\text{Total enzyme}} \cong \frac{L_0}{K_i + L_0}$$

This relationship was determined by Graves and Wu for several values of K_i, and as shown in Figure 4.3, the resulting curves are of the Langmuir isotherm type, which is precisely the type of curve (as has already been shown) that is found in bioselective adsorption. Further, it can easily be concluded that K_i values of 10^{-3} M or less will result in low levels of enzyme retention, except at unreasonably high ligand loadings. Normally, ligand substitution is less than 20 mM, although, as indicated earlier, it is extremely difficult to determine the local surface concentration of ligand, especially on derivatives of cellulose or other heterogeneous matrices.

Graves and Wu have shown that this treatment of affinity chromatographic

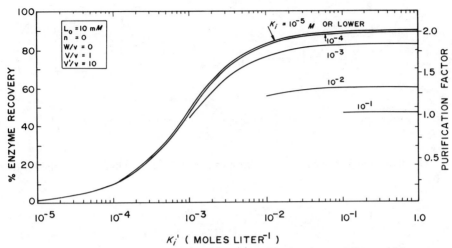

Figure 4.3. Fractional recovery and purification factor for the case of no washing steps, a 1:1 sample:gel volume ratio and a 10:1 eluting solution:gel ratio. Reprinted with permission from Graves and Wu (1).

theory can readily explain the observation that the capacity of a bioselective adsorbent is often much less than the number of ligands bound to the matrix. There is little doubt that steric factors are involved and that an enzyme may well cover several ligand molecules in addition to the one to which it is actually bound; for example, see Figure 4.4. Equilibrium effects are also a significant factor in the actual performance of a bioselective adsorbent. These effects are due to the fact that the enzyme exists in both a bound form and a free form, with no enzyme molecule spending all of its time bound to the same ligand molecule (see Figure 4.5).

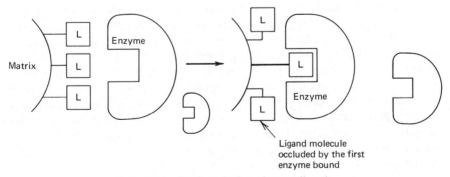

Figure 4.4. Steric occlusion of excess ligand.

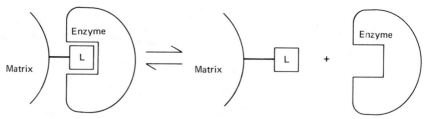

Figure 4.5. Equilibrium of enzyme and matrix-bound enzyme–ligand complex.

Mathematically, the capacity is expressed as $(EL)/(L_0)$, which can be shown to be

$$\text{Capacity} \frac{(EL)}{(L_0)} = \frac{E_0 V}{K_i(V + v) + L_0 v + E_0 v}$$

This, again, is a hyperbolic relationship, and several calculations of adsorbent capacity as determined by Graves and Wu are given in Table 4.1, where ω is the ratio of V to v. Under these conditions, where $\omega = 1$ and reasonable experimental values for K_i, L_0, and E_0 are chosen (e.g., $L_0 = 10$ mM, $E_0 = 10^{-2}$ M, $L_0 = 10^{-2}$ M, and $K_i = 10^{-2}$ M), these authors indicate that the capacity, due merely to equilibrium effects, will be only 83%. The additional 17% of the enzyme applied to the column will fail entirely to be adsorbed to the adsorbent.

The equilibrium model can also be used for the prediction of elution profiles. Graves and Wu have employed three types of elution with this treatment, namely (1) elution using a competitive inhibitor, (2) elution using a change in conditions that will alter the K_i of the original enzyme–inhibitor complex formation, and (3) elution of an enzyme from a weak inhibitor–matrix by way of continuous elution with the original buffer employed during enzyme application.

The latter method is useful only when an enzyme is retarded on a bioselective adsorbent rather than being tightly adsorbed. Such conditions have rarely been useful for enzyme purification. Far more commonly, the enzyme is eluted by changing the buffer, ionic strength, pH, or some similar variable such that the K_i of the enzyme–ligand complex formation is thereby increased. This procedure can be easily misinterpreted, since nonspecific binding also changes under such conditions. Further, the K_i might not be altered significantly under these conditions. Prior determination of K_i for the free ligand under various buffer conditions is necessary to ensure the investigator that the elution (and thus the enzyme purification) is due specifically to enzyme–ligand interactions rather than "nonspecific" forces, such as ion-exchange chromatography.

Mathematically, enzyme elution is caused by adding a volume V' of a new

TABLE 4.1. Percentage Ligand Saturation Under Various Conditions
(for $L_0 = 0.01\ M$)

E_0 (M)	K_i (M)	EL/L_0 (%)		
		$\omega = 1$	$\omega = 10$	$\omega = 100$
10^{-4}	10^{-3}	0.8264	4.5455	8.2465
	10^{-4}	0.9709	8.2645	33.2226
	10^{-5}	0.9881	9.0009	47.5964
	10^{-6}	0.9899	9.0818	49.7488
10^{-5}	10^{-3}	0.0833	0.4739	0.8929
	10^{-4}	0.0979	0.8929	4.7393
	10^{-5}	0.0997	0.9794	8.3264
	10^{-6}	0.0999	0.9890	9.0082
10^{-6}	10^{-3}	0.0083	0.0476	0.0900
	10^{-4}	0.0098	0.0900	0.4950
	10^{-5}	0.0100	0.0988	0.9001
	10^{-6}	0.0100	0.0998	0.9803
10^{-7}	10^{-3}	0.0008	0.0048	0.0090
	10^{-4}	0.0010	0.0090	0.0497
	10^{-5}	0.0010	0.0099	0.0907
	10^{-6}	0.0010	0.0100	0.0989

Source: Reprinted with permission from Graves and Wu (1).

buffer to the gel of volume v, such that a new inhibitor constant K_i' results. This results in the expression

$$\text{Enzyme recovery} = \frac{\text{enzyme in } V'}{E_0 v} =$$

$$\frac{(V')}{(V' + v)} \left\{ 1 - \frac{L_0 v\, (K_i v + L_0 v)}{(K_i\, (V + v) + L_0 v)\, (K_i'\, (V' + v) + L_0 v)} \right.$$

$$\left. - \left(\frac{V}{V + v} \right) \left[1 - \frac{L_0 v}{K_i\, (V + v) + L_0 v} \right] \right\}$$

It can be seen from this expression that the efficiency of a bioselective adsorbent is dependent upon K_i, K_i', V, V', and L_0. Obviously a small sample volume (V/v) and a large elution volume (V'/v) will favor efficient recovery of the enzyme in the eluant. Figure 4.3 graphically illustrates this dependence.

Again, the appropriate choice of a ligand that binds tightly to the enzyme is very important for enzyme purification.

Enzyme elution can also be achieved by utilizing a soluble enzyme inhibitor with either a more favorable K_i and/or a concentration higher than that of the original inhibitor. In practice, this form of elution tends to be the best and, wherever practical, should be the method employed. Nonetheless, when the single step equilibrium model is employed, it is unfortunately also the most difficult to describe accurately. Graves and Wu have shown that the original equations can be greatly simplified with little loss in their utility by assuming that very little soluble enzyme, not complexed to an inhibitor, would be present in a solution of enzyme plus soluble inhibitor. This situation would be true of nearly any system that might be employed.

Under these conditions, when a volume V containing the soluble inhibitor I is added to the enzyme–bioselective adsorbent complex, a new enzyme inhibitor EI will form, such that

$$E + I \underset{k_2'}{\overset{k_1'}{\rightleftharpoons}} EI$$

and

$$K_s = \frac{(E)(I)}{(EI)}$$

where (I) is the concentration of the new, soluble, inhibitor and (EI) is the concentration of the soluble enzyme inhibitor complex. Under these conditions, the enzyme recovery can be illustrated by

$$\frac{\text{Enzyme recovery}}{\substack{\text{Total enzyme in the} \\ \text{washed gel}}} = \frac{V'(EI)}{vE_i} \cong \frac{(p)}{(p+1)} \left[\frac{pI}{2E_i + pI + K_s(1+p) + (K_s(L_0 - E/K_i)} \right]$$

where E_i represents the moles of enzyme initially in the gel and p equals V_1/v.

This approximation, which can be further manipulated to provide such information as the dependence of percent recovery versus K_s, is still rather cumbersome. By assuming that fairly standard conditions exist (e.g., $L_0 \gg E_i$, $L_0/K_i \gg p$), this expression can be further simplified to yield

$$\text{Enzyme recovery} = \frac{V^1(EI)}{vE_1} = \frac{(p)}{(p+1)} \left[\frac{pI_0}{pI_0 + K_s L_0/K_i} \right]$$

From this and the separately determined values of K_s and K_i the enzyme recovery can be predicted for any given elution volume V' assuming an eluant containing a given concentration, I_0, of soluble ligand. To extrapolate these measurements from a single-step equilibrium "batch" model to a chromatographic system, one must assume the column to be a *single* thoroughly mixed system. The danger inherent in this is obvious, despite the claim by Graves and Wu that many chromatographic columns indeed behave in this fashion.

4.2.2. The Staged-Transfer Column Chromatographic Model (2)

Because of the dangers inherent in extrapolating a single-stage model to infinitely staged chromatographic systems, Wankat (2) has derived a bioselective adsorbent model based essentially on the staged transfer and equilibrium step system developed by Martin and Synge (15) as a model for general chromatographic procedures.

In elementary terms, Wankat's approach is to divide the column into small volumes and apply successively a treatment similar to Graves and Wu to each small volume. As these small volume "slices" termed "stages," become smaller, the treatment becomes closer and closer to the actual column operation. (See Figure 4.6.)

In this model, the equilibrium of free enzyme with the immobilized bioligand to form an immobilized enzyme–ligand complex is also considered. As shown before:

$$E + L \underset{k_2}{\overset{k_1}{\rightleftharpoons}} EL$$

and

$$K = \frac{(E)\,(L)}{(EL)}$$

We next consider each stage (or small volume increment) of the column as a volume with a matrix possessing a total surface area A_T, and we define the

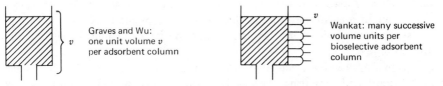

Graves and Wu:
one unit volume v
per adsorbent column

v

Wankat: many successive
volume units per
bioselective adsorbent
column

Figure 4.6. Comparison of single-transfer and equilibrium step models.

area covered by the enzyme–ligand complex as A_E. If L_0, E_0, and (EL) represent, respectively, the concentration of ligand, initial enzyme concentration, and enzyme ligand concentration per stage, in moles per square centimeter, then

$$L_0 \times A_T = \text{total mole of ligand per stage}$$

$$(EL) \times A_T = \text{moles of } EL \text{ per stage}$$

and thus the area of the matrix covered by the enzyme ligand complex is

$$(A_E) \left[(EL) \times A_T \right] = \text{area covered by } EL, \text{ cm}^2/\text{stage}$$

and the area of matrix containing unbound ligand, at equilibrium, in each stage is

$$A_T - \left[(A_E)(EL)A_T \right]$$

which yields, by substitution,

$$\text{Free ligand} = (L) = L_0 \left[1 - A_E(EL) \right]$$

assuming that the concentration of ligand is much greater than the concentration of enzyme in each stage. Substituting the last equation into the definition for K_i, and solving for EL, we obtain the equation bound enzyme

$$(EL) = \frac{(E)(L_0)}{K_i + (E) A_E L_0}$$

Division by (E) produces

$$\frac{\text{Bound enzyme}}{\text{Free enzyme}} = \frac{(EL)}{(E)} = \frac{L_0}{K_i + A_E L_0(E)}$$

The latter two equations are of the Langmuir type and qualitatively resemble the relationships derived by Graves and Wu. However, these relationships are valid for each stage, and the concentrations are expressed in moles per square centimeter of area of stage rather than moles per volume v, as was the case in the single-stage model. This distinction must be kept in mind throughout the discussion.

Elution from the bioselective adsorbent by any of the three methods (change in K_i, continued prolonged washing with the original buffer, or addition of a soluble competing ligand to the eluant buffer) discussed previously can be described by use of either model. The staged equilibrium model lends itself most readily to describing the latter type of elution. This is fortunate because elution with a competing ligand is, in fact, the preferred method of elution since it is the type of elution that is most likely to lead to substantial enzyme purification.

Elution with a soluble ligand is also effected by the formation of a soluble enzyme–ligand complex; thus

$$E + I \underset{k_2}{\overset{k_1}{\rightleftharpoons}} EI$$

and

$$K_i = \frac{(E)\,(I)}{(E)}$$

The concentration of the soluble ligand per stage may be defined as

$$(I) + (EI) = \frac{(I_T)}{V_2}$$

where (I_T) is the total ligand in moles and V_2 is the volume of liquid within each stage. Substituting this into the definition of K_i and solving for (EI), we obtain

$$(EI) = \frac{(E)\,(I_T)/V_2}{K_s + (E)}$$

Assuming the column has a length L and consists of N stages, each of which has a height equivalent to one theoretical plate ($HETP$), then

$$(N)\,(HETP) = L$$

In this model we remove a volume v, from each stage and transfer it to the following stage. At the next stage a new equilibrium between bound and free enzyme is formed followed by a second transfer of volume v into the next stage. This procedure is continued until the enzyme is eluted from the column.

To describe this model in mathematical terms, Wankat first defined the total enzyme in any stage in the column after each transfer step E_T as the moles of enzyme transferred into the stage E_{in} plus the moles of enzyme left behind E_{left}:

$$E_T = E_{in} + E_{left}$$

Similarly, the total ligand I_T is equal to the ligand transferred in plus the ligand left behind during transfer:

$$I_T = I_{in} + I_{left}$$

We can generalize these statements to be applicable to all stages in the column and for any one of the successive transfer steps by defining $E_{Tj,s}$ as the total enzyme at stage j during the transfer step s, and $I_{Tj,s}$ as the total ligand at stage j and transfer step s. It can also be shown that the fraction of enzyme in each stage that is present in the soluble form $(f,_{j,s})$ is

$$(f_{j,s}) = \text{fraction soluble enzyme} = \frac{\text{moles of enzyme as } E \text{ and } EI}{\text{total moles of enzyme, } E, EI, \text{ and } EL}$$

$$= \frac{(E + EI)\,V_2}{(E + EI)\,V_2 + (EL)(AT)}$$

Using this relationship, plus the mass balance equation for each stage, we obtain

$$E_T = (E)\,(V_2) + (EI)\,(V_2) + (EL)\,(A_T)$$

Wankat derived (in much more rigor than given here) the following relationship:

$$E_{Tj,s} = (E)\,(V_2) + \frac{I_{Tj,s}}{K_s + E} + \frac{(E)\,(L_0)\,(A_T)}{K_I + (E)\,(A_E)\,(L_0)}$$

Wankat solved this equation sequentially for (E), (EL), and EI and then calculated $f_{j,s-1}$. Finally he obtained $E_{Tj,s}$ and increased the stage or transfer step by 1 and repeated the process. This procedure has been placed on a computer program and is available for general use (16). Using this procedure one can predict the outcome of any affinity chromatography procedure if K_s and K_i are independently determined.

Initially, this procedure may appear complicated, and, indeed, wherever a simpler, single-stage equilibrium model will adequately describe a bioselective adsorption, it is the method of choice. Whenever, however, the column is very long or when separation of two or more proteins with affinity for the same immobilized ligand is attempted, the multistep process of Wankat is clearly superior to the single-stage description.

The third model, that of Denizot and Delange (3), is certainly the most complex. It is based on the random walk model of Giddings and Eyring (14) and is formulated by a statistical approach. A detailed description is not presented here. The advantage of this theory is that the elution profile for a bioselective chromatographic process in theory can easily yield the values of k_1, k_2, and so on and from these K_s and K_i can be calculated. Unfortunately, this model, like the simpler ones, ignores much of the complexity created by nonspecific adsorption very often encountered in real chromatographic purifications.

4.3. PITFALLS IN BIOSELECTIVE ADSORPTION

Bioselective adsorption is both one of the most useful and one of the most complex methods of protein separation. Far too often the problems inherent in the process have been overlooked by the enthusiastic biochemist who has succeeded in achieving a good—if not excellent—purification by an apparently bioselective method. Much of the time, especially in the earlier literature, the separation achieved was due not entirely to bioselective factors, but rather to combinations of conventional ion exchange, hydrophobic, and affinity factors. On the other hand, there are reports in the literature where purification by way of affinity chromatography did not result in satisfactory purification. Indeed, many such negative results have never been published, and those researchers who fail to achieve the desired end product often have a negative view of affinity chromatography. The fact remains that often neither group of researchers—the overenthused or the pessimist—has a good understanding of exactly what to expect (or not to expect) of bioselective adsorption, nor do they have any understanding of the pitfalls of the process.

The most significant problem in bioselective adsorption to date has been the introduction of charged residues or hydrophobic functions during the derivatization of the matrix to form an "activated" form of the matrix that will react with and immobilize the substrate. We have already (Chapter 3) discussed various methods that can be used to form derivatives that minimize this problem.

The purification of β-galactosidase by Cuatrecasas (17) is probably the best example of the purification of an enzyme by what was reported to be affinity chromatography but where nonspecific factors, rather than bioselective factors, were actually the mode of purification. In this work the investigator activated agarose by way of cyanogen bromide, which, in turn, was coupled to a spacer molecule. The spacer molecule was then attached to the enzyme inhibitor, p-aminophenylthiogalactoside. (See Figure 4.7.)

β-Galactosidase, isolated from E. coli, was retained by this adsorbent, but neither high concentrations of free p-aminophenylthiogalactoside nor other substrates of the enzyme were capable of eluting the enzyme in a good, well-defined peak. However, one could elute the enzyme by changing the pH of the

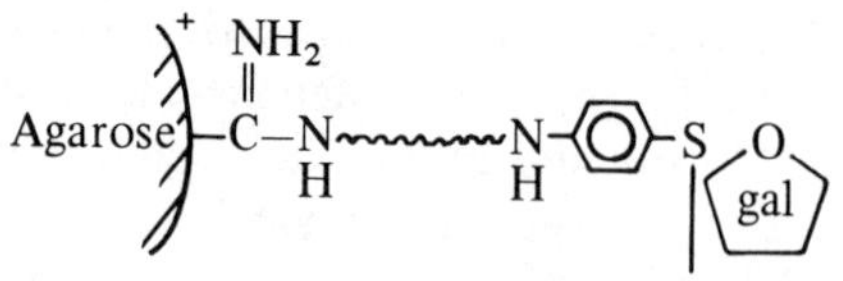

Figure 4.7. *p*-Aminophenylthiogalactoside–agarose.

buffer. This elution resulted in substantial purification of the enzyme, which was incorrectly attributed to affinity chromatography. Unfortunately, Cuatrecasas et al. did not note that under these elution conditions the K_i of the free inhibitor does not change, nor is there any reason to believe that the K_i of the immobilized inhibitor should change. Thus *bioselective* elution *should not* have occurred, but ion exchange could well take place and would explain the results seen.

Later work by Hofstee (18) and by Nishikawa and Bailon (19) demonstrated that a matrix prepared by using aniline instead of the enzyme inhibitor would purify β-galactosidase in the same fashion, even though aniline itself has little or no effect on the enzyme per se. (See Figure 4.8.)

Both the positive charge of the isourea function and the hydrophobic "leash" were required for enzyme retention and purification under these conditions. An isourea function, pK_a about 10.4, is introduced into the matrix whenever an amine group is attached to a cyanogen bromide-activated agarose (20,21). To eliminate this source of nonspecific adsorption, which is a very strong ion-exchange effect, many investigators have begun coupling cyanogen bromide-activated agarose using hydrazides as leashes, since hydrazides yield noncharged derivatives (22). Other investigators eliminate the use of cyanogen bromide completely, coupling the ligand to agarose via bifunctional bisoxirane (23–25). Murphy et al. (26) have shown that cyanogen bromide coupling causes considerably more nonspecific binding than does the bisoxirane method. These investigators coupled glucagon to Sepharose by both the oxirane and CNBr methods. Serum from nonimmunized rabbits and from glucagon-immunized rabbits were both passed through these columns. Some serum protein from the nonimmunized rabbits binds nonspecifically to the oxirane-prepared matrix, but it is far lower than the amount of protein that is nonspecifically adsorbed to the CNBr-prepared matrix. Thus as you might expect, antiglucagon antibodies

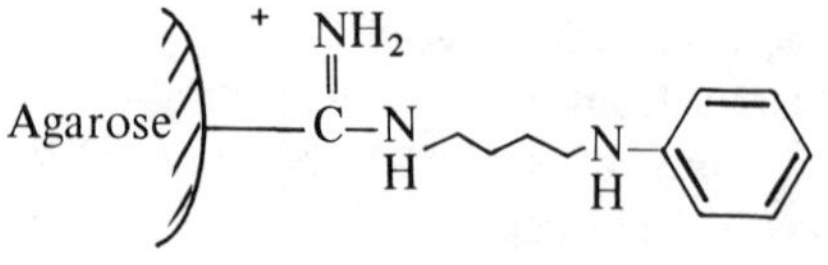

Figure 4.8. Anilinoalkylamine agarose prepared from CNBr activated agarose.

purified on glucagon coupled by way of CNBr require much more drastic elution conditions than do those that are purified by using glucagon coupled to agarose by way of an oxirane bridge.

Unfortunately, the vast majority of commercially prepared bioselective adsorbents are prepared by way of CNBr coupling. To ensure that bioselective adsorbence of the enzyme being purified is actually occurring, an attempt should be made to use a substrate or inhibitor to effect enzyme desorption. Only in this manner can it be ascertained that the affinity of the enzyme for the chromatographic packing actually involves bioselective processes. Pitfalls of bioselective adsorption are listed in Table 4.2.

Occasionally, bioselective elution is not possible because of "compound affinity." This is nonspecific binding that *adds* to true bioselective binding in a nonlinear fashion, such that the final binding is more than the sum of both the specific and the nonspecific binding forces (27). Compound affinity may, in fact, be very useful if the bioselective ligand is normally weakly bound to the enzyme in free solution. Even so, elution from such a compound affinity chromatographic packing should be able to be accomplished in a bioselective fashion. This is usually accomplished by increasing the ionic strength of the buffer and/or adding organic solvents, such as 1,2-propanediol or glycerol, until

TABLE 4.2. Problems of Bioselective Adsorption

Problem	Solution
Nonspecific binding due to CNBr activation	Use high-ionic-strength or alternate activation methods
"Leash" effects, either hydrophobic or ionic	Change binding method or chromatographic solvent
Enzyme loss through covalent attachment to residual reactive functions	Deactivate chromatographic packings using appropriate "dummy" ligands or proteins
Failure to bind to one immobilized substrate of a two-substrate reaction	Determine whether a compulsory ordered mechanism occurs by (a) including the second substrate in the protein application buffer or (b) using the second substrate as the immobilized ligand
Lack of demonstrable bioselective elutability	Try to elute with bioselective eluant dissolved in an enzyme deforming buffer, such as urea or an organic solvent

a concentration just below that required for "nonspecific elution" of the enzyme is achieved. Enzyme desorption is then obtained by the addition of the free substrate or inhibitor to that particular solvent composition. *If* enzyme desorption cannot be obtained through the use of a bioselective eluant and can *only* be obtained by use of strong enzyme deforming solvents, it should be suspected that bioselectivity was not actually involved, or that it contributed to a negligible extent to the purification. Furthermore, bioselective elution should be attempted not only to prove that a given purification is achieved by bioselective adsorption, but also because an enhanced level of purification generally results when bioselective elution is performed.

Failure to ascertain that bioselective forces are involved in a supposed affinity chromatographic purification can be dangerous. As has been stated earlier, purification of β-galactosidase by using *p*-aminophenylthiogalactoside coupled to agarose by way of a hydrophobic bridge was initially considered to be an affinity chromatographic purification. However, the purification was actually an example of hydrophobic chromatography and was largely unrelated to any bioselective property of the enzyme.

4.4. SPACER ARM EFFECTS

Often bioselective ligands are attached to appropriate matrixes by using leashes or spacer arms to hold the ligand away from the matrix and allow the enzyme to sterically reach the ligand. (See Figure 4.9.)

Unfortunately, these spacer arms often contribute nonspecific binding factors. In some instances (28,29) the spacer arm employed contains anionic groups capable of acting in an ion-exchange fashion. More commonly, the problem encountered with the use of spacer arms is the interaction between the hydrophobic nature of the spacer arm and "oily" or "hydrophobic" regions of proteins.

No leash—
enzyme sterically
restricted from
binding the inhibitor

Leash—allows enzyme to bind
these inhibitors without
restrictions

Figure 4.9.   Effect of spacer arm ("leashes") on enzyme binding.

Individual hydrophobic interactions may be considered to be weak, since the chief driving force is entropic and amounts to only a few calories per residue. However, when much of the matrix is hydrophobic there may be appreciable interaction with the protein. . . . indeed, a wholly hydrophobic material, such as polystyrene, may literally turn a protein inside out with the hydrophobic amino acids, which normally are chiefly on the inside of a protein being turned to the outside to form "micellelike" regions on the polystyrene. (See Figure 4.10.) This interaction ultimately denatures the protein, and little or no active enzyme is recoverable from the polystyrene under any eluant conditions. Of course, hydrocarbon spacer arms attached to agarose or any hydrophobic matrix will not react as extremely with a protein as this, but these interactions can be appreciable.

The problem could be compounded by the presence of both ionic groups at the matrix surface and hydrocarbon moieties that serve as spacer arms. This is discussed more thoroughly in Chapter 9 since hydrophobic chromatography is a useful variant of bioselective adsorption, even though it lacks appreciable bioselectivity.

Quite naturally, nonspecific adsorption is not the only problem that may be encountered in bioselective adsorption. The stability of the matrix and the bond between it and the spacer arm or the ligand is equally important. Agarose, for example, is unstable in the presence of denaturing solvents and must be cross-linked with bifunctional reagents before it can be safely used with solutions containing urea, guanidine, detergents, or even certain chaotropic salts, such as LiCl or KSCN. Also the isourea group, formed by the usual cyangen bromide coupling procedures used with agarose, is not stable to prolonged incubation with nucleophilic reagents. β-Mercaptoethanol, hydroxide ion, or

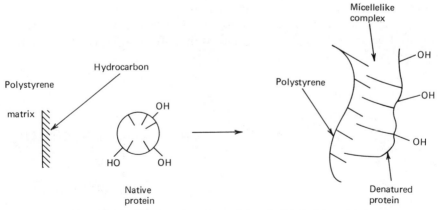

Figure 4.10. Denaturation of proteins on hydrophobic matrices.

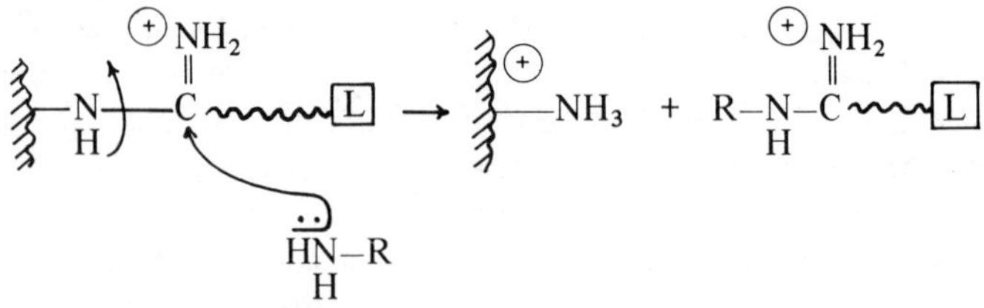

Figure 4.11. Elution by nucleophiles of ligands bound to CNBr activated agarose.

alkaline amines will react with it to remove the spacer arm and ligand and leave behind an ionic residue that is devoid of any bioselective adsorbent activity (30–32). (See Figure 4.11.) The ligand that is solubilized by this reaction may also elute with it any adsorbed enzyme and thereby cause an unpredictable loss of enzyme. The use of bisoxirane coupling with agarose eliminates this problem since the ether bond formed by this procedure is very stable. Likewise, avoidance of the use of alkaline pH levels and the inclusion of nucleophilic reagents in buffers applied to the adsorbent will help to minimize this problem.

A similar stability problem has been reported in the use of porous glass as a matrix, but in this case it appears to be the glass, per se, that is unstable (33). Although glass may appear to be an inert, insoluble substance, it actually possesses appreciable solubility when the surface area is large enough. The hydrolysis seems to involve a front side S_n2 attack by water on the silica utilizing the d orbitals of silicon (34). Similarly, substituents on the silane coating may participate in the hydrolysis. The classic example is the instability of β-halo silanes which spontaneously hydrolyze to yield silyl alcohols plus alkenes (35). (See Figure 4.12 for an example.) The resulting silanol group is acidic, with a pK comparable to that of tyrosine (i.e., 10.5); therefore, any exposed silanol surface has an anionic ion-exchange surface. Furthermore, the highly polarized silanol group will itself form a dipole attraction for positively charged proteins.

Since, in most other respects, porous glass or silica is an ideal support matrix, it would be useful to be able to circumvent this difficulty. Two approaches have

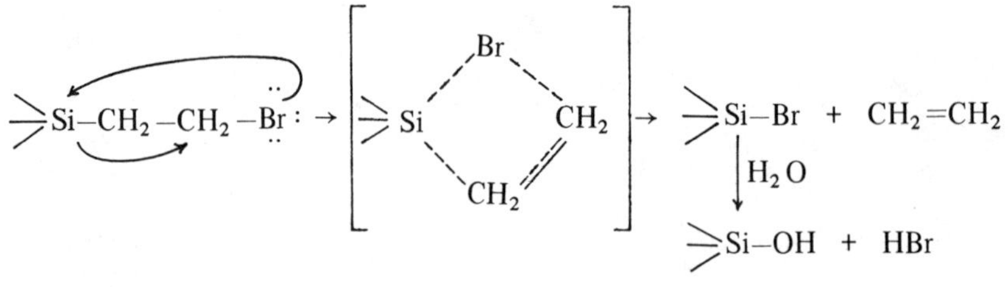

Figure 4.12. Hydrolysis of β-halosilanes.

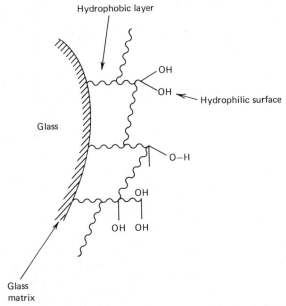

Figure 4.13. Glass coated with a protective double layer of hydrophobic and hydrophilic media.

proved useful: (1) creation of a hydrophobic coating immediately surrounding the glass (36,37) with hydrophilic ends, thus preventing water or other nucleophiles from reaching the silica groups and thereby protecting them from hydrolytic cleavage and (2) coating the glass with zirconia that is both partially positively charged and stable to hydrolysis under normal affinity chromatographic conditions (38,39). (See Figure 4.13.)

$$-\overset{\textstyle |}{\underset{\textstyle |}{Si}}-O-\overset{\textstyle |}{\underset{\textstyle |}{Zr}}-O-\overset{\textstyle |}{\underset{\textstyle |}{Si}}-CH_2\text{\wedge\wedge\wedge}L$$

The partial positive charge helps hold the two silica layers together, ultimately forming a very stable matrix.

There are many other potential difficulties in bioselective adsorption. Some of these, plus methods of evaluating the difficulty, are listed in Table 4.3. Certainly this list should not be considered exhaustive. Bioselective adsorption is a useful tool for the enzymologist, but by no means is it a replacement for classical purification methods nor a "one-step cure all." Rather, it is a valuable new technique that compliments classical methods and, far from being a magic one-step process, it requires thoughtful, careful laboratory application and analysis.

TABLE 4.3. Methods of Evaluating Bioselective Adsorption

Problem	Effect	Method of Evaluation
1. Ion-exchange type nonspecific adsorption	Poor purification: nonspecific purification process	Check effect of ionic strength and change of pH on purification; also attempt bioselective elution—if this isn't possible, nonspecific purification is suspected
2. Ligand–matrix bond instability	Adsorbent has poor and diminishing capacity; enzyme "tails" and is slowly eluted throughout the chromatography	Check eluate for free ligand
3. Ligand binds enzyme too tightly	Enzyme is not elutable from the column unless deforming buffers are used	Bind ligand to matrix by a cleavable linkage; when the linkage is cleaved, the enzyme should elute
4. Excess "activated-matrix" functions that covalently couple enzyme to the matrix	Enzyme is not elutable from the column, even when deforming buffers or denaturing solvents are employed as the eluant	Bind ligand to matrix as before; cleavage of ligand will not elute the enzyme—also, prior treatment of affinant with bovine serum albumin will prevent the problem
5. Hydrophobic adsorption (nonspecific)	As in (1)	Check the effect of glycerol or propylene glycol on purification
6. Hydrophobic–ionic combined non-specific adsorption ("detergent chromatography")	As in (1)	Neither propylene glycol nor high ionic strength alone will cause elution, but both together will

References

1. P. J. Graves and Y. T. Wu (1974), in *Methods in Enzymology* (W. B. Jakoby and M. Wilchek, eds.) **34**, 140.
2. P. C. Wankat (1974), *Anal. Chem.* **46**, 1400.

3. F. C. Denizot and M. A. Delange (1975), *Proc. Nat. Acad. Sci. (USA)* **72**, 4840.

4. B. M. Dunn and I. M. Chaiken (1974), *Proc. Nat. Acad. Sci. (USA)* **71**, 2382.

5. B. M. Dunn and I. M. Chaiken (1975), *Biochemistry* **14**, 2343.

6. T. C. J. Gribnau (1977), Ph.D. thesis, Nijmegen University, Nijmegen, The Netherlands.

7. T. C. J. Gribnau (1979), *J. Solid Phase Biochem.* (in press).

8. K. Kasai and S. Ishii (1975), *J. Biochem. (Tokyo)* **77**, 261.

9. A. H. Nishikawa, P. Bailon, and A. H. Ramel (1974), *Adv. Exp. Med. Biol.* **42**, 33.

10. C. R. Lowe, M. J. Harvey, and P. D. G. Dean (1974), *Eur. J. Biochem.* **42**, 1.

11. S. Okada, Y. Husimi, S. Tanabe, and A. Wada (1975), *Biopolymers* **14**, 33.

12. J. Turkova (1978), *Affinity Chromatography*, Elsevier, Amsterdam, p. 13.

13. For a brief discussion, see D. E. Metzler (1977), *Biochemistry: The Chemical Reactions of Living Cells*, Academic, New York, p. 305.

14. J. C. Giddings and H. Eyring (1955), *J. Phys. Chem.* **59**, 416.

15. A. J. P. Martin and R. L. M. Synge (1941), *Biochem. J.* **35**, 1358.

16. Program, adapted for IBM 7094, can be obtained from Wankat (2).

17. P. Cuatrecasas (1970), *J. Biol. Chem.* **245**, 3059.

18. B. H. J. Hofstee (1973), *Biochem. Biophys. Res. Commun.* **50**, 751.

19. A. H. Nishikawa and P. Bailon (1975), *Arch. Biochem. Biophys.* **168**, 576.

20. S. Kohn and M. Wilchek (1978), *Biochem. Biophys. Res. Commun.* **84**, 7.

21. K. Broström, S. Ekman, L. Kagedal, and S. Akerström (1974), *Acta Chem. Scand. B.* **28**, 102.

22. M. Wilchek and T. Miron (1974), in *Methods in Enzymology* (W. B. Jakoby and M. Wilchek, eds.) **34**, 72.

23. L. Sundberg and J. Porath (1974), *J. Chromatogr.* **90**, 87.

24. J. Porath and L. Sundberg (1970), *Protides Biol. Fluids, Proc. Colloq.* **18**, 401.

25. J. Porath (1974), in *Methods in Enzymology* (W. B. Jakoby and M. Wilchek, eds.) **34**, 13.

26. R. F. Murphy, J. M. Conlon, A. Imam, and G. J. C. Kelly (1977), *J. Chromatogr.* **135**, 427.

27. P. O'Carra, S. Barry, and T. Griffin (1974), in *Methods in Enzymology* (W. B. Jakoby and M. Wilchek, eds.) **34**, 108.

28. S. Barry and P. O'Carra (1973), *Biochem. J.* **134**, 595.

29. C. C. Huang and D. Aminoff (1974), *J. Chromatogr.* **371**, 462.

30. E. Janowicz and S. E. Charm (1976), *Biochim. Biophys. Acta* **428**, 157.

31. T. C. J. Gribnau and G. I. Tesser (1974), *Experientia* **30**, 1228.

32. G. I. Tesser, H.-U. Fisch, and R. Schwyzer (1974), *Helv. Chim. Acta* **57**, 1718.

33. H. H. Weetall and A. M. Filbert (1974), in *Methods in Enzymology* (W. B. Jakoby and M. Wilchek, eds.) **34**, 59.

34. R. D. Smith and P. E. Corbin (1949), *J. Am. Cer. Soc.* **32**, 195.

35. E. C. Rochow (1951), *Chemistry of the Silanes*, 2nd ed., Wiley, New York.

36. W. H. Scouten and A. Bergold (1978), unpublished results.

37. D. C. Locke, J. T. Schmermund, and B. Banner (1972), *Anal. Chem.* **44**, 90.

38. H. H. Weetall (1976), in *Methods in Enzymology* (K. Mosbach, ed.) **44**, 134.

39. R. A. Messing (1976), in *Methods in Enzymology* (K. Mosbach, ed.) **44**, 148.

GENERAL LIGANDS

Often separate affinity chromatography matrices have to be prepared for each protein to be purified, but for a large number of proteins, a bioselective adsorbent that has an affinity for an entire *class* of proteins, such as nucleoproteins, glycoproteins, dehydrogenases, and other similar groups or classes of proteins can be employed. Once a class of proteins has been adsorbed to a class specific ligand, termed a *general ligand*, classical purification techniques will permit separation of these proteins to purify one, or several, members of the group. These class- or group-specific bioselective adsorbents are generally termed "general" ligands to differentiate them from "specific" ligands that are recognized by only one or two enzymes.

Almost any classification scheme for proteins will suggest a potential "class" or group of general ligands. Enzymes that require coenzymes can be divided into NAD^+, biotin, coenzyme A, and other binding enzymes. Or glycoproteins may interact specifically with lectins, proteins that bind tightly to the carbohydrate portion of the glycoprotein. There are also receptor proteins on cell surfaces that recognize certain hormones, such as steroid receptors, cAMP receptors, insulin receptors, and antibody receptors. Then there are the nucleotide recognizing proteins, including those that recognize mononucleotides (ATP, AMP, etc.) and those that bind nucleic acids (RNA, DNA, and poly dAT). A large number of proteases will recognize and bind with varying affinity to a broad number of protein "substrates." Many proteinases are also inhibited by broad-spectrum proteinase "inhibitors," and thus these also form a group of "general ligand"-selective enzymes.

While the types of group-specific ligands are numerous, we concentrate on certain enzyme–cofactor "groups" and on glycoprotein–lectin complexes. Separate chapters are devoted to nucleic acid binding proteins, antigen–antibody complexes, and other protein–protein interactions.

5.1. IMMOBILIZED COENZYMES

Many enzymes require the presence of low-molecular-weight cofactors for enzymatic activity. These cofactors may bind to the active site and actually serve as cosubstrates, or they may bind to a different site, the allosteric site, where they function to regulate enzymatic activity. If the cofactor is very tightly bound to the enzyme, it is termed a *prosthetic group*. The enzyme plus the tightly bound prosthetic group is called a *holo enzyme*. If the prosthetic group is removed, the remaining protein portion of the enzyme is named the *apoenzyme*.

Very often the cofactor is not tightly bound to the enzyme, but serves as a soluble cofactor for many enzymes. Such a cofactor is termed a *coenzyme*. Among these, NAD^+ and $NADP^+$ are coenzymes that carry electrons between various parts of the cell; coenzyme A is a carrier for acyl groups, whereas folic acid and its derivatives serve to transport one-carbon units (methyl groups, methylene groups, etc.) throughout the cell.

The most widely studied application of immobilized cofactors in affinity chromatography is the use of immobilized analogues of NAD^+ and $NADP^+$. This is because most dehydrogenases require one or both of these cofactors (a few utilize other cofactors instead) and because dehydrogenases are enzymes of wide scientific and medical interest and broad distribution. For example, a substantial number of clinical tests are based on the presence of various dehydrogenases. In one medical application, the five isoenzymes of lactate dehydrogenase (an *isoenzyme* is an isomeric form of an enzyme) from human blood plasma can be separated and quantified by electrophoresis. The results of the electrophoresis will allow the physician to differentiate between cardiac difficulty or gall bladder problems, both of which may produce similar clinical symptoms.

5.1.1. Affinity Chromatography on NAD^+ ($NADP^+$) Analogues

Both NAD^+ and $NADP^+$ are multifunctional molecules that interact in a complex fashion with various enzymes. (See Figure 5.1)

Many dehydrogenases have a nucleotide-binding fold that is a folded region or domain of the enzyme that recognizes the adenine portion of the coenzyme. A similar domain is also present in many enzymes that utilize AMP, ATP, and other nucleotides. Thus AMP-based bioselective adsorbents are applicable not only for isolation of NAD^+ binding enzymes, but also for isolating a wide variety of other enzymes. Each recognizes the bioselective adsorbent and binds to it, although the degree of affinity to each adsorbent will be different for each enzyme. By utilizing bioselective elution procedures, each dehydrogenase that can bind to the adsorbent can be eluted separately and isolated in high purity.

One factor that permits bioselective elution of dehydrogenases from immobilized NAD^+ analogues is the fact that these enzymes generally form ternary complexes. A ternary complex occurs if the enzyme binds both NAD^+ and its other substrate at the same time in the same enzyme–substrate complex. Very

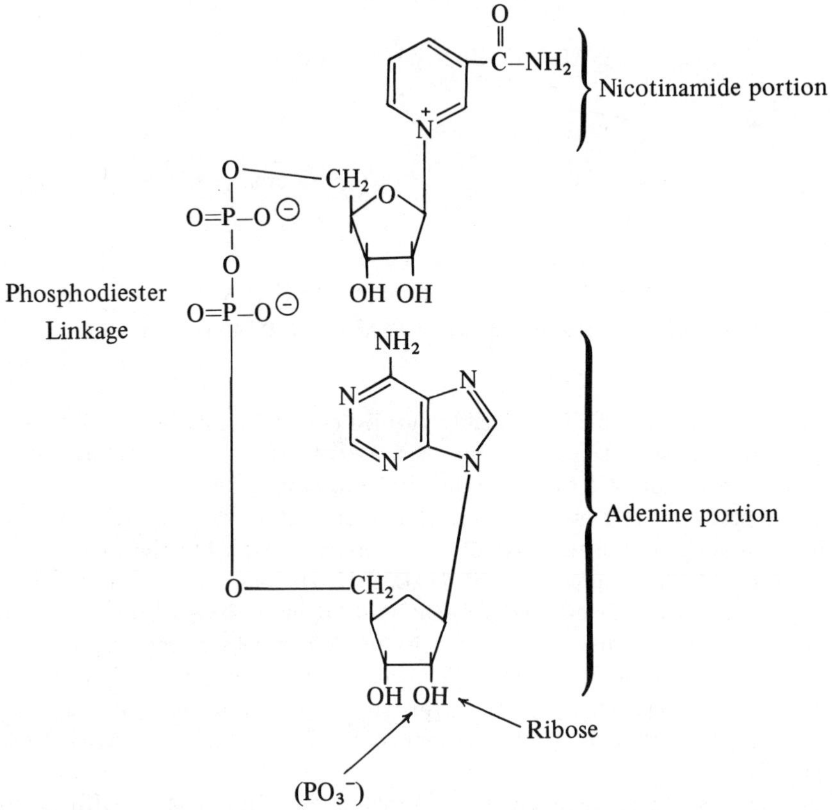

Figure 5.1. The structure of NAD^+ ($NADP^+$ contains PO_3^- at the point indicated.).

often NAD^+ must bind first (a compulsory ordered mechanism), and then the second substrate binds to the NAD^+-enzyme complex to form the ternary complex. Usually NAD^+ is bound much more tightly in the ternary complex than in the initial binary complex between enzyme and NAD^+. Finally, in the binary complex NADH is often more tightly bound than NAD^+. (See Figure 5.2.)

The immobilization of NAD^+ or $NADP^+$ to support matrices is often difficult because of the instability of NAD^+ or its analogues. In addition, many of the reactive portions of the coenzymes cannot be changed without altering the ability of the enzyme to recognize and bind to the immobilized NAD^+ analogue. Many such analogues (with various degrees of usefulness) have been synthesized, but often the synthesis is tedious, and whenever such analogues are available commercially, they are apt to be quite expensive. This may not be a significant problem for the investigator who wishes to use small columns of immobilized

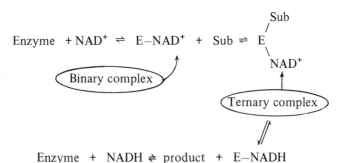

Figure 5.2. Compulsory ordered mechanism of NAD^+ utilizing enzymes.

ligand to "clean up" an already highly purified enzyme in final stages of purification, but it does not permit large-scale use of affinity chromatography for routine purification of NAD^+- ($NADP^+$)-utilizing enzymes.

There are, however, two methods of immobilization that are both easy and have lead to good purifications of some enzymes. Probably the method that should first be attempted for any NAD^+- ($NADP^+$)-bioselective adsorption is the preparation of diazo-NAD^+ ($NADP^+$) adsorbents. These adsorbents can be readily prepared and have a high capacity for certain dehydrogenases; for example, lactate dehydrogenases (1) and glyceraldehyde-3-phosphate dehydrogenase (2) have been purified by utilizing this adsorbent. (Note that the diazonium group attacks carbon 8 of the adenine portion of NAD^+ or $NADH^+$.) (See Figure 5.3.)

In the case of glyceraldehyde-3-phosphate dehydrogenase, Chaffotte and co-workers (3) were able to purify substantial quantities of the enzyme from sturgeon muscle with 90 to 95% recovery using this method. Conversely, yeast glyceraldehyde-3-phosphate dehydrogenase did not bind significantly to this bioselective adsorbent.

To purify sturgeon muscle glyceraldehyde-3-phosphate dehydrogenase, NAD^+ was coupled to a diazotized p-aminobenzoylated amino derivative of Sepharose. A six-carbon hydrophobic spacer arm and a three-carbon hydrophilic spacer arm were each employed in separate experiments. (See Figure 5.4.) Chromatographic supports prepared with either spacer arm was effective in adsorbing large amounts of enzyme (~ 60 mg/100 g of matrix). Only about 65% of the bound enzyme could be eluted from the material prepared from the hydrophobic packing, whereas over 95% could be recovered from the hydrophilic spacer arm containing matrix. The lower recovery from the hydrophobic spacer arm containing bioselective adsorbent was attributed to the previous observation (4) that muscle glyceraldehyde-3-phosphate dehydrogenase possesses a high degree of hydrophobicity. The enzyme was eluted from either adsorbent by using a linear gradient of NAD^+ (~ 20 to 100 mM), albeit the enzyme eluted as

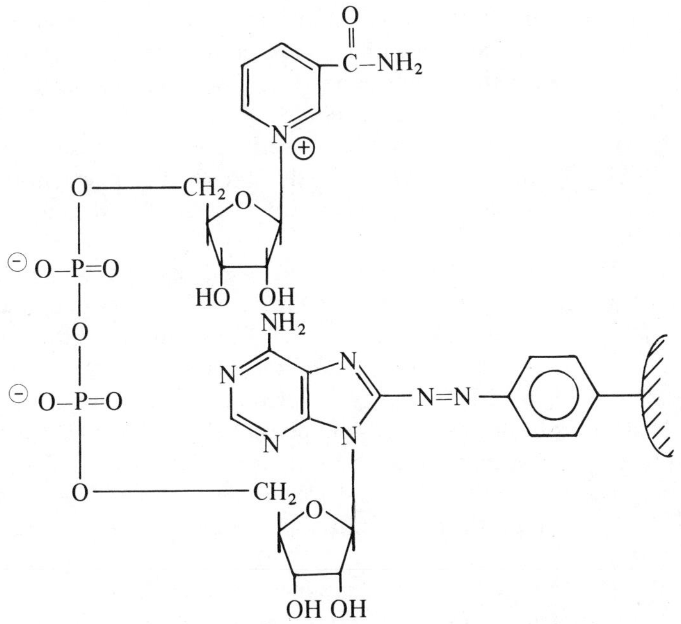

Figure 5.3. Structure of azo – NAD$^+$.

a sharp, well-defined peak from the matrix with the hydrophilic spacer and as a very diffuse, broad peak from the matrix prepared with the hydrophobic spacer arm.

Another readily prepared NAD$^+$ or NADP$^+$ affinity ligand is produced by coupling NAD$^+$ (NADP$^+$) to a carboxylate matrix by way of dicyclohexyl-carbodiimide or a similar condensing agent. This yields a matrix with NAD$^+$

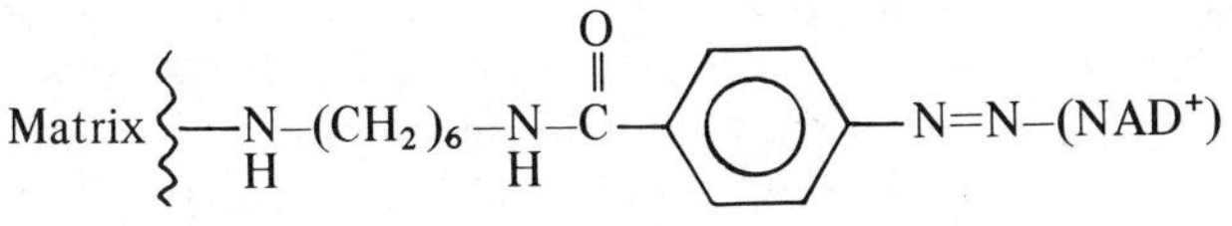

Six-carbon hydrophobic spacer arm

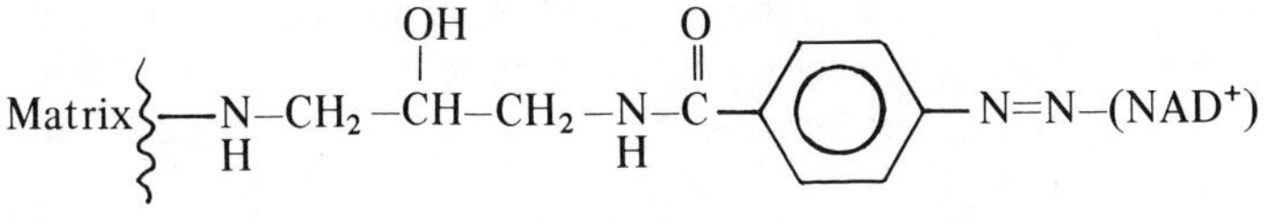

Three-carbon hydrophilic spacer arm

Figure 5.4. NAD$^+$ bound to a matrix by hydrophilic or hydrophobic spacer arms.

(NADP$^+$) bound through both the N^6 amino group and the ribosyl hydroxyls (5); such a heterologous and ill-defined bioselective adsorbent would be of very little use in quantitative applications of affinity chromatography, as for example, the determination of K_m or the demonstration of the existence of a ternary complex. Nonetheless, it could be a very useful matrix for purification of NAD$^+$- (NADP$^+$)-utilizing enzymes, chiefly because it can be synthesized in large quantities (up to 1 1 of adsorbent) at a reasonable cost and with little expense of time or effort.

Grover and Hammes (6) were able to purify β-hydroxybutyrate dehydrogenase by using a combination of affinity chromatography on ϵ-aminocaproyl-NAD agarose (prepared by this method) and hydrophobic chromatography (see page 241). This enzyme requires phospholipids for maximal activity, and as can be seen in Figure 5.5, the enzyme–lecithin complex binds much more tightly to NAD–agarose than does the non-phospholipid-containing apoenzyme.

The combination of ϵ-aminocapryl-NAD agarose and hydrophobic chromatography procedures produced a yield of 75% active, phospholipid-free enzyme. That is significantly more than the "classical" method of purification of this enzyme yields.

The utility of ϵ-aminocaroyl-NAD$^+$, produced by the carbodiimide method, has been demonstrated by Lowe et al. (5), who isolated a dozen different dehydrogenases using this method. Most were eluted by increasing the ionic strength of the eluant, although they could also be eluted with NAD$^+$-containing buffers, thus demonstrating that the purification was based on bioselective adsorption of these dehydrogenases.

The basic difficulty with this method of preparation of an NAD$^+$- or NADP$^+$-containing bioselective adsorbent is its heterogeneity. Frontal analysis chromatography by Lowe et al. (5,7) demonstrated that at least two different forms of immobilized NAD$^+$ existed in this adsorbent and that only about 0.1% of the bound ligand was responsible for the tight binding of the enzyme to the support packing. The remaining portion of the ligand bound the enzyme rather ineffectively. It was suggested that the tight-binding NAD$^+$ was attached to the matrix through the N^6 position of the adenine portion of the molecules, whereas the bulk of the enzyme was bound through the hydroxyls (8). In another method NAD$^+$ (NADP$^+$) has also been coupled to various support media by way of periodate cleavage of the ribosyl hydroxyls, followed by reaction of the resulting aldehydes with an amine-containing matrix and subsequent reduction of the Schiff base. Bioselective adsorbents prepared by this method failed to bind most dehydrogenases effectively (9). Similar results were seen with other methods in which NAD$^+$ or NADP$^+$ was bound by way of the ribosyl hydroxyls (10,11).

Although carbodiimide-coupled NAD$^+$ produces a heterogeneous bioselective adsorbent, its usefulness should not be underestimated. Whenever a fairly large

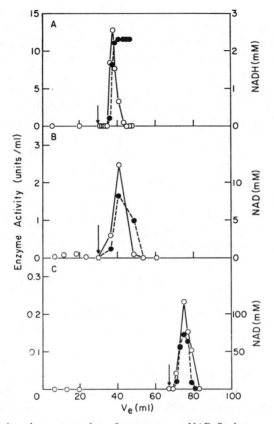

Figure 5.5. Affinity chromatography of enzymes on NAD–Sepharose. Arrows indicate when nucleotide was applied on column; solid lines and open circles represent enzyme activity; dotted lines and filled circles indicate nucleotide concentration. (*A*) Lactate dehydrogenase, 1 ml of 0.1-*M* sodium phosphate buffer (pH 7.5) containing 59 units of lactate dehydrogenase was applied on a 10-ml NAD–Sepharose column, which was washed with the same buffer, and the enzyme was eluted with 3-m*M* NAD in the same buffer. (*B*) Apo-β-hydroxybutyrate dehydrogenase, 2 ml of the β-hydrobutyrate dehydrogenase containing 14 units of the enzyme, specific activity 0.96 units/mg protein in 0.05-*M* sodium phosphate buffer (pH 7.1), 5-m*M* dithiothreitol, was applied to a 10-ml NAD–Sepharose column. The column then was washed with the application buffer, and the enzyme was eluted with 5 ml of 200-m*M* NAD prepared in the same buffer, readjusted to pH 7.2, followed by the application buffer. (*C*) β-Hydroxybutyrate dehydrogenase–lecithin complex, β-hydroxybutyrate dehydrogenase–soybean lecithin complex; NAD was removed by passage through two Sephadex G-25 columns equilibrated with 0.05-*M* sodium phosphate buffer (pH 7.8), 1-m*M* EDTA, 5-m*M* dithiothreitol, and 4.5 ml of this preparation containing 4.6 units of the enzyme were applied on the gel in the same buffer. The column was washed with the application buffer and eluted as in (*B*). Reprinted with permission from Grover and Hammes (6).

scale purification is attempted, this method is a likely first choice, if for no other reason than the minimal time and expense involved. Most dehydrogenase-binding bioselective adsorbents other than diazo-NAD^+ or carbodiimide-coupled NAD^+ are prepared from expensive, but homogeneous, derivatives of 5'-AMP. The only other reasonably priced bioselective adsorbent for dehydrogenases as a group is Blue Dextrin agarose and similar adsorbents containing Cibacron Blue F3GA. This is a dye that binds to most nucelotide-binding proteins, including kinases, dehydrogenases, and a number of nonnucleotide binding enzymes. The wide use of "blue matrices" is due to its ease of application, its commercial availability, and the fact that it has been widely investigated; thus a large body of literature exists concerning its use. Because of its broad application to many types of protein, a separate section is devoted to "Blue-Dextrin" chromatography.

Chemically defined NAD^+ ($NADP^+$) analogues, although somewhat expensive, can be very valuable in both quantitative and qualitative applications. These matrices consist primarily of NAD^+, NADP, 5'-AMP, or 5',2'-AMP coupled to a spacer arm through the N^6 or C^8 positions of the adenine rings. N^6-Carboxymethyl-NAD^+, for example, has been immobilized to agarose and is reported to have a high affinity for many dehydrogenases (5,12).

While sturgeon muscle glyceraldehyde dehydrogenase is bioselectively adsorbed to diazo-NAD^+ supports, the yeast enzyme is not retained on this type of adsorbent. However, the yeast glyceraldehyde dehydrogenase can be purified by using N^6-(6-aminohexyl)-AMP coupled to CNBr-activated agarose (3). The yeast glyceraldehyde-3-phosphate dehydrogenase is bioselectively eluted from this material with 1-mM NAD^+, resulting in an 8.7-fold purification over the crude cell extract. Since this enzyme constitutes a large percentage of the protein in yeast grown under the conditions employed in this experiment, the 8.7 purification resulted in a nearly homogeneous enzyme product. Moreover, Chaffotte et al. (3) adapted the method to employ a fairly large column (34 cm $\times$ 4 cm) with a yield of approximately 160 mg and a recovery of 22%. Such a large column would be prohibitively expensive if purchased commercially. It can, however, be easily prepared in a laboratory. Scaleup of the synthesis of most bioselective adsorbents from 1-ml to 1000-ml volumes is usually easily effected.

Interestingly, even when agarose was attached in a chemically defined fashion to the N^6-adenine portion of the 5'-AMP, the adsorbent (like ϵ-aminocaproyl-NAD^+) exhibited a biphasic adsorption of enzyme activity, suggesting the existence of two or more types of immobilized ligand. This could be due to chemical differences, but it would be just as likely that these differences are geometric, rather than chemical, and that a portion of the immobilized N^6-(6-amionhexyl)-5'-AMP are much more available for binding than a second class of sterically hindered ligand. This situation would result in two *apparent*

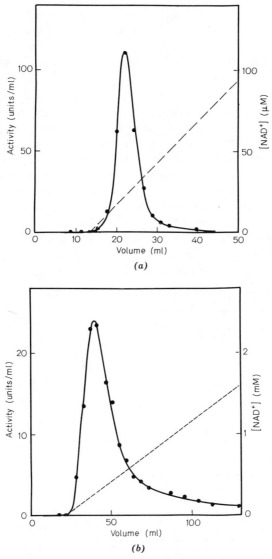

Figure 5.6. (*A*) Interaction of sturgeon apoglyceraldehyde-3-phosphate dehydrogenase with the hydrophilic azo-linked NAD'–Sepharose derivative (1a); 2.55 mg of enzyme in 2.5 ml (specific activity 298 units mg) was applied to a column of derivative 1a (13 cm × 0.9 cm). The enzyme was eluted at a flow rate of 25 ml/hour with a linear gradient of NAD. (*B*) Interaction of sturgeon apoglyceraldehyde-3-phosphate dehydrogenase with the hydrophobic azo-linked NAD'–Sepharose derivative (1b); 1 mg of enzyme in 2 ml specific activity 250 units/mg) was applied to a column of derivatives 1b (11.8 cm × 0.9 cm). The enzyme was eluted at a flow rate of 30 ml/hour with a linear gradient of NAD. Reprinted with permission from A. F. Chaffotte, C. Roucous, and F. Seydoux (1977), *Eur. J. Biochem.* 78, 309.

binding constants for the formation of the same enzyme-immobilized ligand complex.

Andersson et al. (12) used N^6-(6-aminohexyl)-5'-AMP to purify and preparatively separate the isoenzymes of horse liver alcohol dehydrogenase. It was essential to first partially purify the enzyme by chromatography of the crude cell extract on a carboxymethyl cellulose column prior to affinity chromatography. This procedure was apparently necessary because of the relatively large number of AMP-recognizing enzymes present in the crude extract that competed with the alcohol dehydrogenase for the immobilized ligand. After partial purification, the extract was applied to a column of N^6-(6-aminohexyl)-5'-AMP-agarose and eluted with a mixture of 1.5-mM cholic acid plus 0.2-mM NAD$^+$. The resulting alcohol dehydrogenase peak fractions were pooled, dialyzed, and rechromatographed on a second immobilized 5'-AMP agarose column. In this case the enzyme was eluted by utilizing a gradient of NAD$^+$ (0 to 8 mM) in 1.5-mM cholic acid. The steroid-binding isoenzyme of alcohol dehydrogenase forms a tight ternary complex with soluble NAD$^+$ and cholic

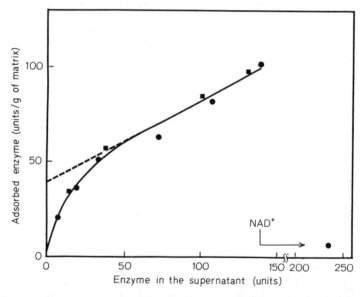

Figure 5.7. Batchwise adsorption of yeast glyceraldehyde-3-phosphate dehydrogenase in a crude extract on immobilized AMP derivative (11). The specific activity of the enzyme in the crude extract was 15.7 units/mg; (■) 1 g of derivative II suspended in 5 ml of Tris sulfonate buffer, (●) 3 g of derivative II in 15 ml of Tris sulfonate buffer. Desorption of the enzyme on addition of NAD (final concentration 0.5 mM) in incubation mixture is indicated by arrow. Reprinted with permission from A. F. Chaffotte, C. Roucous, and F. Seydoux (1977), *Eur. J. Biochem.* **78**, 309.

acid in the eluant and is thus eluted early in the gradient. The nonsteroid-recognizing isoenzyme does not form a ternary complex of NAD$^+$ and cholate and thus was only eluted at higher concentrations of NAD$^+$. In this case 29 mg of the steroid-utilizing isoenzyme was purified from 23 g of crude liver protein. The yield overall was 18%, but most of the loss occurred during the carboxy-methyl cellulose chromatography. Both affinity chromatographic steps resulted in essentially quantitative recoveries. (See Figure 5.8 and Table 5.1.)

Immobilized AMP analogues can be modified for application in purification of NADP$^+$-binding proteins by employing immobilized 2'-AMP or 2'5'-AMP derivatives. Wermuth and Kaplan (13) employed 8-(6-aminohexyl)amino 2'-AMP-agarose to obtain a homogeneous preparation of pyridine nucleotide transhydrogenase from *Pseudomonas aeruginosa,* with a yield of 15 mg of enzyme from 800 g of cell paste. The overall yield was 22%, but the recovery in the affinity chromatography step was 71%, with most of the loss once again in the classical purification steps, namely, ion-exchange and gel-permeation chromatography.

In this purification affinity chromatography could be employed as the initial step since there are far fewer enzymes that recognize 2'-AMP than bind to 5'-AMP. Kinases, NAD$^+$ dehydrogenases, and a number of enzymes with AMP regulatory sites bind to 5'-AMP, but only NADP$^+$-utilizing enzymes bind 2'-AMP. In this preparation 50 ml of 2'-AMP-agarose was added in batch mode to 2 l of a crude sonicate prepared from 800 g of cells and the mixture stirred overnight at 4°C. The adsorbent was collected the following day by suction filtration; it was washed with buffer and the enzyme was then eluted with 10-mM 2'-AMP (NADP$^+$ and NADPH also effected elution, but neither 5'-AMP, NAD$^+$, nor NADH would do so). This batch absorption effected an approximately 150-fold purification. Subsequently the enzyme was purified another fivefold by the combination of ion-exchange and gel-permeation chromatography.

Proper maintenance of pH in this procedure was critical, as the capacity of the bioselective adsorbent at pH 9.0 was only half that at pH 7.0. Conversely, elution with 2'-AMP was effective from pH 8 to 9, but below pH 7.0 the enzyme was bound so tightly to the adsorbent that elution with 10-mM 2'-AMP was impossible.

These investigators prepared 2'-AMP-agarose by the method of Lee et al. (14), but commercial preparations could have been employed equally well, but at a cost, based on current prices, of about $600.00 for 40 ml of reusable adsorbent. Hopefully, new synthetic procedures and increased use of general ligands in affinity chromatography will decrease the cost of preparative quantities of these adsorbents.

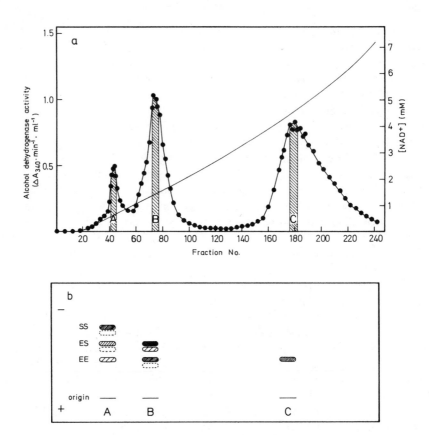

Figure 5.8. Separation of the isozymes of alcohol dehydrogenase from a horse liver crude extract on an AMP–Sepharose column by elution with NAD plus cholic acid, (*a*) elution profile from the column (1 × 14 cm containing 9.6 ml wet gel) loaded with 7 ml of a crude extract containing 340 mg of protein and 26 units of ethanol dehydrogenase activity. Prior to elution with a gradient of NAD (0 to 8 m*M*, 1.5 m*M* in cholic acid, total volume 200 ml) the column was washed with 5 volumes of 0.1-*M* sodium phosphate, pH 7.5, 1 m*M*, in glutathione followed by 3 volumes of the same solution but also 1.5 m*M* in cholic acid. Fractions of 0.8 ml were collected at a flow rate of 3.0 ml/hour (● ●). Horse liver ethanol dehydrogenase activity. (*b*) Agarose–gel electrophoresis in 0.05-*M* Tris-HCl, pH 8.5, followed by staining for enzymatic activity. Hatched areas indicate less intense bands and dotted outlines hardly visible bands. The relative intensity of the bands varies somewhat with concentration of samples. Even the pattern of bands may vary slightly with formation of subfractions. This can, however, be depressed by the presence of thiols. Forms just below the main isozymes are subfractions and the letters refer to the hatched areas in (*a*). Reprinted with permission from Andersson et al. (12).

120

TABLE 5.1. Purification of Homogeneous Steroid-Active Isozyme of Horse Liver Alcohol Dehydrogenase by Affinity Chromatography on a Sepharose-Bound AMP-Analogue[a]

Step	Enzyme Activity Units/ml	Enzyme Activity Total Units	Total Protein, mg	Specific Activity, Units/mg	Purification n-fold	Yield, %
Crude extract	0.67 (4.5)	321 (2160)	23,000	0.014 (0.09)	1 (1)	100 (100)
CM-Cellulose	0.13 (1.6)	115 (146)	1110	0.10 (0.13)	7.1 (1.4)	36 (6.8)
Ammonium sulfate precipitation	0.67 (0.89)	108 (144)	930	0.12 (0.15)	8.6 (1.6)	34 (6.7)
First-affinity chromatography	1.0 (0.62)	71 (43)	40	1.8 (1.1)	129 (11)	22 (2.0)
Second-affinity chromatography	0.84 (0.46)	58 (31)	29	2.0 (1.1)	143 (11)	18 (1.4)

Source: Reprinted with permission from L. Anderson, H. Jörnvall, and K. Mosbach (1975), Anal. Biochem. 69, 401.

[a] The first figures for specific activity, purification, and yield apply to alcohol dehydrogenase activity tested against 5β-DHT as substrate to estimate values most closely related to the SS isozyme. Because of the presence of the ES isozyme in the crude extract, the value for the yield is still a minimum value. Figures for alcohol dehydrogenase activity tested against ethanol as substrate are given within parentheses and give little indication about purification or yield of the SS isozyme.

5.2. ANALOGUES OF ATP, ADP, AND AMP
FOR THE PURIFICATION OF KINASES

Until recently no simple method, other than Cibacron Blue chromatographic packings, existed to prepare general ligand adsorbents for kinases and most other ATP-AMP-ADP (ADP-adenosine diphosphate; ATP-adenosine triphospate) utilizing enzymes. The N^6 derivatives of 5'-AMP, which are very useful for purification of dehydrogenase, give poor results with kinases since relatively few kinases bind to these adsorbents. Recently, however, 8-azo-ATP, 8-azo-ADP, 8-(6-aminohexylamino)ATP, and 3-aminopyridine-adenine dinucleotide derivatives have been employed for the purification of a substantial number of kinases from such diverse animal tissue as fish muscle and human brain (15). Just as significantly, Lee et al. have developed a rapid synthesis of 8-(6-aminohexylamino)-ATP and a synthesis of 3-aminopyridine adenine dinucleotide–agarose bioselective adsorbents that is completed in a single 24-hour procedure, starting with the commercially available precursor, 3-aminopyridine adenine dinucleotide.

By using a combination of either isoelectric focusing or ammonium sulfate precipitation, plus chromatography on 8-azo-ATP-agarose homogeneous preparations of creatine kinase, adenylate kinase, pyruvate kinase, and aldolase from a *single* rabbit muscle extract could be obtained in two steps. The crude extract was applied to a 60-ml column of 8-azo-ATP that was subsequently eluted with a gradient of 0- to 10-mM ATP. As shown in Figure 5.9, adenylate kinase

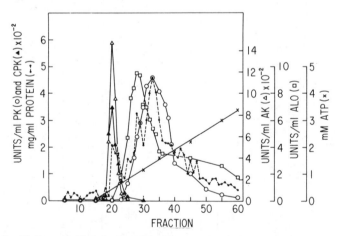

Figure 5.9. Elution profile of adsorbed kinases and aldolase from 8-azo-ATP–Sepharose column (1 × 15 cm). An ATP gradient of 0 to 10 mM (X) was employed to elute adenylate kinase (△), creatine kinase (▲), pyruvate kinase (0), and aldolase (■). Dashed line represents protein concentration. Reprinted with permission from Lee et al (15).

and creatinase kinase are eluted together in a sharp peak at about 0.5-mM ATP, whereas pyruvate kinase and aldolase are slightly separated, eluting at about 1.0- and 1.5-mM ATP, respectively. The creatinine and adenylate kinases are then separated by isoelectric focusing while pyruvate kinase and aldolase are separated by "salting out" with ammonium sulfate. Homogeneous, crystalline enzymes result from these procedures. (See Figure 5.10 and Table 5.2.)

Whereas 8-azo-ATP-agarose was the ligand used in most of these studies, a number of immobilized ATP analogues were used with excellent results in several cases. For example, human brain adenylate kinase was retained by 8-azo-ATP agarose, whereas human brain creatinine kinase was not adsorbed. The retained adenylate kinase could be eluted free of creatinine kinase activity in 80% purity by a gradient of ATP. Conversely, 3-aminopyridine adenine dinucleotide-agarose retained creatinine kinase, but not adenylate kinase. The retained creatinine kinase was elutable with ATP in approximately 30% purity. Both adenylate kinase and creatinine kinase were retained on 8-(6-aminohexyl-amino)-ATP-agarose and coeluted in an ATP gradient.

Utilizing a much more laborious procedure, Trayer et al. (16,17) also synthesized 8-(6-aminohexylamino)-ATP-agarose and demonstrated its usefulness in the purification of glucokinase and the ATP-ase fragments derived from myosin. They also synthesized P^1-(6-aminohex-1-yl)-P^2-5^1-(adenosyl)pyrophosphate-agarose and demonstrated its utility as a bioselective adsorbent for the same enzymes. Conversely, N^6-(6-aminohexyl)-AMP-agarose and its ATP analogue were both ineffective in retaining these enzymes. It would therefore, appear that many kinases required a free N^6 amino function. (See Figure 5.11.)

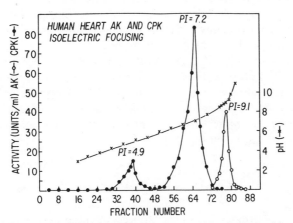

Figure 5.10. Separation of adenylate kinase and isoenzymes of creatine kinase from human heart by isoelectric focusing. Reprinted with permission from Lee et al. (15).

**TABLE 5.2. Purification Procedure for Kinases
from Rabbit Muscle Extract[a]**

Purification Step	Specific Activity (IU/mg)			
	Creatine Kinase	Adenylate Kinase	Pyruvate Kinase	Aldolase
Crude extract	20	7	7.4	0.24
8-azo-ATP-Sepharose	185	625	100	4.5
Isoelectric focusing	650	1400	—	—
Ammonium sulfate				
fractionation	—	—	160	20

Source: Reprinted with permission from Lee et al. (15).

[a]Homogeneous enzymes were obtained after the second step. The data represent the average results of three complete experiments.

Very successful results have been reported by Ramadoss et al. (18) on the purification of phosphofructokinase using N^6-(6-aminohexyl)-carbamoyl-methyl)-ATP–agarose. This easily synthesized bioligand was minimally 90 times more effective in adsorbing this enzyme than was N^6-(6-aminohexyl)-ATP-agarose. The bound phosphofructokinase is eluted with a gradient of fructose-diphosphate and ADP. This step, followed by a single ultracentrifugation, yielded a homogeneous enzyme. Yields were often quantitative, with a minimum of 50% recovery, depending on the source of the crude extract. The major loss of protein was in the "classical" ultracentrifugation step, and the recovery from the bioselective adsorption step was usually quantitative.

Lindberg and Mosbach (19) first demonstrated the usefulness of N^6-[(6-aminohexyl)carbamoyl-methyl]-ATP as a bioligand for affinity chromatography, using it to purify citrate synthetase from pig heart.

Many other derivatives of ATP, AMP, and ADP have been prepared, with mixed results. Derivatives of the C_8 position are useful primarily for the purification of kinases. Analogues with modification of the phosphate groups have not been generally useful bioligands, although the *p*-aminophenyl derivative of dATP is an effective bioselective adsorbent for T_4 and *E. coli* ribonucleotide reductase (20,21). (See Figure 5.12.)

5.3. OTHER NUCLEOTIDE-UTILIZING ENZYMES

Many other nucleotides have been utilized in the bioselective adsorption of various proteins. Whereas ATP functions physiologically as a general energy

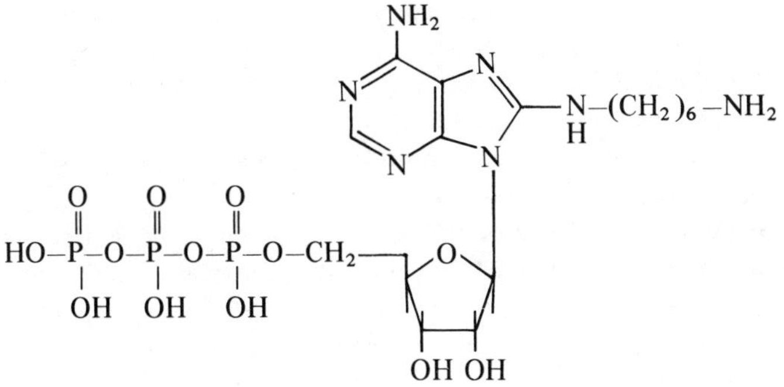

8-(6 Aminohexylamino)—ATP

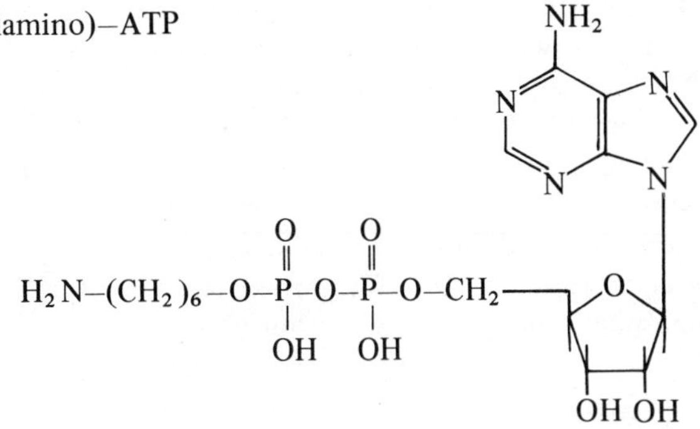

P^1-(6-Aminohex-1-yl)-P^2-5^1-(adenosyl)pyrophosphate (HADP)

Figure 5.11. Amino–ATP derivatives used as bioligands.

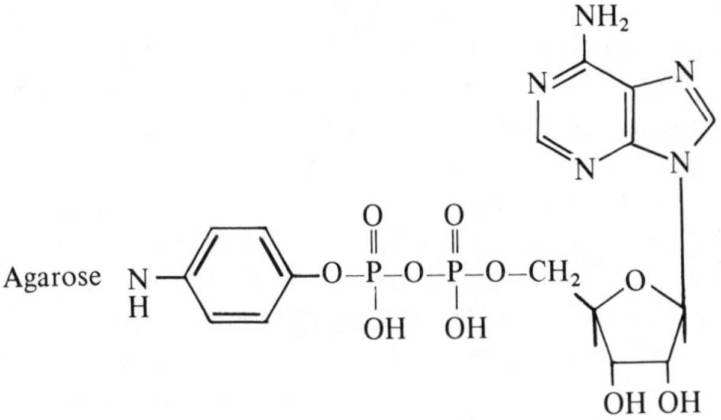

Figure 5.12. P^2-(p-Aminophenyl)-P^1-adenyl phosphate agarose.

125

source, other nucleotides have very restricted usages. Uridine derivatives are involved as cofactors in many sugar transferase reactions, cytidine moieties are involved with phospholipid and glycoprotein sysnthesis, and GTP functions in protein synthesis reactions. Because only a small group of enzymes recognize these nucleotides, a high degree of purification is obtainable by using bioselective adsorption on immobilized GTP, CTP, UTP.

One striking example of enzyme purification on nonadenine nucleotides is the recent purification by Paulson et al. (22) of the sialyltransferase from bovine colostrum. A 440,000-fold overall purification was effected in six steps that included two affinity chromatographic steps on CDP-agarose. The largest purification in a single step was approximately 1500 fold in the first of the two bioselective adsorption procedures. A 28-fold purification was achieved by the second. In the first affinity chromatography step, the enzyme was eluted by an NaCl gradient, whereas the second step included bioselective elution by using CDP. The recovery for each was only 50 to 75%, which is somewhat lower than in many other affinity chromatographic procedures. This could possibly be caused by the hydrophobic nature of the spacer arm used in this preparation. Several workers have reported increased recovery occur when hydrophobic spacer arms were replaced with hydrophilic ones; for example, Chaffotte et al. (3) obtained approximately three-fold improvement in the recovery of glyceraldehyde-3-phosphate dehydrogenase. Conversely, Lowe and Mosbach (23) were able to increase the recovery of lactate dehydrogenase from a medium consisting of NAD^+ coupled to agarose by way of a hydrophobic spacer, by increasing the hydrophobicity of the buffer.

Dihydroneopterin triphosphate synthetase has been purified from dialyzed crude extracts of *Lactobacillis plantarium* by using GTP that was coupled to agarose by the periodate method of Gilham (24) (page 230). The enzyme was eluted with a recovery of 60 to 70% and a 350-fold purification by using GTP solutions. The presence of phosphatases in the crude extract produced complications but these were eliminated by initial partial purification of the enzyme by classical methods. Immobilized GDP, GMP, or guanosine analogues were ineffective for the preparation of a useful bioselective adsorbent.

Paulson et al. (22,25) also partially purified the UDP galactosyltransferase from colostrum with excellent results by chromatography on UDP-hexylamine-agarose (see Figure 5.8). Previously, UDP-galactose galactosyl transferase had been purified from bovine milk and milk whey by using the same bioselective adsorbent (26-28). By chromatography on UDP-agarose, a 163-fold purification of the enzyme could be effected from crude whey. Further purification (ca. fivefold) was accomplished by affinity chromatography on α-lactalbumin-agarose to yield an electrophoretically homogeneous protein. α-Lactalbumin is a protein component of milk that combines with UDP–galactose-galactosyltransferase to form the lactose synthetase complex responsible for the formation of milk lactose.

Elution of the enzyme could be performed by using a "deforming" agent (e.g., urea), by using EDTA to remove the Mn^{++} ions needed for UDP–galactose binding, or by competitive elution with N-acetylgalactoseamine. A combination of both EDTA and N-acetylgalactoseamine proved to be the most effective method with recoveries of 60 to 80%. The entire purification from whey (based on the use of two bioselective adsorbents) gave a 6209-fold overall purification with a 40% yield.

Uridine diphosphate-agarose has also been used by Barker et al. (26) for the purification of glycogen synthetase from rabbit muscle. As shown in Figure 5.13, most of the glycogen synthetase activity from the rabbit muscle extract (which was pretreated with α-amylase to remove endogenous glycogen) was bound to the bioselective adsorbent. The enzyme was not eluted with wash buffer, with glycerol phosphate, nor with 10-mM UDP, but it was recovered free of glycogen phosphorylase by elution with a buffer containing its other substrate, namely, glycogen.

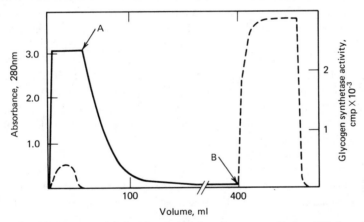

Figure 5.13. The glycogen-rich fraction containing glycogen synthetase (20) from 600 g of rabbit muscle was incubated with pancreatic α-amylase (2mg) in 50 ml of 0.05-M glycerol phosphate buffer, pH 7.8, containing 0.004-M EDTA, 0.04-M mercaptoethanol, and 0.005-M NaCl. After incubation at 37°C for 2 hours the digest was centrifuged at 30,000g, and the resulting supernatant solution was applied to a column (1 X 10 cm) of UDP-Sepharose at 25°C. After the sample was applied the column was washed at A (arrow) with the glycerol phosphate buffer (500 ml) and then with the same buffer containing glycogen (20 mg per ml) at B (arrow). Solid line indicates adsorbence and dashed line, glycogen synthetase activity. Reprinted with permission of the American Society of Biological Chemists, Inc. from Barker et al., (26).

5.4. CIBACRON BLUE–AGAROSE
AND BLUE DEXTRAN–AGAROSE

Blue Dextran 2000 was initially synthesized for use as a void volume marker for gel-permeation chromatography. This was accomplished by coupling the reactive dye, Cibacron Blue F3GA, to a high-molecular-weight (2,000,000) polydextran (29). Occasionally when the Blue Dextran was mixed with certain enzymes, the Blue Dextran and the enzyme eluted together in the void volume during gel-permeation chromatography, even if the protein, in the absence of Blue Dextran, would normally be found in the inclusion volume of the column (31–34). Such initial experimental suggestions that proteins bind to Blue Dextran were followed by investigations of the nature of the protein–Blue Dextran interaction (35–36) and by various applications of the protein–Blue Dextran interaction to the purification of several enzymes (32, 35–38). A soluble Blue Dextran polymer was initially used, perhaps because of its ready commercial availability, after which methods were later developed to immobilize Blue Dextran 2000 to form a new chromatographic media for enzyme purification.

Koppenschlager et al. (35, 391) were the first to immobilize Blue-Dextran 2000 by trapping the macromolecule within the pores of a polyacrylamide gel, thereby leaving the chromophore unaltered by the immobilization. (See Figure 5.14.) They were able to demonstrate that yeast phosphofructokinase can be purified by chromatography on the immobilized dye by using ATP (2 to 3 mM) as the eluant. Elution with AMP or ITP was not possible. The second substrate, fructose-6-phosphate, effected the elution of the enzyme minimally at alkaline pH and actually increased the strength of the enzyme dye complex formation at pH below 7.0. Kinetic analysis indicated that Blue Dextran was a competitive inhibitor with respect to ATP but that it did not show competitive inhibition with regard to fructose-6-phosphate. This fact strongly suggests that Blue Dextran binds to the same site as ATP.

Ryan and Vestling (41) coupled Blue Dextran to CNBr-activated agarose and obtained a useful affinity chromatography packing that could be readily prepared in large quantities. For example, lactate dehydrogenase is (42,43)

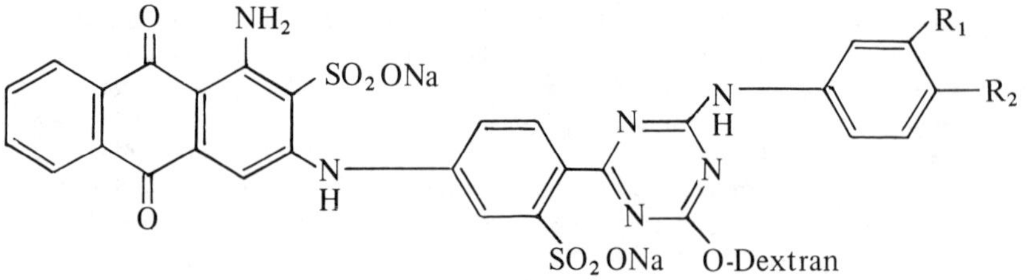

Figure 5.14.   Blue Dextran 2000.

adsorbed to Blue Dextran agarose and can be bioselectively eluted with 1-mM NADH. Malate dehydrogenase (43), adenylate cyclase (44), phosphoglycerate kinase (42), ribonuclease (45), alcohol dehydrogenase (46), and pyruvate kinase (42), among others, have been purified by use of this type of bioselective adsorbent. In later investigations the dye, Cibacron Blue F3GA, was attached through the triazine function directly to the matrix. Bohme et al. (47) and Roschlau and Hess (48) coupled the dye to Sephadex whereas Esterday and Esterday (45) and Heyns and DeMoor (49) coupled it to agarose. Both materials were effective as bioselective adsorbents, but the agarose-based material was generally more useful for reasons discussed previously (see pages 21 through 24).

Immobilized Cibacron Blue has a very broad affinity for a wide range of proteins that includes both those possessing a nucleotide-binding site and some without such a site. A type of hydrophobic chromatography, sometimes termed "detergent chromatography," is the probable explanation for some of the binding. This results from a combination of ionic and hydrophobic interactions (50). To visualize this situation, a protein should be viewed as chiefly a hydrophilic ion-bearing surface surrounding a primarily hydrophobic core. On this hydrophilic surface, however, there are crevices reaching into the core. Some of the crevices are enzyme active sites; others are receptor sites or places for membrane-protein contact. Still others may have no biological function. These crevices may possess positive or negative groups near their surfaces. An immobilized amphipathic compound with both ionic and hydrophobic groups will "fit" into these crevices, thereby causing the proteins to bind to the adsorbent. Such binding *may* be bioselective if the amphipathic compound "fits" a particular set of ionic and hydrophobic residues at the active site (see Chapter 9). (See Figure 5-15.)

A similar type of nonbiospecific binding was proposed by Wilson (51) to account for the observation that Blue-Agarose and Blue Dextran bind serum albumin (52,53) and interferon (54). However, the concept that Cibacron Blue F3GA binds in a specific way to the nucleotide binding sites of kinases and

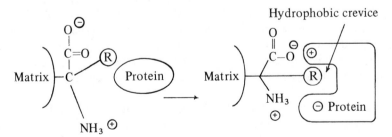

Figure 5.15. Binding of an amphipathic adsorbent to a protein.

dehydrogenases has been rather fully substantiated. First, ATP and/or other nucleotides will compete with Cibacron Blue for the active site of these enzymes as shown by kinetic analysis (55). Second, nucleotides that do *not* bind to a specific enzyme will not elute the enzyme from a Cibacron Blue bioselective adsorbent. Conversely, those nucleotides that bind to the enzyme are very effective bioselective eluants. Finally, Bohme et al. (47) immobilized the related dye, Cibacron Brilliant Blue FBR-P, on dextran and showed that despite the very close similarity between the two dyes, and in spite of the fact that they have identical numbers and types of ionic and hydrophobic groups that are merely arranged differently on the two dyes, the Blue FBR-P dye complex was incapable of adsorbing phosphofructokinase under conditions in which Cibacron Blue F3GA bound the enzyme tightly.

In addition to these observations, Edwards and Woody (56) have studied the binding of Cibacron Blue F3GA and a similar dye, Congo Red, to various dehydrogenases using circular dichoism (CD) of the chromophore absorbence bands. Both Cibacron Blue and Congo Red yielded strong CD spectra when bound to M_4-lactate dehydrogenase, H_4-lactate dehydrogenase, glyceraldehyde-3-phosphate dehydrogenase, or malate dehydrogenase. The induced circular dichroism progressively disappeared as increasing concentrations of NAD^+ (or $NADP^+$, if appropriate) were added again, indicating that the dye and NAD^+ ($NADP^+$) competed for the same site. Congo Red appears to bind these dehydrogenases more tightly than does Cibacron Blue F3GA and thus might be a better bioligand for general ligand chromatography. (See Figures 5.16 and 5.17.) No doubt future investigations will yield other amphipathic molecules with even more useful characteristics.

Over 30 different enzymes have been purified by use of Cibacron Blue F3GA, among the most recent of which was the purification of two deoxynucleoside kinases from *Lactobacillius acidophilus*. A 1225-fold overall purification was effected using a combination of streptomycin and ammonium sulfate precipitations plus bioselective adsorption on Cibracon Blue F3GA-agarose (57). When an ammonium sulfate-fractionated extract was applied to the bioselective adsorbent and eluted with a gradient of Mg^{++} and ATP in 15% glycerol, the deoxynucleoside kinase activity emerged in two separate peaks. The first peak showed activity toward deoxycytosine and deoxyadenosine, and the second had activity toward deoxyguanosine and deoxyadenosine. Overall yields were between 10 and 20% and 4 to 19 mg of enzyme from 400 gms of cells. The most substantial loss of enzyme occurred during the separation of the two different deoxynucleotide kinases, which were initially eluted from the Blue–agarose as overlapping peaks.

Lamkin and King (46) superbly illustrated the usefulness of Blue–Agarose for isolating dehydrogenases in a relatively pure state from microamounts of starting material. They obtained 73 μg of electrophoretically homogeneous

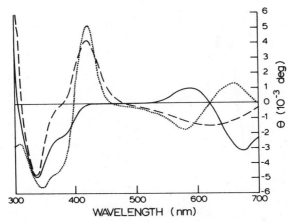

Figure 5.16. Circular dichroism spectra of beef heart lactic dehydrogenase (LDH) plus Congo Red:

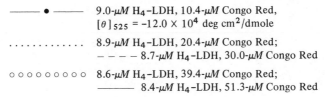

————— • ————— 9.0-μM H_4-LDH, 10.4-μM Congo Red, $[\theta]_{525} = -12.0 \times 10^4$ deg cm^2/dmole

. . . . 8.9-μM H_4-LDH, 20.4-μM Congo Red; — — — — 8.7-μM H_4-LDH, 30.0-μM Congo Red

o o o o o o o o o 8.6-μM H_4-LDH, 39.4-μM Congo Red; ————— 8.4-μM H_4-LDH, 51.3-μM Congo Red

Reprinted with permission from R. A. Edwards and R. W. Woody, *Biochem. Biophys. Res. Commun.* 79, 470 (1977).

alcohol dehydrogenase from 500 mg of cotton seed mill by using blue–agarose chromatography as the sole purification step. The purified enzyme produced one electrophoresis band when stained for protein and two bands, a major one and a very faint one, when stained for alcohol dehydrogenase activity by way of the tetrazolium blue method. Many other examples of the purification of dehydrogenases and kinases could be cited; the procedure has failed in very few instances. Although some kinases and dehydrogenases are not retained by blue-dextran-agarose, they are retained by Blue Agarose (51), possibly because blue-dextran sterically prevents maximal binding. On the other hand, immobilization of blue-dextran on CNBr-activated agarose also gives a product with additional positive charges, namely, isourea functions, which could be a source of the lowered affinity of many enzymes for blue-dextran-agarose. This lower affinity could be useful when the binding of an enzyme to blue-agarose is estraordinarily tight. The presence of various amounts of minor impurities in Cibacron Blue F3GA could also account for differences between Blue Dextran and Blue Agarose (58). Abreu et al. (59) used chromatography on Blue Dextran-

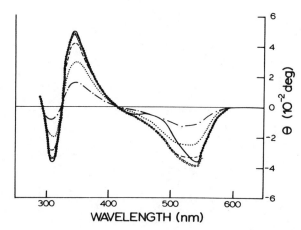

Figure 5.17. Circular dichroism spectra of rabbit muscle glyceraldehyde-3-phosphate dehydrogenase:

———————————	30-μM GAPDH, 120-μM Cibacron Blue;
	$[\theta]_{675} = -3 \times 10^3$ deg cm^2/dmole
——— ———	24.8-μM LADH, 10-μM Cibacron Blue;
	$[\theta]_{620} = -15 \times 10^3$ deg cm^2/dmole
. . . .	28-μM cMDH, 27-μM Cibacron Blue;
	$[\theta]_{660} = 5.2 \times 10^3$ deg cm^2/dmole

Reprinted with permission from R. A. Edwards and R. W. Woody, *Biochem. Biophys. Res. Commun.* 79, 470 (1977).

agarose to separate two forms of pyruvate dehydrogenase from *Azotobacter vinelandii* (Table 5.3). The four-component pyruvate dehydrogenase complex appeared to bind irreversibly while the three-component complex bound reversibly and was eluted with 0.6-M potassium chloride. The complex was not eluted with NAD$^+$, CoA, or AMP, thus suggesting that the binding is not bioselective. Conversely, the pyruvate dehydrogenase complex from *E. coli* was not retained at all on Blue Dextran–agarose. Quite possibly, chromatography on blue–agarose would give a bioselective method for purification of the complex since the large size of the complex (molecular weight over 10^6) makes steric restrictions a very significant factor.

Several important purification procedures involving nonnucleotide binding proteins have been performed with the use of Cibacron Blue F3GA-containing adsorbents, most notably that of serum albumin and of interferon. One of the most difficult problems encountered with the electrophoresis of serum is the presence of large quantities of serum albumin, which masks the appearance of many other proteins. The removal of serum albumin from serum samples significantly contributes to the purification and/or electrophoresis of the non-

TABLE 5.3. Partial and Overall Activities of 4- and 3-Component Complexes from *Azobacter vinelandii*, Before and After Blue Dextran–Sepharose Chromatography, Respectively

Enzyme	Four-Component Complex			Three-Component Complex		
	Units/mg Protein	Units/μmol FAD	Total Units	Units/mg Protein	Units/μmol FAD	Total Units
Pyruvate-NAD$^+$ reductase	7.0	4100	105	12.0	4450	109
Pyruvate-Fe(CN)$_6$ reductase	0.12	70	1.8	0.13	40	1.2
Lipoyl-trans-acetylase	3.4	2000	51	7.4	2700	67
Lipoamide dehydrogenase	1.9	1100	28.5	9.5	3500	86.5

Source: Reprinted with permission from Abreu et al. (59).

albumin proteins. Chromatography of serum on Blue Dextran-agarose will remove 96% of the serum albumin with little loss of other serum protein. For example, 90 to 99% of rabbit serum IgG passes unretained through a column of Cibacron Blue, which shows the potential of Blue–Agarose chromatography as a very valuable preliminary step in the purification of immunoglobulins. Burgett and Greenley (60), in an excellent review of Cibacron Blue F3GA chromatography, have detailed such a method for utilization of blue–agarose for the removal of albumin from serum. Basically, a serum sample that was previously dialyzed against 0.02-M phosphate buffer, pH 7.1, is applied to a column containing five times its volume of blue–agarose that has been pre-equilibrated with the same buffer. The column is then eluted with two volumes of the same buffer, yielding a diluted protein solution containing most serum proteins minus the albumin that is retained on the column. The albumin is then eluted with 1.4-M NaCl, and the column is regenerated by washing sequentially with 2 volumes of 8-M urea and 2 volumes of the starting buffer. The Blue Dextran column is reusable for a nearly infinite number of such albumin separations.

Interferon, another of the nonnucleotide-binding proteins isolated with the use of Cibacron Blue F3GA has considerable potential application as an antiviral agent (61). Jankowski et al. (53) have extensively investigated the affinity of human fibroblast and human leukocyte interferons for various immobilized aromatic compounds including Cibacron Blue F3GA, benzene, napththalene, anthracene, and their monoaminoderivatives immobilized on agarose. Leukocyte interferon binds to anthracene agarose, is retarded by napthalene agarose, and is unaffected by benzene or any of the monoamino aromatic compounds. It also emerges unretarded from Cibacron Blue-agarose. Conversely, fibroblast interferon is retained on these chromatographic packings and is eluable with a gradient of ethylene glycol. This illustrates the basically hydrophobic nature of this chromatographic process. In one step, human fibrocyte interferon is purified 800 fold with 95% recovery with the use of "hydrophobic" chromatography of Blue–Agarose. (See Figure 5.18.)

5.5. COENZYME A-BINDING PROTEINS

Although many enzymes recognize coenzyme A as a cofactor or cosubstrate, relatively few affinity chromatographic purification procedures that are based on coenzyme A affinity have been reported. This may be due to the extremely high cost of coenzyme A for use either as a bioligand or as a bioselective eluant. [In fact, a "reverse" affinity chromatography procedure has been designed by Chibata et al. (62,63) to purify coenzyme A. These authors purified coenzyme A by chromatography of a protein-free cell extract on a column of coenzyme A

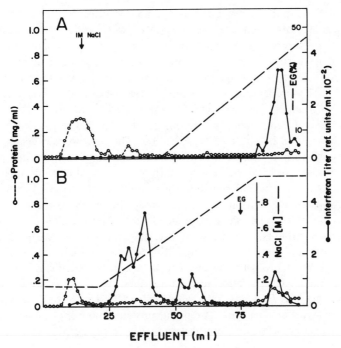

Figure 5.18. Chromatography of human fibroblast and leukocyte interferons on Blue Dextran agarose. (*A*) Human fibroblast interferon (9 ml), 500 reference units per ml, was applied on a column (0.9 × 15 cm); the column was washed with 1-*M* NaCl in 0.02-*M* sodium phosphate, pH 7.4 (↓), and then a linear concentration gradient of ethylene glycol (1:G) was developed by mixing 30 ml of equilibrating solvent with 30 ml of 50% (v/v) ethylene glycol in the equilibrating solvent. (*B*) Human leukocyte interferon (2 ml, 5400 reference units/ml) was applied on a column (0.9 × 15 cm); the column was washed with equilibrating buffer, and then a linear concentration gradient of sodium chloride was developed by mixing 30 ml of 0.15-*M* NaCl in 0.02-*M* sodium phosphate, pH 7.4, and 30 ml of 1.0-*M* NaCl in 0.02-*M* sodium phosphate, pH 7.4. Finally, the column was developed with 50% (v/v) ethylene glycol in 1.0-*M* NaCl, 0.02-*M* in sodium phosphate, pH 7.4 (EG). Reprinted with permission from Jankowski et al. (53).

binding proteins.] A few enzyme purifications have been effected by using columns of immobilized coenzyme A. For example, Kleinsek et al. (64) have purified 3-hydroxy-3-methyl glutaryl-coenzyme A reductase using a commercial thioester linked coenzyme A agarose. The affinity chromatography column was the last step in the purification, which ultimately yielded a homogeneous protein with a specific activity 20 fold above that ever found previously. The bioselective adsorption step had a recovery of 95 to 100%. However, an extremely large loss of activity occurred in the classical purification steps that preceded the affinity chromatography, and the overall yield was only 3%.

To avoid the problems that occur in the immobilization of coenzyme A, the "half-molecule" of coenzyme A, 3'5'-AMP, was synthesized in Mosbach's laboratory and subsequently employed as a coenzyme A analogue in the purification of pig heart succinate thiokinase in much the same manner as 5'-AMP and 2'5'-AMP have been employed as analogues of NAD$^+$ and NADP$^+$ respectively (65). The preparation of the bioselective adsorbent was simple and inexpensive. Commercially available 3'5'-AMP (0.148 mmole) was added to 0.2-M NaHCO$_3$ (3 ml) and the resulting solution mixed with 3 ml of epoxy-activated agarose in 0.4-M NaHCO$_3$, pH 11.7. The mixture was gently agitated for 20 hours, washed, and then packed into a column. From a crude pig heart extract a 50-fold purification of succinic thiokinase could be achieved by applying the extract to the column and then washing the column with 0.5-mM NADP$^+$ followed by elution of the enzyme with an 8-ml pulse of 0.2-mM CoA. Succinic thiokinase was not retained on 5'-AMP or 2'5'-AMP containing bioselective adsorbents, thereby demonstrating the bioselective nature of the adsorbent. It is not yet known whether this will be a widely useful bioselective adsorbent as some problems occurred in its use in the purification of succinyl thiokinase, chiefly (1) the enzyme is not retained at moderate buffer concentrations (100-mM Tris–HCl, pH 7.0) and (2) the K_I for 3'5'-AMP was only 6 × 10^{-4} M. Two other coenzyme A-binding proteins, phosphotransacetylase and carnitine acetyl transferase, were also shown by these authors to bind to 3'5'-AMP–agarose.

5.6. FOLATE-BINDING PROTEINS

Folic acid is readily available, is relatively inexpensive, and possesses reactive functional groups (carboxylic acid residues) that can easily be coupled to amino-alkylagarose or other amine-containing matrices by using the carbodiimide method. Moreover, methotrexate and aminopterin (two analogues of folic acid) are of considerable medical values as chemotherapeutic agents. Their mode of action is to bind to folic acid-binding proteins, especially dihydrofolate reductase (K_a = 10^{-9} M), much more tightly than folate binds to these enzymes (K_a = 10^{-7} M). For these reasons, and because of the important role that folic acid plays in metabolism, numerous bioselective adsorbents have been synthesized for use in the purification of those enzymes that use folic acid or its derivatives as cofactors.

Several problems occurred in initial attempts to prepare immobilized folate or its analogues. Folate could be readily bound to cellulose by use of the carbodiimide method (66) to produce a bright yellow-colored product that bound little, if any, dihydrofolate reductase, which was undoubtedly due to the steric hindrance of the cellulose structure itself (see page 36). Next, methotrexate was coupled to aminoethyl starch. The resulting product bound dihydrofolate reductase very tightly, and it could then be isolated by gel-permeation chromatography as an enzyme–methotrexate–starch complex (67). Erickson and Mathews (68) purified dihydrofolate reductase 6000 fold starting from crude extracts of bacteriophage T_4-infected *E. coli* by chromatography on N^{10}-formylaminopterin coupled to aminoethyl polyacylamide.

The preparation of Erickson and Mathews yielded a homogeneous protein with about 80% recovery. However, such good results were not produced when other methods of immobilizing folate, aminopterin, or methotrexate were used. Two different problems appeared to be involved. First, an adequate concentration of the expensive aminopterin or methotrexate dyes often was not used during the ligand coupling step, thereby leaving many aminoalkyl groups underivatized. These groups consequently served as ion-exchange centers and cause a great deal of nonspecific binding of proteins to the bioselective adsorbent (69). Second, in some experiments such high concentrations of ligand were bound, and the protein–ligand binding was so tight that the conditions required for elution produced considerable denaturation of the enzyme desired.

Recently, Whiteley et al. (70) have synthesized an improved bioselective adsorbent for dihydrofolate reductase by initially coupling 1 mole of folate to 1 mole of 1,6-diaminohexane by use of a carbodiimide mediated condensation. This yielded a mixture of α- and γ-linked 1-aminohexyl-6-aminofolate. The mixture was not separated but was added directly to CNBr-activated agarose. By this procedure no underivatized amino groups remained on the matrix to form ion-exchange centers.

The resulting folate–agarose column had excellent bioselective adsorption properties for several species of dihydrofolate reductase. The dihydrofolate reductase from aminopterin-resistant murine leukemia cells was retained even with 1-M KCl but were sharply eluted in 2-mM folic acid (see Figure 5.19). Crude extracts of the cells were applied to the column with 60 to 80% recovery of electrophoretically homogeneous enzyme. When *L. caseii* extracts were used as the enzyme source, the enzyme was eluted with 0.5-M KCl, thereby demonstrating the lower binding constant between the enzyme from this source and folic acid. The eluted enzyme was not homogeneous, and two additional steps using DEAE–cellulose and gel-permeation chromatography were required to obtain a homogeneous preparation. Two forms of the enzyme were separated by this procedure, an NADPH-containing form that bound tightly the folate–agarose and an NADPH-free form that had very low affinity for folate–agarose.

Folate-binding protein from bovine milk has been purified on folate–agarose by using a combination of pH extremes and deforming buffers (71) to effect elution. First, the pH was lowered to 3.6 to dissociate the naturally bound folate. The pH was readjusted to 7.0, and the solution was applied to the bioselective adsorbent. The folate-binding protein was bound tightly to the column and could be eluted only by 0.8-M urea, pH 5.

Recently, Hansen et al. (72) have purified bovine milk folate-binding protein

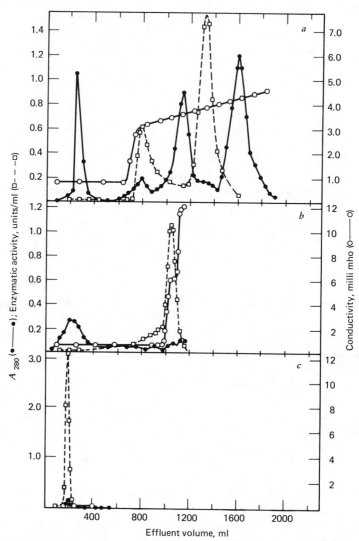

Figure 5.19. Chromatography of dihydrofolate reductase from an amethopterin-resistant strain of *L. casei* on DEAE–cellulose. The enzymatically active fractions comprising peaks 1 and 2 were combined. (*b*) Affinity chromatography of peak 2 activity on folate–Sepharose. (*c*) Application of the combined affinity column peak activity extract in a column of Sephadex G-50. Reprinted with permission from Whiteley et al. (70).

in a different and novel application of bioselective elution. They accidentally discovered that folate-binding protein (from which folate has been removed) has a great affinity for Sephadex G-100 and is completely adsorbed on such a column. If, however, folate is applied to the column containing adsorbed folate-binding protein, the protein is desorbed in a sharp peak. Ultragel may be substituted for Sephadex with identical results. The nature of the interaction between folate-binding protein and the polysaccharides that comprise the gel-permeation matrix is as yet unknown.

5.7. COENZYME B$_{12}$

Some of the most strikingly successful examples of affinity chromatography are in the purification of B$_{12}$-binding proteins. This is partially because there are very few B$_{12}$-binding proteins and partially because classical techniques have worked so very poorly for their purification. Thus far there are 10 enzymes (73) identified that utilize B$_{12}$ in the so-called coenzyme form and a number of other proteins that bind vitamin B$_{12}$, possibly for transport or storage. Among the latter are the intrinsic factors required for vitamin B$_{12}$ uptake during digestion, transcobalamin I and II, which are serum proteins, and the so called "R-type" proteins, which bind B$_{12}$ in milk, white blood cells, and various tissues and fluids. (See Figure 5.20.)

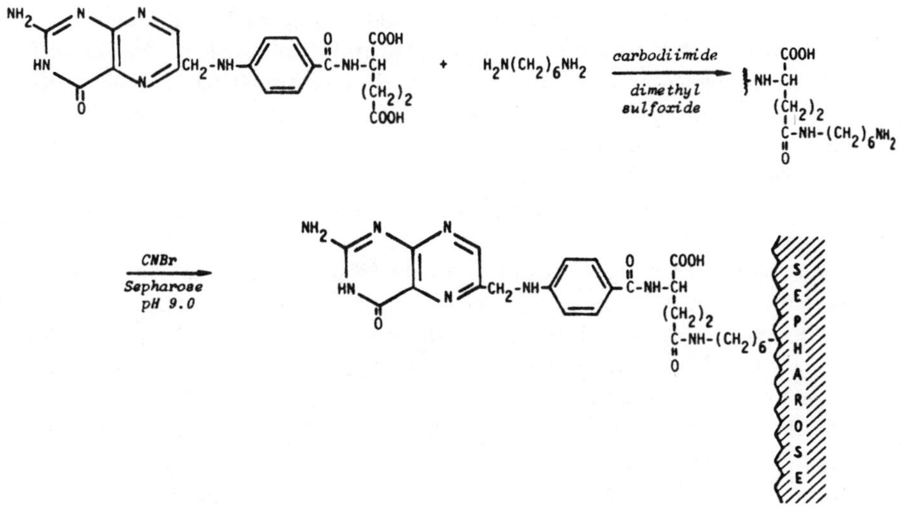

Figure 5.20. Reaction sequence for the preparation of folate–Sepharose. Reprinted with permission from Whiteley et al. (70).

A very stable ligand, for B_{12}-utilizing enzymes, and one that is easy to synthesize, is the monocarboxylate derivate of vitamin B_{12} (74). Murthy et al. (75) used this ligand attached to aminoalkyl agarose by the carbodiimide method to purify methylamalonyl–coenzyme A mutase from *Propionibacterium shermanii* by over 300 fold. A two-step method was employed. First, the crude cell extract was applied to a DEAE–cellulose column and eluted with a pH-6.8 phosphate buffer, 0.3 M, thereby resulting in a two-fold purification and eliminating all endogenous B_{12}. This eluate was, in turn, applied to the vitamin B_{12}-agarose column. After the column was extensively washed, the enzyme was bioselectively eluted with 0.2 M of the coenzyme form of B_{12} (5′-deoxyadenosylcobalamin). The resulting enzyme was twice as active as any previous preparation and considerably more stable, perhaps because of the presence of B_{12} during purification and storage (*Note:* all steps in purification procedures involving B_{12} and B_{12}-affinity chromatography were done *in the dark* since B_{12} and its analogues are light sensitive.)

Similar purification methods for vitamin B_{12}-binding proteins from milk (76), saliva (76,77), serum (78,79), and gastric juices (80) have been performed with a high degree of success. Gastric secretions contain two vitamin B_{12}-binding proteins: (1) *intrinsic factor,* needed for B_{12} absorption from the gut, and (2) *nonintrinsic factor protein,* a vitamin B_{12}-binding protein unrelated to B_{12} adsorption. Intrinsic factor is lacking in persons with pernicious anemia. Francis et al. (80) have isolated porcine intrinsic factor and employed it in the treatment of pernicious anemia. Administration of 54 μg of intrinsic factor resulted in a 10-fold improvement in the daily adsorption of vitamin B_{12}.

To purify intrinsic factor, crude gastric fluid was applied to a vitamin B_{12}-agarose column that was then washed extensively, followed by elution with a gradient of guanadine HCl. Intrinsic factor eluted in the 0.4-M guanadine HCl fraction, nonintrinsic factor in the 4- to 7.5-M fractions. Both were further purified by dialysis and hydroxylapatite chromatography. Intrinsic factor was purified by about 1000 fold by this procedure (see Table 5.3). Burger et al. (76) used a similar procedure that employed two affinity steps and produced a somewhat lower yield.

Undoubtedly, the most successful purification procedure employing vitamin B_{12}-agarose chromatography is a four-step preparation of transcobalamin II (a serum B_{12} transport protein) that utilizes affinity chromatography, carboxymethyl cellulose, DEAE–cellulose, and Sephadex G-100 gel-permeation steps. A total purification of 2,000,000 fold and a yield of 12.8% resulted from an initial volume of 1400 l of plasma (78). The final protein was shown to be homogeneous by gel electrophoresis.

Olesen et al. (81) have prepared a somewhat different type of bioselective matrix by coupling 5′-deoxy-adenosyl-cobalamin to albumin–cellulose or succinylated γ-globulin. The resulting material was useful in purifying various

B_{12}-binding proteins. Similarly, Yamada and Hogenkamp (82) coupled a coenzyme-B_{12} to agarose to yield a B_{12}-agarose bioselective adsorbent that was very efficient in the purification of ribonucleotide reductase and could be used as a raw material to synthesize other immobilized B_{12}-agarose analogues by reduction or by substitution of the axial ligands.

5.8. THIAMINE-BINDING PROTEINS

There are two types of thiamine-binding protein: (1) those that are involved in thiamine synthesis, transport, or phosphorylation and that recognize of the thiamine group per se and (2) those that utilize thiamine pyrophosphate as a coenzyme and require both the thiamine and the pyrophosphate groups for binding. To specifically isolate the first type of protein, for example, the thiamine transport protein from *E. coli* cell membranes, Matsuura et al. (83–85) coupled thiamine pyrophosphate to aminoalkylagarose by way of a water-soluble carbodiimide. Using this bioselective matrix, these authors were able to

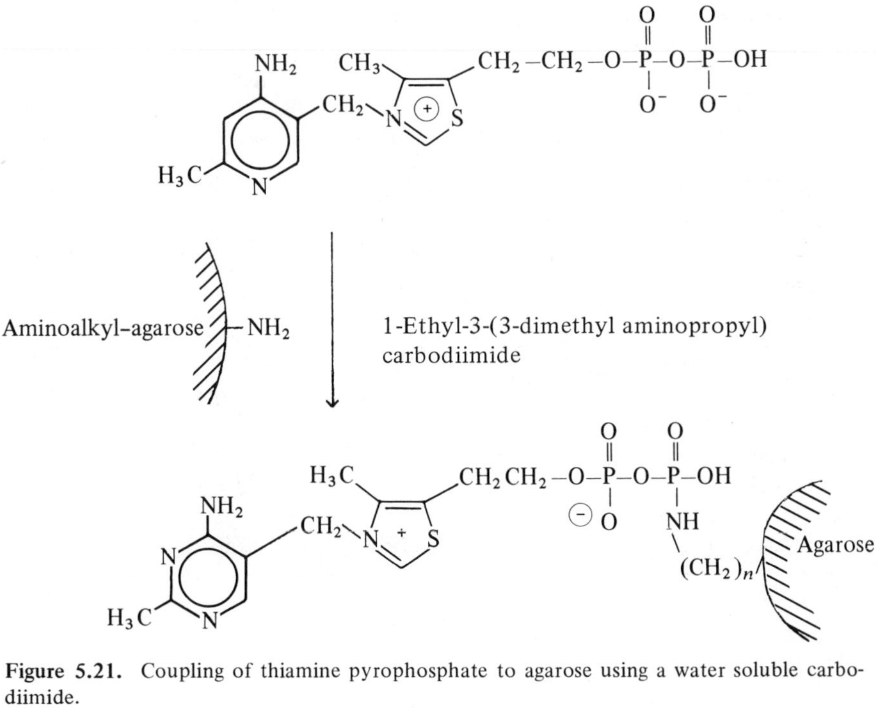

Figure 5.21. Coupling of thiamine pyrophosphate to agarose using a water soluble carbodiimide.

isolate homogeneous thiamine binding protein from either an *E. coli* cell shock fluid or from a partially purified *E. coli* extract. Even so, this method has a significant disadvantage in that the thiamine pyrophosphate–agarose is very unstable, and 70% of the thiamine is lost from the column within a week at 4°C.

A much more stable thiamine pyrophosphate–agarose has been prepared by O'Brien et al. (86) by the diazonium procedure. *p*-Aminobenzoyl agarose (100 ml) was diazotized (87), and 15 g of thiamine pyrophosphate in saturated sodium borate was added and the pH adjusted to 8.6. The resin was kept in this solution for 8 hours at 4°C and then thoroughly washed. A reddish-colored agarose adsorbent with a ligand concentration of 6 μmoles of thiamine pyrophosphate per ml resulted. This thiamine pyrophosphate agarose was used as the last step in the purification of pyruvate oxidase from *E. coli*. The partially purified enzyme was applied to the bioselective adsorbent in a pH-5.7 buffer containing 10% glycerol and 10-mM MgCl$_2$. The column was extensively washed and the enzyme bioselectively eluted with 10-mM thiamine phyrophosphate. Alternatively, the enzyme could be eluted with 0.1-M KCl. (See Table 5.4.)

TABLE 5.4.   Purification of Pyruvate Oxidase[a]

Step	Protein mg	Total Activity[b]	Specific Activity[b] Units/mg Protein	Recovery %
Crude extract	—	405,600	—	100
Protamine sulfate precipitation	26,600	352,800	13	87.0
Ammonium sulfate fractionation	8,500	317,400	37	78
Heat step supernatant	4,500	268,800	59	66
Concentrated DEAE– Sephadex fractions	307	239,300	781	59
Concentrated affinity chromatography fractions	56	185,200	3325	46
Low-ionic-strength precipitation	29	151,000	5189	37

Source: Reprinted with permission from O'Brien et al. (86).

[a] Starting with 300-g *E. coli* cell paste.

[b] Enzymatic activity is expressed in decarboxylase units.

5.9. FLAVIN-BINDING PROTEINS

Arsenis and McCormick (88,89) were among the pioneers of affinity chromatography, using FMN derivatives covalently linked to ion exchangers, for example, carboxymethyl and DEAE-cellulose. These derivatives were surprisingly successful, considering that a large amount of ion-exchange capacity remained after derivatization of the ion exchanger.

Much more successful purifications have been made by chromatography on FMN coupled to aminoalkyl-agarose by use of the carbodiimide method (90). Waters et al. (91) were able to isolate bacterial luciferase from several luminous bacteria by utilizing bioselective adsorption on FMN-agarose. The immobilized FMN could also be utilized as a substrate in the luciferase reaction. In a typical enzyme preparation, the luciferase obtained from a crude cell extract was adsorbed onto DEAE-cellulose in a batch procedure. The enzyme was eluted with $0.35 M$ phosphate buffer, pH 7.0, and the eluate was dialyzed against $0.02 M$ phosphate, pH 7, containing $0.5 mM$ dithiothreitol. The dialyzed, partially purified luciferase was then passed through a Sepharose-hexanoate column to remove any proteins that might hydrophobically bind to the spacer arm. Luciferase was unretained by this column and was applied directly to a FMN-aminohexyl-agarose bioselective adsorbent. Luciferase (about 40% in purity) was eluted with either an FMN or NaCl gradient (see Figure 5.22).

Blankenhorn et al. (92) have coupled 3-carboxymethyl-FMN to agarose and

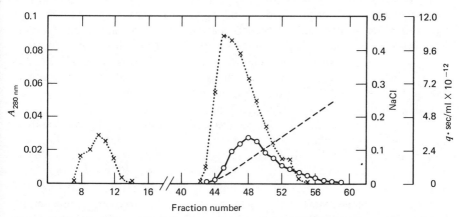

Figure 5.22. Partially purified luciferase was applied to FMN–Sepharose column. Some contaminating proteins came through in the wash (tubes 8 through 12). Luciferase, as well as some other unidentified proteins, was eluted with NaCl. Reprinted with permission from C. A. Waters, J. R. Murphy, and J. W. Hastings (1974), *Biochem. Biophys. Res. Commun.* **57**, 1152.

have employed it in the purification of chicken egg white flavin binding protein, which is the major flavin storage protein in eggs. The 3-carboxymethyl–FMN was chosen since the N-3 position appears to be relatively unnecessary for FMN binding, whereas the ribitol phosphate chain (which is altered in carbodiimide method) (90) is necessary for flavin binding. The major·drawback in the 3-carboxymethyl–FMN method, however, is that it is not yet commercially available (unlike the materials for the carbodiimide procedure).

To isolate ovoflavoprotein by affinity chromatography, protein-bound FMN is first removed either by dialysis against KBr or by DEAE chromatography. The resulting crude apoovoflavoprotein was applied to the bioselective adsorbent in 0.1-M acetate buffer, pH 5.0, and 0.2 M in NaCl. (This procedure and all subsequent FMN-agarose affinity chromatographic steps were done in the dark.) The column was then extensively washed with the same buffer and the apoflavoprotein was eluted with 0.1-M ammonium acetate, pH 3.2, containing 0.5-M NaCl. The apoflavoprotein prepared in this fashion was essentially homogeneous and could be recovered in large amounts with good yields (see Table 5.5).

5.10. PYRIDOXAL ENZYMES

Pyridoxal phosphate is the major cofactor associated with amino acid metabolism. It and its amine derivative, pyridoxamine phosphate, possess many possible sites for immobilization or modification. Unfortunately, most of these sites are also necessary for the binding of pyridoxal phosphate to pyridoxal phosphate-requiring enzymes. The aldehyde (or amine) group appears to be the least necessary in this regard and thus has been used as the binding site in preparing several pyridoxal–agarose bioselective adsorbents (93,94). To prepare the adsorbent, succinyl-aminoethyl-agarose is added to an aqueous solution of pyridoxamine phosphate and a water-soluble carbodiimide at pH 4.7 (95). Pyridoxamine phosphate that is directly attached to agarose without an intervening spacer arm is ineffective as a bioselective adsorbent.

To purify tyrosine aminotransferase by affinity chromatography, a crude liver extract is incubated with 1-mM tyrosine for at least 15 minutes at 37°C, followed by dialysis at 4°C to remove the tyrosine. (Without preincubation with tyrosine only a· very low percent (20 to 40%) of the enzyme is retained.) Next, the dialyzed extract is applied to a column of the adsorbent in 50-mM Tris·HCl, pH 7.6, containing 25-mM KCl and 5-mM MgCl$_2$. The enzyme is subsequently eluted with a gradient consisting of the same buffer containing 0.5 M NaCl and from 10- to 100-mM pyridoxal phosphate. The NaCl is needed in the pyridoxal phosphate gradient to elute the enzyme to weaken the apparently large affinity of the enzyme for pyridoxamine phosphate (10^{-7} M).

One of the more striking applications of pyridoxal phosphate–agarose was

TABLE 5.5. Purification of Ovoflavoprotein by Affinity Adsorption on FMN–Agarose

Fraction	Volume, ml	Total Protein		Apoprotein		Specific Binding Capacity, μg Riboflavin/mg Protein
		mg	%	mg	%	
Egg white first purified by DEAE–cellulose chromatography						
Original	1200	176 000	100	852	100	0.057
After DEAE–cellulose	195	878	0.5	722	85	8.5
After affinity chromatography	420	557	0.32	549	65	11.6
Egg white first purified by KBr dialysis						
Original	300	44 000	100	213	100	0.058
After KBr treatment	835	28 900	66	188	87	0.077
After affinity chromatography	198	173	0.4	161	75	10.9

Source: Reprinted with permission from Blankenhorn et al. (92).

in the purification of polyribosomes enriched in mRNA coding for tyrosine aminotransferase (95,96). A homogenate was prepared from hepatoma cells that were induced to increase their synthesis of the enzyme with 1-μM dexamethasone phosphate. This homogenate was then differentially centrifuged to produce a crude polysomal fraction. This fraction was applied in the Tris $\cdot$KCl$\cdot$Mg^{++} buffer previously used in the affinity chromatography of the enzyme. The column was extensively washed to remove unbound material, and the polysomes are subsequently eluted with 50-M Tris$\cdot$HCl, pH 4.0, containing 25-mM KCl, 10-mM MgCl$_2$, 500-mM NaCl, 10-mM pyridoxal phosphate, and 1% bovine serum albumin. The polysomes, collected from the eluant by centrifugation at 100,000 g, are very enriched in tyrosine aminotransferase m-RNA as shown by way of protein synthesis. The pH-4.0 elution procedure is necessary since the polysomes are unstable at even slightly alkaline pH.

Another enzyme purified using pyridoxal phosphate is glutamate oxaloacetate transaminase (93), which was purified from pig heart extract using N'-(6-aminohexylamino)pyridoxamine–agarose. Pyridoxamine directly attached to agarose again proved to be an ineffective adsorbent. Ryan and Fottrell (97) also purified aspartate aminotransferase from *Rhizopus japonicum* by way of bioselective adsorption on pyridoxal phosphate–agarose with a 170-fold purification and a 60% yield.

5.11.   BIOTIN-BINDING PROTEINS

Biotin was first used as a bioselective adsorbent in 1965 by McCormick, who coupled biotin to cellulose by way of an ester linkage (98) and employed it to purify avidin, a biotin-binding protein from egg white. Many improvements in the process have been made by attaching biotin to Sephadex (99) or agarose (100,101). Most of these procedures have been employed to purify avidin, and few biotinyl enzymes have been studied by use of this technique. For example, N-biotinyl polylysine coupled to agarose was used for the purification of egg white avidin with a 90% yield and a 4000-fold purification in a single bioselective adsorption step (101). Avidin bound so tightly to this material that elution was best effected with 6-M guanadine hydrochloride at pH 1.5. Biotin coupled to aminoalkyl agarose has also been used to purify acetyl coenzyme A carboxylate (a biotin requiring enzyme). The enzyme was not strongly bound, but only retarded by the bioselective adsorbent (102).

Hofmann et al. (103) have utilized biotin–avidin interactions in a unique application of bioselective adsorption. They coupled biotin to the β-N-terminal amino function of insulin by incubating the N-hydroxysuccinimide ester of biotin with insulin. Simultaneously, avidin was bound to agarose which had been activated with 2,4,6-trifluoro-5-chloropyrimidine (104). The biotinyl insulin was now used as a bifunctional reagent, part with high affinity for avidin–agarose and part with affinity for adipose cells through their insulin receptor sites. The biotin-containing hormone can be chemically modified in free solution and the modified insulin–biotin complex attached to the avidin–agarose. The effect of the modification can subsequently be tested by using bioselective adsorption of adipose cells. The coupling of insulin to agarose in this process does not modify the insulin to form a "superinsulin" as does direct attachment of insulin to CNBr-activated agarose (see page 325). Moreover, the biotinyl–insulin is more tightly bound and is more stable when noncovalently coupled to avidin–agarose than when covalently bound directly to CNBr-activated agarose.

5.12.   LIPOIC ACID

Relatively few enzymes recognize lipoic acid, and thus one would expect that it could be very useful in the purification of those proteins that do, for example, lipoamide dehydrogenase, pyruvate dehydrogenase, and α-ketoglutarate dehydrogenase. Scouten et al. (105) have employed lipoic acid coupled to aminoalkyl glass or silica gel to purify lipoamide dehydrogenase from pig heart, yeast, or *E. coli.* Cox and Lowe (106) and Visser and Strating (107) have reported the separation of pig heart lipoamide dehydrogenase isoenzymes by chromatography on lipoic acid coupled to aminoalkyl agarose or on aminoalkylagarose per se. This separation is essentially identical to results obtained when lipoamide dehydrogenase is chromatographed on DEAE–cellulose and thus may be the result of the ion-exchange effect of the positively charged isourea function known to exist in aminoalkyl agarose prepared from CNBr-activated agarose and diaminoalkanes, rather than by affinity chromatography per se. Visser and Strating have also demonstrated hydrophobic chromatography of lipoamide dehydrogenase and have suggested that chromatography of lipoamide dehydrogenase on lipoate–glass could be hydrophobic chromatography rather than bioselective adsorption. Since the active site of lipoamide dehydrogenase is very hydrophobic (108), it may be difficult to differentiate between hydrophobic and affinity interactions.

5.13.   BIOSELECTIVE ADSORPTION OF GLYCOPROTEINS ON IMMOBILIZED LECTINS

Lectins are proteins (chiefly but not exclusively of plant origin) that bind tightly to specific carbohydrate residues. The best known lectin is concanavalin A, which is one of the lectins isolated from jack beans (*Canavalia ensiformis*). It, like other lectins, will bind to cell-surface carbohydrates. Concanavalin A recognizes α-D-mannopyranosyl and α-D-glucopyranosyl residues and will agglutinate cells that have these carbohydrates on their surfaces. Lectins have been used to type blood cells since certain lectins agglutinate type B red blood cells and others, type A or O. Other serological classifications may also be related to cell surface glycoproteins and may ultimately be identified by lectin agglutination. For this reason, plant lectins are also known as *phytohemagglutinins.*

Because lectins recognize sugars in a bioselective fashion and because they recognize those specific types of glycoprotein that contain these sugars, lectins have a twofold relationship to affinity chromatography, in that they can be (1) purified by using bioselective adsorption on immobilized carbohydrates and (2) immobilized to produce a bioselective adsorbent for glycoproteins. Some of the better-known lectins along with their carbohydrate specificity are listed in Table 5.6.

5.14. PURIFICATION OF GLYCOPROTEINS BY WAY OF IMMOBILIZED LECTINS

The recent increase in the number of glycoproteins that have been purified by use of bioselective adsorption on immobilized lectins is in general due to two factors: (1) advances in methodology in affinity chromatography, including the commercial availability of many lectins, such as concanavalin-A agarose, and (2) to the current intense interest in membranes and membrane proteins from cell surfaces.

Among the proteins and biomolecules purified by affinity chromatography on lectin agarose are enzymes, transport proteins, virus, and even whole cells. The ability to separate glycoproteins from nonglycoproteins is a very useful addition to the chromatographic tools of the biochemist since the process overlaps none of the enzyme-isolation techniques that are in general use, for example, ion-exchange chromatography, gel-permeation chromatography, or isopycnic centrifugation.

Hemopexin is a serum protein responsible for transporting free heme from the blood stream to the liver. Its concentration is dependent on the concentration of free heme in the circulatory system and thus can be used in the monitoring of internal hemolysis. Unfortunately, prior to the development of affinity chromatography the methods for purifying hemopexin had proved to be unreliable. Vretblad and Hjorth (110) have purified hemopexin from human serum 114 fold with a yield of 29 mg/100 ml of serum and a 37% recovery by using bioselective adsorption on wheat lectin–agarose. The entire purification process could be completed in 2 days. Initially, serum was fractionated with polyethylene glycol, and the 15- to 25% polyethylene glycol precipitate was

TABLE 5.6. Properties of Lectins (Lectins Listed by Group (109))

Group	Sugar Residue(s) Recognized	Source of Lectin[a]
Group I	L-Fucose	*Lotus tetragonolobus*[5-10]; weakly inhibited by L-galactose *Ulex europaeus* (gorse)[11]; contains another lectin belonging to group VII *Ulex parviflorus*[12]; weakly inhibited by other sugars
Group II	N-Acetyl-D-glucosamine Group inhibited by N-acetylated chitodextrins	*Triticum vulgare* (wheat germ)[1,2,13]; also inhibited by N-acetylneuraminic acid (NANA)[14] *Solanum tuberosum* (potato tuber)[15,16]; also inhibited by muramic acid
Group III	N-Acetyl-D-galactosamine	*Dolichos biflorus* (horse gram)[7,17,18] *Phaseolus lunatus* (lima bean, also called *P. limensis*)[5,7,17] *Phaseolus vulgaris* (red kidney bean, black kidney bean, yellow wax bean; bean meal is source)[1,2] *Vicia cracca*[5,7,19]; also contains a nonspecific lectin in group VIII *Euonymus europaeus*[20] *Helix pomatia* (a snail)[21,22]
Group IV	D-Galactose Group also inhibited by L-arabinose, D-fucose, lactose, raffinose, and melibiose	*Crotalaria juncea* (sunn hemp)[7,23,25] β-specific *Ricinus communis* (castor bean)[6,15,16] *Abrus precatorius*[26] *Griffonia simplicifolia*[2,27]
Group V	N-Acetyl-D-galactosamine and D-galactose These lectins are inhibited almost equally by both sugars	*Sophora japonica* (japanese pagoda tree)[6,20] *Glycine max* (soybean)[1] α-specific *Caragana arborescens*[7] *Bandaeirea simplicifolia*[23] α-specific *Bauhinia variegata*, var. *candida*[7]

(Continued)

149

TABLE 5.6. (*Continued*)

Group	Sugar Residue(s) Recognized	Source of Lectin[a]
		Momordia charantia[15]
		Erythrina subrosa[24]
		Coronilla varia,[6,20] α-specific
		Crotalaria zanzibarica[24]
		Arachis hypogea,[28,29] β-specific
Group VI	D-Glucose	*Sesamum indicum*[15]
		Pisum sativum (garden pea); inhibited about four times better by D-mannose[27]
Group VII	β-Glycosides and β-N-acetylglucosaminides Group inhibited most strongly by *N*,*N*′-diacetylchitobiose, but by salicin [(2-hydroxy-methyl)-phenyl-β-D-glucopyranoside)], phenyl-β-D-glucopyranoside, and cellobiose	*Ulex europaeus* (gorse)[11]
		Ulex galli[7]
		Ulex nanus[7]
		Cytisus sessilifolius[5,7,8,23,30]; inhibited also by lactose
		Laburnum alpinum[5,7,8,30], inhibited also by lactose
		Clerodendrum viscosum[31]; pulp is source
Group VIII	Methyl-α-D-mannoside, D-mannose, D-glucose, L-sorbose, and N-acetyl-D-glucosamide	*Pisum sativum* (garden pea)[6,7,32]; also inhibited by D-glucose, but about one-fourth as efficiently[27]
		Lens culinaris (common lentil)[1]
		Canavalia ensiformis (jack bean; gives concanavalin A)[1];
		Vicia cracca[2,7]; also contains a lectin in group III[19]
		Lathyrus sativus L.[27,33,34]
Group IX	N-Acetylneuraminic acid (NANA)	*Limulus polyphemus* (hemolymph of horseshoe crab[35])

150

Source: Adapted with permission from Kristiansen (109).

[a]References for Table 5.6:

1. N. Sharon and H. Lis (1972) *Science* **177**, 949; H. Lis and N. Sharon (1973), *Annu. Rev. Biochem.* **42**, 541.
2. J. Tobiška (1964), "Die Phythämagglutinine," p. 191. Akademie-Verlag, Berlin, 1964; O. Prokop and G. Uhlenbruck (1969), "Human Blood and Serum Groups," p. 72. Elsevier, Amsterdam.
3. M. M. Burger (1973), *Fed. Proc., Fed. Amer. Soc. Exp. Biol.* **32**, 91.
4. A. R. Oseroff, P. W. Robbins, and M. M. Burger (1973), *Annu. Rev. Biochem.*, p. 647.
5. W. T. J. Morgan and W. M. Watkins (1953), *Bri. J. Exp. Pathol.* **34**, 94.
6. M. Krüpe (1955), *Hoppe-Seyler's Z. Physiol. Chem.* **299**, 277.
7. O. Mäkelä,(1957), *Ann. Med. Exp. Biol. Fenn.* **35**, Supplement 11.
8. W. M. Watkins and W. T. J. Morgan (1962), *Vox Sang.* **7**, 129
9. G. F. Springer and P. Williamson (1962), *Biochem. J.* **85**, 282.
10. J. Yariv, A. J. Kalb, and E. Katchalski (1967), *Nature (Lond.)* **215**, 890.
11. I. Matsumoto and T. Osawa (1969), *Biochim. Biophys. Acta* **194**, 180.
12. M. Lalaurie, B. Marty, and J.-P. Fabre (1966), *Trav. Soc. Pharm. Montpellier* **26**, 215.
13. Pattern of inhibition studied by A. K. Allen, A. Neuberger, and N. Sharon (1973), *Biochem. J.* **31**, 155.
14. P. J. Greenaway and D. LeVine (1973), *Nature (Lond.) New Biol.* **241**, 191.
15. M. Tomita, T. Osawa, Y. Sakurai, and T. Ukita (1970), *Internat. J. Cancer* **6**, 283.
16. G. I. Pardoe, G. W. G. Bird, and G. Uhlenbruck (1969), *Z. Immunitaetsforsch. Allerg. Klin. Immunol.* **137**, 442.
17. Z. Yosisawa and T. Miki (1963), *Proc. Jap. Acad.* **39**, 187.
18. M. E. Etzler and E. A. Kabat (1970), *Biochemistry* **9**, 869.
19. K. Aspberg, H. Holmén, and J. Porath (1968), *Biochim. Biophys. Acta* **160**, 116.
20. O. Mäkelä, P. Mäkelä, and M. Krüpe (1959), *Z. Immunitaetsforsch. Exp. Ther.* **117**, 220.
21. S. Hammarström and E. A. Kabat (1969), *Biochemistry* **8**, 2696.
22. S. Hammarström, *Methods in Enzymology*, Vol. 28, p. 368.

(Continued)

151

TABLE 5.6. *(Continued)*

23. O. Mäkelä (1959), *Nature (Lond.)* **184**, 111.
24. H. M. Bhatia and W. C. Boyd (1962), *Transfusion* **2**, 106.
25. B. Ersson, K. Aspberg, and J. Porath (1973), *Biochim. Biophys. Acta* **310**, 446.
26. M. Krüpe and A. Ensgraber (1962), *Z. Immunitaetsforsch. Exp. Ther.* **123**, 355.
27. V. Hořejší and J. Kocourek (1973), *Biochim. Biophys. Acta* **297**, 346.
28. W. C. Boyd, D. M. Green, D. M. Fujinaga, J. S. Drabik, and E. Waszczenko-Zacharczenko (1959), *Vox Sang.* **6**, 456.
29. G. Uhlenbruck, G. I. Pardoe, and G. W. G. Bird (1969), *Z. Immunitaetsforsch. Allerg. Klin. Immunol.* **138**, 423.
30. T. Osawa (1966), *Biochim. Biophys. Acta* **115**, 507.
31. G. W. G. Bird (1961), *Nature (Lond.)* **191**, 292.
32. M. Tichá, G. Entlicher, J. V. Kóstir, and J. Kocourek (1969), *Experientia* **25**, 17.
33. G. Louženský (1970), Isolation of Hemagglutinins from the Seeds of the Sweet Pea (*Lathyrus sativus* L. var. *Karcagi*), thesis, Charles University, Prague, 1970.
34. J. Kocourek and G. Louženský (1971), *Abstr. 7th FEBS Meeting, Varna*, p. 134.
35. J. J. Marchalonis and G. M. Edelman (1968), *J. Molec. Biol.* **32**, 453.

chromatographed on three consecutive columns: DEAE–agarose, wheat germ lectin–agarose, and Sephadex G-25. The DEAE–cellulose and the wheat lectin–agarose steps provided the most substantial degree of purification. The yield on the ion-exchange column was only 50%, whereas the recovery from the wheat-germ lectin–agarose was essentially quantitative. Elution from the lectin-agarose column was effected by a pulse of 10% N–acetyl-D-glucosamine. glucosamine or N-acetyl-neuraminic acid.

This procedure could also be used to purify nonglycoprotein fractions of serum proteins containing trace contaminates of serum glycoprotein. Figure 5.23 shows the results that occur when transferrin contaminated with hemopexin was passed through a wheat germ lectin–agarose column. Transferrin passes directly through the bioselective adsorbent, whereas hemopexin is adsorbed and can be bioselectively eluted later with N-acetylglucosamine.

Tay-Sachs disease results from the absence of the enzyme N-acetylhexosaminidase. Its purification is necessary for study of the disease, and, if a proper method could be developed for placing the enzyme in a useful position in the cell, it could potentially be utilized for enzyme replacement therapy. This possibility (although remote for this enzyme) is one logical future application of enzymes isolated by way of bioselective adsorption. Brattain et al. (111) have demonstrated that N-acetylhexosaminidase can be purified 35 fold from human placental crude extracts with a yield of 76%. Using low concentrations

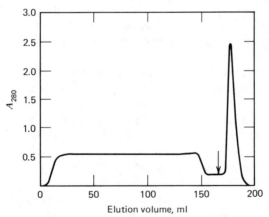

Figure 5.23. Chromatography of hemopexin transferrin mixture on wheat germ lectin Sepharose 6MB. A wheat germ lectin Sepharose 6MB column (16 mm × 45 mm: total volume 9 ml) was equilibrated at 20°C with 0.050-M sodium phosphate buffer containing 0.2-M NaCl and 0.02% (w/v) NaN$_3$, pH 7.0. Crude hemopexin (140 ml) was applied to the column at 22 ml/hour and 5.5 ml fractions were collected. The column was washed with equilibration buffer. Arrow indicates the application of buffer (10 ml) containing N-acetyl-D-glucosamine (100 mg/ml). Reprinted with permission from Vretblad and Hjörth (110).

of α-methylmannoside as the eluant and operating at 4 to 9°C, Cohen et al. (112) and Bishayee and Bachhawat (113) had previously used concanavalin-A-agarose chromatography for N-hexosaminidase purification. Brattain et al. (111) improved the yield by eluting at 37°C with 0.5-M α-methylmannoside. Takahashi et al. (114) purified rat kidney γ-glutamyltransferase on concanavalin A-agarose by using high concentrations (1 M) of α-methylmannoside as the eluant. Poor yields were produced with lower levels of a α-methylmannoside or when poorer inhibitors of the lectin (e.g., D-mannose) were employed in the eluant.

Kinzel et al. (115,116) employed lectin–agarose chromatography to separate intact tissue culture cells. Cells had been previously adsorbed to concanavalin A-agarose (117) and wheat germ lectin–agarose (118), but the adsorbed cells were not easily eluted from these bioselective adsorbents. To alleviate this difficulty, Kinzel et al. (115,116) utilized *Lens culinaris* lectin immobilized on agarose. *Lens culinaris* lectin has the same bioselectivity as concanavalin A but possesses binding constants that are 50 times lower (119). For this reason immobilized *Lens culinaris* lectin will bind Hela cells and will differentiate between transformed and nontransformed cells. The adsorbed cells can be bioselectively eluted with methyl-α-D-mannopyranoside. The results are dependent on the amount of *Lens culinaris* lectin bound to the agarose, and a lectin concentration of about 2.5 mg/ml of beads appear to be optimal. Above this concentration the adsorbed cells are poorly recovered in the eluant; below this optimum the adsorption is slow and the beads have a very low capacity. The binding is also dependent on the cell types employed and the temperature at which the binding is performed. Membrane glycoproteins from normal and transformed cells can also be separated in a similar fashion by way of bioselective adsorption on immobilized lectin (120).

Occasionally various unexpected factors complicate lectin-affinity chromatography just as they do any form of bioselective adsorption. For example, Turbeenniemi et al. (121) determined that lysyl hydroxylose bound very tightly to concanavalin A that had been coupled to agarose. The adsorbed enzyme could not, however, be eluted with 1.0-M α-methyl mannoside, with salt gradients or deforming buffers. Elution was effected by using 0.3-M α-methyl mannoside plus 50% ethylene glycol and a 3000-fold purification of the enzyme could be effected with this procedure. Apparently a "compound affinity" existed between lysyl hydroxylase and concanavalin A–agarose. This compound affinity was a synergistic combination of hydrophobic and bioselective adsorption.

Considerable interest in artifically glycosylated proteins has recently been generated. Proteins coated with polydextrins (122) or with polyethylene glycol (123,135,136) appear to not be recognized as foreign proteins when injected into an animal different from the type from which the proteins were originally isolated and thus do not elicit the formation of antibodies. Marshall and Humphreys (124) have used concanavalin A–agarose chromatography to separate dextrinized enzymes from those that were not derivatized during the preparation of artificial enzyme–dextrin conjugates. Dextrin conjugates of catalase, α-amylase, and trypsin were separated from excess unreacted enzyme through chromatography on concanavalin A–agarose. Prior to the use of concanavalin A–agarose, dextrinized proteins were separated from starting material by gel filtration. Lower recoveries (60 to 80%) of enzyme were found by using lectin chromatography compared to gel filtration (90% yields), but the lectin procedure was considerably easier and produced a more highly derivatized product (the latter was an absolute necessity in preventing cross reactivity of the protein in immunological experiments).

5.15. LECTIN PURIFICATION ON IMMOBILIZED SUGAR DERIVATIVES

Undoubtedly the easiest way to purify lectins is by bioselective absorption on immobilized sugar derivatives. Lis and Sharon (125), in a review of plant lectins, have described the successful purification of 16 different hemoagglutinins using this technique. All of the over 800 known lectins could quite probably be successfully prepared in this fashion. Those that recognize α-D-glucose residues may be bioselectivity adsorbed to the polydextrin Sephadex (109) and subsequently eluted with α-methyl-D-glucoside. Likewise, lectins that bind D-galactose can be purified by using the polygalactan agarose as the adsorbent. Special adsorbents have been synthesized for the bioselective adsorption of many other lectins that recognize L-fructose (126), N-acetyl glucosamine (127,128), or a wide variety of other carbohydrates (109, 129–131). Some of these bioselective adsorbents are synthesized by way of lengthy chemical procedures, whereas others are made by immobilizing naturally occurring polysaccharides. For example, Sutoh et al. (132) purified peanut lectin, a galactose binding anti-T agglutinin, on a bioselective adsorbent prepared by trapping guar gum (a naturally occurring galactomannan) within a polyacrylamide matrix. This preparation method of a bioselective adsorbent was very simple and consumed very little time. Moreover, the resulting lectin was electrophoretically homogeneous with a high anti-T titer. Previously this lectin had been isolated by affinity chromatography by using either unmodified agarose (133) or N-(ε-aminocaproyl)-β-D-galactopyranosylamine-agarose (134). The latter was a much better adsorbent than unmodified agarose, but the ligand, N-(ε-aminocaproyl)-β-D-galactopyranosylamine, is not commercially available and requires a lengthy synthesis. Entrapped guar gum was just as effective as the ligand agarose matrix in adsorbing peanut lectin and was almost as easy to use as unmodified agarose. Preparation of the adsorbent requires 3 to 4 hours and the entire purification can be done in 2 to 3 days with a yield of 410 mg of homogeneous lectin from 250 g of raw peanuts. (See Figure 5.24.)

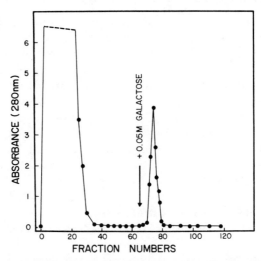

Figure 5.24. Elution profile of affinity chromatography of peanut lectin on a column (2.5 cm × 8 cm) of fresh guar-polyacrylamide beads. The column was equilibrated in 0.15-M NaCl and was washed with the same solution to remove nonlectin proteins. Arrow indicates position at which elution of the lectin with 0.05-M D-galactose in 0.15-M NaCl was started. Reprinted with permission from Sutoh et al. (132). Copyright © Academic Press, 1977.

References

1. C. R. Lowe and P. D. G. Dean (1971), *FEBS Lett.* **14**, 313.

2. J. D. Hocking and J. I. Harris (1973), *FEBS Lett.* **34**, 280.

3. A. F. Chaffotte, C. Roucous, and F. Seydoux (1977), *Eur. J. Biochem.* **78**, 309.

4. P. O'Carra, S. Barry, and T. Griffin (1974), *FEBS Lett.* **43**, 169.

5. C. R. Lowe, M. J. Harvey, D. B. Craven, and P. D. G. Dean (1973), *Biochem. J.* **133**: 499.

6. A. K. Grover and G. G. Hammes (1974), *Biochim. Biophys. Acta* **356**, 309.

7. C. R. Lowe and P. D. G. Dean (1974), *Affinity Chromatography*, Wiley-Interscience, New York, pp. 96–98.

8. C. R. Lowe, K. Mosbach, and P. D. G. Dean (1972), *Biochem. Biophys. Res. Commun.* **48**, 1004.

9. M. Wilchek and R. Lamed (1974), in *Methods in Enzymology* (W. B. Jakoby and M. Wilchek, eds.) **34**, 475.

10. F. M. Novais (1971), Ph.D. thesis, University of Birmingham, Birmingham, UK.

11. P. O'Carra, S. Barry, and T. Griffin (1974), in *Methods in Enzymology* (W. B. Jakoby and M. Wilchek, eds.) **34**, 108.

12. L. Andersson, H. Jörnvall, and K. Mosbach (1975), *Anal. Biochem.* **69**, 401.

13. B. Wermuth and N. O. Kaplan (1976), *Arch. Biochem. Biophys.* **176**, 136.

14. C.-Y. Lee, D. A. Lappi, B. Wermuth, J. Everse, and N. O. Kaplan (1974), *Arch. Biochem. Biophys.* **163**, 561.

15. C.-Y. Lee, L. H. Lazarus, D. S. Kabakoff, P. J. Russell, Jr., M. Laver, and N. O. Kaplan (1977), *Arch. Biochem. Biophys.* **178**, 8.

16. I. P. Trayer, H. R. Trayer, D. A. P. Small, and R. C. Bottomley (1974), *Biochem. J.* **139**, 609.

17. R. Barker, I. P. Trayer, and R. L. Hill (1974), in *Methods in Enzymology*, (W. B. Jakoby and M. Wilchek, eds.) **34**, 479.

18. C. S. Ramadoss, L. J. Luby, and K. Uyeda (1976), *Arch. Biochem. Biophys.* **175**, 487.

19. M. Lindberg and K. Mosbach (1975), *Eur. J. Biochem.* **53**, 481.

20. O. Berglund and F. Eckstein (1972), *Eur. J. Biochem.* **28**, 492.

21. O. Berglund, O. Karlström, and P. Reichard (1969), *Proc. Nat. Acad. Sci. (USA)* **62**, 820.

22. J. C. Paulson, W. E. Beranek, and R. L. Hill (1977), *J. Biol. Chem.*, **252**, 2356.

23. C. R. Lowe and K. Mosbach (1974), *Eur. J. Biochem.* **49**, 511.

24. P. T. Gilham (1971) in *Methods in Enzymology* (L. Grossman and K. Moldave, eds.) **21**, 191.

25. R. L. Hill, H. D. Hill, Jr., V. R. Naik, J. C. Paulson, J. I. Rearick, and M. Schwyzer (1976), *Tenth Internat. Congress of Biochem.*, Hamburg, Germany.

26. R. Barker, K. W. Olsen, J. H. Shaper, and R. L. Hill (1972), *J. Biol. Chem.* **247**, 7135.

27. R. Barker, C. K. Chiang, I. P. Trayer, and R. L. Hill (1974), in *Methods in Enzymology* (W. B. Jakoby and M. Wilchek, eds.) **34**, 317.

28. J. H. Shaper, R. Barker, and R. L. Hill (1971), *Fed. Proc.* **30**, 1265.

29. P. Andrews (1965), *Biochem. J.* **96**, 595.

30. J. K. Baird, R. F. Sherwood, R. J. G. Carr, and A. Atkinson (1976), *FEBS Lett.* **70**, 61.

31. G. E. Staal, J. F. Koster, H. Kamp, L. Van Milligen-Boersma, and C. Veeger (1971), *Biochim. Biophys. Acta* **227**, 86.

32. G. E. Staal, J. Visser, and C. Veeger (1969), *Biochim. Biophys. Acta* **185**, 39.

33. H. D. White and W. R. Jenks (1970) *Am. Chem. Soc. Meeting Abstr.* **43**.

34. H. Haeckel, B. Hess, W. Lauterborn, and K. H. Wusler (1968), *Hoppe-Seyler's Z. Physiol. Chem.* **349**, 699.

35. G. Kopperschlager, R. Freyer, W. Diezel, and E. Hofmann (1968), *FEBS Lett.* **1**, 137.

36. K. G. Blume, R. W. Hoffbauer, D. Busch, H. Arnold, and G. W. Lohr (1971), *Biochim. Biophys. Acta* **277**, 364.

37. A. C. W. Swart and H. C. Hemker (1970), *Biochim. Biophys Acta* **222**, 692.

38. M. Silink, R. Reddel, M. Bethel, and P. B. Rowe (1975), *J. Biol. Chem.* **250**, 5982.

39. G. Kopperschlager, W. Diezel, R. Freyer, S. Liebe, and E. Hofmann (1971), *Eur. J. Biochem.* **22**, 40.

40. M. F. Meldoleis, V. Macchia, and P. Lacetti (1976), *J. Biol. Chem.* **251**, 6244.

41. L. D. Ryan and C. S. Vestling (1974), *Arch. Biochem. Biophys.* **160**, 279.

42. S. T. Thompson, R. Cass, and E. Stellwagen (1975), *Proc. Nat. Acad. Sci. (USA)* **72**, 669.

43. S. T. Thompson, R. Cass, and E. Stellwagen (1976), *Anal. Biochem.* **72**, 293.

44. E. Stellwagen and B. Baker (1976), *Nature* **261**, 719.

45. R. L. Esterday, and I. M. Esterday (1974), in *Immobilized Biochemicals and Affinity Chromatography* (R. B. Dunlop, ed.) Plenum, New York, pp. 122, 123.

46. G. E. Lamkin and E. E. King (1976), *Biochem. Biophys. Res. Commun.* **72**, 560.

47. H. J. Bohme, G. Kopperschlager, J. Schulz, and E. Hofmann (1972), *J. Chromatogr.* **69**, 209.

48. P. Roschlau and B. Hess (1972), *Hoppe-Syeler's Z. Physiol. Chem.* **353**, 441.

49. W. Heyns and P. DeMoor (1974), *Biochim. Biophys. Acta* **358**, 1.

50. S. Shaltiel (1974), in *Methods in Enzymology* (W. B. Jakoby and M. Wilchek, eds.) **34**, 126.

51. J. E. Wilson (1976), *Biochem. Biophys. Res. Commun.* **72**, 816.

52. L. E. Wille (1976), *Clin. Chim. Acta* **71**, 355.

53. W. J. Jankowski, W. Von Muenchhawson, E. Sulkowsk, and M. A. Carter (1976), *Biochemistry* **15**, 5182.

54. J. DeMaeyer-Guignard and E. DeMaeyer (1976), *C. R. Acad. Sci. (Paris)* **283**, 709.

55. S. T. Thompson and E. Stellwagen (1976), *Proc. Nat. Acad. Sci. (USA)* **73**, 361.

56. R. A. Edwards and R. W. Woody (1977), *Biochem. Biophys. Res. Commun.* **79**, 470.

57. M. R. Diebel, Jr. and D. H. Ives (1977), *J. Biol. Chem.* **252**, 8235.

58. B. H. Weber, K. Willeford, J. G. Moe, and D. Piszkiewicz (1979), *Biochem. Biophys. Res. Commun.* **86**, 252.

59. R. A. Abreu, A. deKok, and C. Veeger (1978), *FEBS Lett.* **82**, 89.

60. M. W. Burgett and L. V. Greenley (1977), *Am. Lab.* **9**, 74.

61. N. B. Finter (1972), *Interferon and Interferon Inducers,* North Holland, Amsterdam.

62. I. Chibata, T. Tosa, and Y. Matuo (1974), in *Methods in Enzymology* (W. B. Jakoby and M. Wilchek, eds.) **34**, 267.

63. Y. Matuo, T. Tosa, and I. Chibata (1974), *Biochim. Biophys. Acta* **338**, 520.

64. D. A. Kleinsek, S. Ranganathan, and J. W. Porter (1977), *Proc. Nat. Acad. Sci. (USA)* **74**, 1431.

65. S. Barry, P. Brodelius, and K. Mosbach (1976), *FEBS Lett.* **70**, 261.

66. B. T. Kaufman and J. V. Pierce (1971), *Biochem. Biophys. Res. Commun.* **44**, 608.

67. G. P. Mell, J. M. Whiteley, and F. M. Huennekens (1968), *J. Biol. Chem.* **243**, 6074.

68. J. S. Erickson and C. K. Mathews (1971), *Biochem. Biophys. Res. Commun.* **43**, 1164.

69. J. M. Whiteley, G. P. Mell, J. H. Drais, and F. M. Heunnekens, in Ref. 67.

70. J. M. Whiteley, G. B. Henderson, A. Russell, P. Singh, and E. M. Zevely (1977), *Anal. Biochem.* **79**, 42.

71. D. N. Salter, J. E. Ford, K. J. Scott, and P. Andrews (1972), *FEBS Lett.* **20**, 302.

72. S. I. Hansen, J. Holm, and J. Lyngbye (1977), *J. Chromatogr.* **134**, 517.

73. C. A. Hall and A. E. Finkler (1971), in *Methods in Enzymology* (D. B. McCormick and C. D. Wright, eds.) **18** C, 108.

74. R. H. Allen and P. W. Majerus (1972), *J. Biol. Chem.* **247**, 7695.

75. V. V. Murthy, E. Jones, T. W. Cole, Jr., and J. Johnson, Jr. (1977), *Biochim. Biophys. Acta* **483**, 487.

76. R. L. Burger and R. H. Allen (1974), *J. Biol. Chem.* **249**, 7220.

77. R. Gräsbeck (1969), *Progr. Hematol.* **6**, 233.

78. R. H. Allen, and P. W. Majerus (1972), *J. Biol. Chem.* **247**, 7695.

79. G. Marcoullis, E–M. Salonen, and R. Gräsbeck (1977), *Biochim. Biophys. Acta* **495**, 336.

80. G. L. Francis, G. W. Smith, P. P. Toskes, and E. G. Sanders (1977), *Gastronenterology* **72**, 1304.

81. K. Olesen, E. Hippe, and E. Haber (1971), *Biochim. Biophys. Acta* **243**, 66.

82. R. H. Yamada and H. P. C. Hogenkamp (1972), *J. Biol. Chem.* **247**, 6266.

83. A. Matsuura, A. Iwashima, and Y. Nose (1972), *J. Vitaminol. (Koyoto)* **18**, 29.

84. A. Matsuura, A. Iwashima, and Y. Nose. (1974), in *Methods in Enzymology* (W. B. Jakoby and M. Wilchek, eds.) **34**, 303.

85. A. Matsuura, A. Iwashima, and Y. Nose (1973), *Biochem. Biophys. Res. Commun.* **51**, 241.

86. T. A. O'Brien, H. L. Schrock, P. Russell, R. Blake, II, and R. B. Gennis (1976), *Biochim. Biophys. Acta* **452**, 13.

87. L. A. Cohen (1974), in *Methods in Enzymology* (W. B. Jakoby and M. Wilchek, eds.) **34**, 102.

88. C. Arsenis and D. B. McCormick (1964), *J. Biol. Chem.* **239**, 3093.

89. C. Arsenis and D. B. McCormick (1966), *J. Biol. Chem.* **241**, 330.

90. M. N. Kazarinoff, C. Arsenis and D. B. McCormick (1974), in *Methods in Enzymology* (W. B. Jakoby and M. Wilchek, eds.) **34**, 300.

91. C. A. Waters, J. R. Murphy, and J. W. Hastings (1974), *Biochem. Biophys. Res. Commun.* **57**, 1152.

92. G. Blankenhorn, D. T. Osuga, H. S. Lee, and R. E. Feeney (1975), *Biochim. Biophys. Acta* **386**, 470.

93. R. Collier and G. Kohlaw (1971), *Anal. Biochem.* **42**, 48.

94. J. V. Miller, Jr. and E. B. Thompson (1971), *Fed. Proc.* **30**, 516.

95. E. B. Thompson (1974), in *Methods in Enzymology* (W. B. Jakoby and M. Wilchek, eds.) **34**, 294.

96. E. B. Thompson and J. V. Miller (1975), in *Methods in Enzymology* (B. W. O'Malley and J. G. Hardman, eds.) **40**, 266.

97. E. Ryan and P. F. Fottrell (1972), *FEBS Lett.* **23**, 73.

98. D. B. McCormick (1965), *Anal. Biochem.* **13**, 194.

99. J. Porath (1967), in *Nobel Symposium 3, Gamma Globulins* (J. Killander, ed.), Almquist and Wiksell, Stockholm and Wiley-Interscience, New York, p. 287.

100. E. Bayer and M. Wilchek (1974), in *Methods in Enzymology* (W. B. Jakoby and M. Wilchek, eds.) **34**, 265.

101. P. Cuatrecasas and M. Wilchek (1968), *Biochem. Biophys. Res. Commun.* **33**, 235.

102. J. S. Wolpert and M. L. Ernst-Fonberg (1973), *Anal. Biochem.* **52**, 111.

103. K. Hofmann, F. Finn, H.-J Friesen, C. Diaconescu, and H. Zahn (1977), *Proc. Nat. Acad. Sci. (USA)* **74**, 2697.

104. T. C. J. Gribnau (1977), *Proceedings of the Symposium "Advances in the Chromatographic Separation of Macromolecules,"* Birmingham, England, The Chemical Society, in press.

105. W. H. Scouten, F. Torok, and W. Gitomer (1973), *Biochim. Biophys. Acta* **309**, 521.

106. E. H. Cox and C. R. Lowe (1973), *FEBS Lett.*

107. J. Visser and M. Strating (1975), *Biochim. Biophys. Acta* **384**, 69.

108. U. Schmidt, P. Graffen, K. Altland, and H. W. Goedde (1969), in *Advances in Enzymology* (F. F. Nord, ed.) **32**, 423.

109. T. Kristiansen (1974), in *Methods in Enzymology* (W. B. Jakoby and M. Wilchek, eds.) **34**, 331.

110. P. Vretblad and R. Hjorth (1977), *Biochem. J.* **167**, 759.

111. M. G. Brattain, P. M. Kimball, T. G. Pretlow (III), and M. E. Marks (1977), *Biochem. J.* **163**, 247.

112. C. M. Cohen, G. Weissman, S. Hoffstein, Y. C. Awasthi, and S. K. Srivastava (1976), *Biochemsitry* **15**, 452.

113. S. Bishayee and B. K. Bachhawat (1974), *Neurobiology* **4**, 48.

114. S. Takahashi, J. Pollack, and S. Seifter (1974), *Biochim. Biophys. Acta* **371**, 71.

115. V. Kinzel, D. Kübler, J. Richards, and M. Stöhr (1976), in *Conconavalin A as a Tool* (H. Bittiger and H. P. Schnebli, eds.), Wiley, Chichester, UK, p. 467.

116. V. Kinzel, D. Kübler, J. Richards, and M. Stöhr (1976), *Science* **192**, 487.

117. G. Edelman, U. Rutishauer, and C. F. Millette (1971), *Proc. Nat. Acad. Sci. (USA)* **68**, 2153.

118. I. Kahane, H. Furthmayr, and V. T. Marchesi (1976), *Biochim. Biophys. Acta* **426**, 464.

119. M. D. Stein, I. K. Howard, and H. J. Sage (1971), *Arch. Biochem. Biophys.* **146**, 353.

120. E. Pearlstein (1977), *Exp. Cell Res.* **109**, 95.

121. T. M. Turpeenniemi, U. Puistola, H. Anttinen, and K. I. Kivirikko (1977), *Biochim. Biophys. Acta* **483**, 215.

122. J. J. Marshall and M. L. Rabinowitz (1976), *J. Biol. Chem.* **251**, 1081.

123. Y. Ashihara, T. Kono, S. Yamazaki, and Y. Inda (1978), *Biochem. Biophys. Res. Commun.* **83**, 385.

124. J. J. Marshall and J. D. Humphreys (1977), *J. Chromatogr.* **137**, 468.

125. H. Lis and N. Sharon (1973), *Annu. Rev. Biochem.* **42**, 541.

126. S. Blumberg, J. Hildesheim, J. Yariv, and K. J. Wilson (1972), *Biochim. Biophys. Acta* **264**, 171.

127. H. Lis, R. Lotan and N. Sharon (1974), in *Methods in Enzymology* (W. B. Jakoby and M. Wilchek, eds.) **34**, 341.

128. R. Lotan, A. E. S. Gussin, H. Lis, and W. Sharon (1973), *Biochem. Biophys. Res. Commun.* **52**, 656.

129. V. Horejsi and J. Kocourek (1974), in *Methods in Enzymology* (W. B. Jakoby and M. Wilchek, eds.) **34**, 361.

130. V. Horejsi and J. Kocourek (1973), *Biochim. Biophys. Acta* **297**, 346.

131. V. Horejsi and J. Kocourek (1974), *Biochim. Biophys. Acta* **336**, 329.

132. K. Sutoh, L. Rosenfield, and Y. C. Lee (1977), *Anal. Biochem.* **79**, 329.

133. T. Terao, T. Irimura, and T. Osawa (1975), *Hoppe-Seyler's Z. Physiol. Chem.* **356**, 1685.

134. R. Lotan, E. Skultelsky, D. Danon, and N. Sharon (1975), *J. Biol. Chem.* **250**, 8518.

135. A. Abuchowski, J. R. McCoy, N. C. Palczuk, T. Van Es, and F. F. Davis (1977), *J. Biol. Chem.* **252**, 3582.

136. A. Abuchowski, J. R. McCoy, N. C. Palczuk, and F. F. Davis (1977), *J. Biol. Chem.* **252**, 3578.

CHAPTER

6

COVALENT CHROMATOGRAPHY

Covalent chromatography is a special type of bioselective adsorption in which the protein being isolated becomes covalently bound to the adsorbent. The type of covalent bond employed is readily cleavable under mild conditions to yield the native protein. The most commonly employed covalent bonds are disulfide bonds and Hg^{++}-sulfur bonds. Not infrequently, the principal bioselective property of the protein used in the purification is that it possesses a particularly reactive and readily accessible thiol function, as, for example, the thiol proteases. These proteases can bind to either thiol or mercurial derivatives of agarose or acrylamide to form a covalent chromatography medium. Other covalent chromatography support materials include immobilized quasisubstrates that form moderately stable covalent enzyme–quasisubstrate complexes that are cleaved either by slow hydrolysis in buffer or more rapidly with an added nucleophile. Serine esterases, for example, can be isolated with an immobilized analogue of the quasisubstrate diisopropylfluorophosphate.

6.1. ACTIVATED THIOLS AS CHROMATOGRAPHIC MATERIALS

Numerous immobilized thiols have been proposed for potential use in biological systems; however, only three have been frequently used in affinity chromatography. These thiols are glutathione derivatives of agarose (1,2), N-acetyl homocysteine derivatives of aminoalkylated agarose (3), and thiol propyl agarose (4). The first possess ionically charged residues and thus may have ion-exchange properties. The latter is a neutral molecule with little, if any, hydrophobicity and thus should be considered as the appropriate matrix whenever ionic charges can interfere with the covalent chromatographic process. (See Figure 6.1.)

The thiolated agarose, regardless of which type is used, must be first "activated"; that is, it must be converting it to a reactive unsymmetrical disulfide, usually by coupling it to 2-thiopyridine or to nitrothiobenzoic acid. The purpose of this procedure is to place a "good leaving group" on the thiol agarose such that it will readily couple to protein thiols through thiol exchange (see Figure 6.2a).

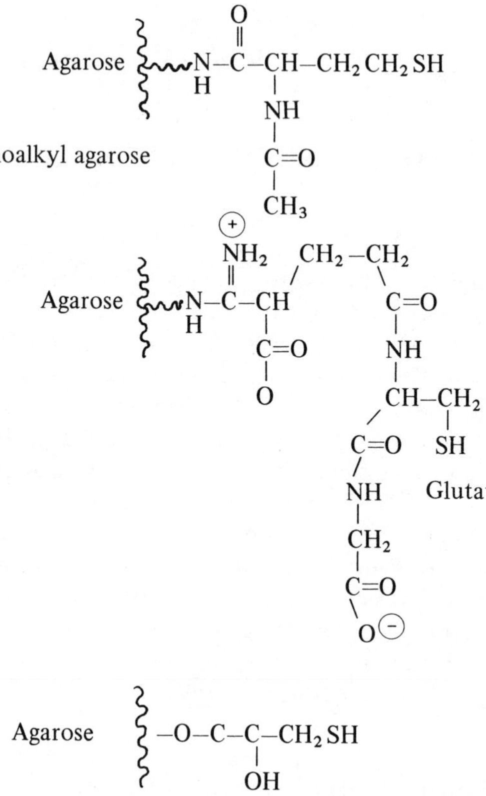

Homocysteine-aminoalkyl agarose

Glutathione-agarose

Thiopropyl-agarose

Figure 6.1. Thiol containing agarose matrices.

The protein(s) that are covalently coupled to the thiolated matrix are washed free of nonsulfhydryl-containing proteins, and the covalently coupled proteins are then subsequently eluted with thiol-containing (e.g., mercaptoethanol, dithiothreitol, or cysteine) buffers (see Figure 6.2*b*).

Unfortunately, the isourea function found in CNBr-activated gels is unstable in the presence of nucleophilic agents (5); therefore, glutathione coupled to CNBr-activated agarose is ultimately released on treatment with β-mercaptoethanol, or dithiothreitol (6,7). Glutathione that is freed from the matrix by this reaction may thus ultimately contaminate the proteins being purified. Moreover, the reusability of glutathione-agarose is severely hampered by its lability to sulfhydryl agents.

Even so, this product has been widely employed for the purification of a vast

(a)

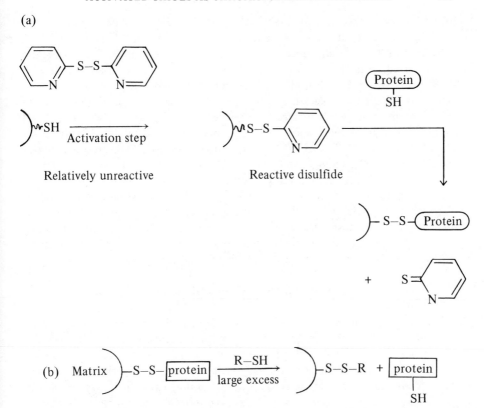

Figure 6.2. Schematic representation of the purification of thiol containing proteins.

number of sulfhydryl-containing proteins, including many of the proteases that contain essential active site thiols (1,2, 8–11). However, since thiol propyl agarose prepared by the method of Axen et al. (4) is not labile to these nucleophilic reagents and is stable under normal conditions of use, thiol propyl agarose may well replace activated glutathione agarose for these purposes.

6.1.1. Synthesis of Activated Glutathione–Agarose (1,2)

Agarose, for example, Sepharose 2B, is activated by any of the classical methods (see pages 42 through 51). The activated agarose is immediately washed with 10 volumes of 0.1-M NaHCO$_3$, pH 8.5, which contains 20 mg of glutathione per gram of agarose. This mixture is shaken for 20 hours at 22°C. The gel is subsequently washed with approximately 10 volumes of 0.1-M NaHCO$_3$, in a sintered glass funnel and then washed in a 1.8-cm × 30 cm column (for 50 g of agarose) sequentially with 240 ml each of (1) 0.1-M

NaHCO$_3$, pH 9.5, containing 1-M NaCl with 1-mM EDTA, (2) 0.1-M sodium acetate, pH 4.3, containing NaCl and EDTA as before, and (3) 1-mM EDTA. A flow rate of 10 ml/hour for each wash solution is recommended. To activate the immobilized glutathione, 50 g of this gel (wet weight) is further washed with 450 ml of 0.1-M Tris-HCl, pH 8.0, containing 0.3-mM NaCl and 1-mM EDTA. The washed gel is suspended in 60 ml of the same buffer containing 30-mM dithiothreitol and incubated for 10 minutes to reduce all available sulfhydryl residues. The gel is then washed with 150 ml each of 1-M NaCl, 0.5-M NaCHO$_3$, 0.3-M NaCl, 0.1-M Tris-HCl (pH 9.0), and 0.3-M NaCl. Each buffer contains 1-mM EDTA. The reduced and washed gel is activated by incubating it for 30 minutes in 1.5-mM 2,2'-dipyridyldisulfide at room temperature with gentle shaking. The activated glutathione-2-pyridyl disulfide agarose is washed to free it from excess 2,2'-pyridyl disulfide by using the same sequence of wash buffers as before. The final wash with 0.1-M Tris-HCl and 0.3-M NaCl, pH 8.0, containing 1-mM EDTA is monitored at 280 nm, and the washing is continued until no further absorbance at 280 nm is detectable in the wash solution.

6.1.2. Application of Activated Glutathione–Agarose

Papain was purified from crude papaya latex to homogeneity by use of covalent chromatography on glutathione-2-pyridyl disulfide–agarose prepared by this procedure. It is very difficult to obtain a homogeneous active papain preparation because papain (as it is normally isolated) consists of variable amounts of three components: (1) an active papain containing a free thiol; (2) an inactive but activable papain whose essential thiol is oxidized to form a protein–cysteine disulfide, and (3) a protein that appears to be papain but cannot be activated by reduction (12,13). To obtain homogeneous active papain, Brocklehurst et al. (1) extracted dried papaya latex by grinding it with sand and celite in 20-mM cysteine, pH 5.7. The crude latex extract was clarified and then subjected to two successive ammonium sulfate precipitations. The first precipitation occurred at 0 to 40% saturation of ammonium sulfate, pH 6.0, and the second at 0 to 35% saturation by utilizing solid ammonium sulfate. The twice precipitated papaya extract was dissolved in 0.1-M Tris-HCl, pH 8.0, or 0.1-M sodium acetate, pH 4.0, each containing 0.3-M NaCl and 1-mM EDTA. The dissolved precipitate was applied to a column of activated glutathione–agarose (a 1.8-cm × 30 cm column for 100 g of crude papaya latex). The column was washed with the application buffer until the absorbence at both 280 nm and 343 nm became less than 0.03 OD units. The column was next equilibrated with 0.1-M Tris-HCl, pH 8, containing 0.3-M NaCl and 1-mM EDTA. Finally, the covalently bound protein was eluted with 50-mM cysteine in the same buffer. The enzymically active fractions were pooled, concentrated, and applied to a Sephadex G-25 column that was previously equilibrated with 0.1-M KCl-1-mM EDTA. Dithiothreitol at a final concentration of 9.5 mM was added to the papain solution immediately prior to gel-permeation chromatography. Partially purified commercial papain could also be purified to homogeneity by covalent chromatography on activated glutathione–agarose.

Ficin is a sulfhydryl protease, similar to papain, which is isolated from the latex of the fig, *Ficus glabrata.* Like papain, it normally consists of an active form, a portion that can be activated by reduction, and an inactivatable form. Unlike papain, ficin consists of at least five active isoenzyme forms that can be separated chromatographically on carboxymethyl cellulose. Ficin can be prepared from crude ficin latex by covalent chromatography by the same method as described earlier for papain (8). Ficin prepared in this fashion is fully active and can be separated into five active isoenzyme forms on carboxymethyl cellulose chromatography. Approximately 10-fold purification is effected by covalent chromatography and the resulting ficin possesses higher specific activity than does enzyme prepared by more arduous classical methods (14).

Brocklehurst and Kierstan (10) and Freidenson and Liener (15) have suggested that the inactivatable forms of both papain and ficin may be "proenzyme" forms that differ from the active enzyme by a disulfide interchange. The isoenzymes of ficin may also differ through similar disulfide interchange.

Bovine serum albumin has a single free thiol group (16,17) that, like papain, may be bound to a free cysteine through a disulfide linkage. Alternatively, bovine serum albumin may form dimers through disulfide linkages. When bovine serum albumin is isolated by classical techniques it contains about 0.6 moles of free thiol per mole of polypeptide. The portion of the protein containing a free sulfhydryl group, often called *mercaptalbumin,* can be isolated from the other forms of serum albumin by classical enzyme isolation techniques, but the yields are low and the procedure is tedious (18). Carlsson and Svenson (19) isolated electrophoretically homogeneous mercaptalbumin from commercial bovine serum albumin in a single covalent chromatography procedure using activated glutathione–agarose. Interestingly, when the purified bovine mercaptalbumin was incubated for 5 hours at pH 9.1, a new electrophoretic species appeared. This probably resulted by thiol–disulfide interchange in the same fashion as reported for the interconversion of the active and inactive forms of papain (10,11).

Urease is among the enzymes successfully purified by covalent chromatography activated glutathione–agarose. High-purity urease is classically obtained from a crude extract of jack bean meal through multiple cycles of gel-permeation chromatography on Sephadex G-200. Norris and Brocklehurst (30) have been able to purify urease from a crude enzyme preparation by a single covalent chromatographic procedure. Urease, with 24 highly reactive free sulfhydryl groups and a total of 84 free SH groups of varying reactivity, is an ideal candidate for activated thiol chromatography.

Egorov et al. (6) used activated glutathione to isolate thiol-containing peptides from thiol proteins. First, the proteins were immobilized on glutathione agarose, after which the immobilized protein was proteolytically degraded. The nonthiol peptides that resulted were easily washed from the column with buffer. Subsequently the covalently bonded thiol peptides were released by elution with a reducing agent, such as β-mercaptoethanol. The coupling yield for the thiol-containing proteins was usually excellent (90 to 97%), although a few proteins, notably ceruloplasm, gave poor binding (20 to 30%). (This may be because ceruloplasm has three free thiols, and, once the first has reacted, the next two may be sterically hindered from further reaction.) The overall yield of thiol-containing peptides released by elution with β-mercaptoethanol was likewise excellent. This method yields thiol peptides only from exposed free sulfhydryl groups and is particularly useful since most proteins have few reactive free sulfhydryls.

The major problem in this procedure is slow leakage of glutathione from the matrix on washing with nucleophilic reducing agents. From 0.1 to 1.0% of the glutathione was removed with each cycle of reduction. This situation is most probably due to a nucleophilic attack by the reducing agent (β-mercaptoethanol) on the isourea linkage between the glutathione and the agarose. If the problem is indeed due to this cause, it might be circumvented by using sodium boro-hydride as the reductant or by coupling the glutathione to oxirane activated agarose instead of cyanogen bromide-activated agarose.

In a typical case, bovine serum mercaptoalbumin (0.5 to 1.0 μmole) in sodium phosphate, pH 7.8, was added to an excess of activated glutathione agarose. This mixture was purged with nitrogen and then incubated with agitation by rotating the tube at room temperature for 6 hours. Ninety seven percent of the protein added was coupled to the beads. Next, the immobilized protein was collected by suction filtration and washed with buffer. The washed immobilized protein was digested with pepsin. The proteolysis was performed by agitating the beads at 37°C for 15 hours. The beads were then collected and sequentially

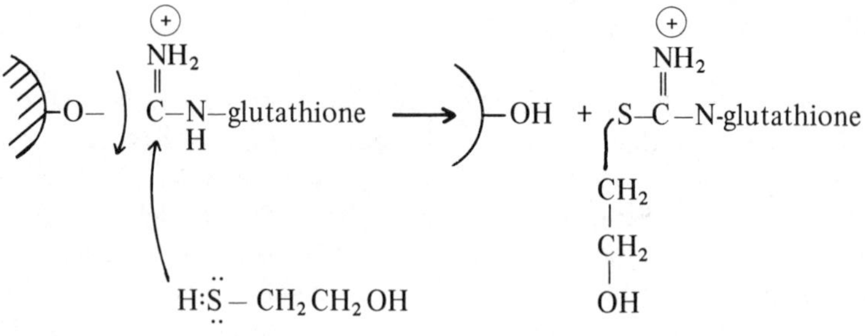

Figure 6.3. Proposed nucleophilic hydrolysis of glutathione coupled to CNBr activated agarose.

washed with several volumes each of 0.2-M ammonium acetate, pH 8.6, and 1.0-M NaCl. The thiol peptides were then detached by incubating the beads for 30 minutes at 25°C in one volume of 0.2-M ammonium acetate, pH 8.6, containing 20-mM β-mercaptoethanol. The freed thiol peptides were then collected and analyzed. In the case of bovine serum albumin, the resulting octapeptide was sequenced and found to be identical in composition to the thiol peptide previously identified (21,22) through use of classical techniques.

Laurell et al. (23) used activated glutathione–agarose, activated cysteine–agarose, and four related thiol agarose matrices for separation of plasma proteins by "thiol–disulfide interchange," as they term this form of covalent chromatography. The elution profiles varied widely according to the type of thiol agarose material used. 3-Thio-2-hydroxypropyl (thiopropyl) agarose preferably bound albumin at pH 8.1 while cysteine selectively bound the α, AT, and IgA fractions. Using various types of thiol materials, they were able to partially fractionate all of the free thiol-containing plasma proteins and to separate non-thiol-containing proteins from those that have reactive sulfhydryl groups. These authors activated the immobilized thiol matrix by using 5,5′-dithiobis-(2-nitrobenzoic acid) (Ellman's reagent, NBs, or DTNB). This is a negatively charged activating agent that forms a yellow monothiol anion that can be used to determine the completeness of the thiol–disulfide interchange. There seems to be little basic difference between DTNB and the 2′-thiopyridine as the activating group; thus the choice of the best agent must be made individually for each protein to be purified.

Lowe (24–26) used DTNB to covalently attach lipoamide dehydrogenase to thiolagarose, which was created by using N-acetyl homocysteine agarose as depicted in Figure 6.4.

The properties of the immobilized enzyme were studied extensively with a significant increase in the thermal stability of the enzyme upon immobilization.

6.1.3. Synthesis and Use of Thiopropyl Agarose

Thiopropyl agarose (or, more properly, 3-thio-2-hydroxypropyl-agarose) is readily synthesized (4) and is also commercially available (Pharmacia). Since the commercial material has a fixed concentration of thiol group, it may be preferable to synthesize your own thiopropyl agarose and thus control its SH-group concentration. The synthesis also offers considerable economic advantage over the commercial material.

Porath and Axen (27) have reported that the following method yields about 50 μmoles of SH per gram of dried covalent chromatography packing. Initially, 30 g of 6% agarose is washed with water and collected by suction filtration. The beads are freed of any interstitial water by suction filtration and are then suspended in 24 ml of 1-M NaOH.

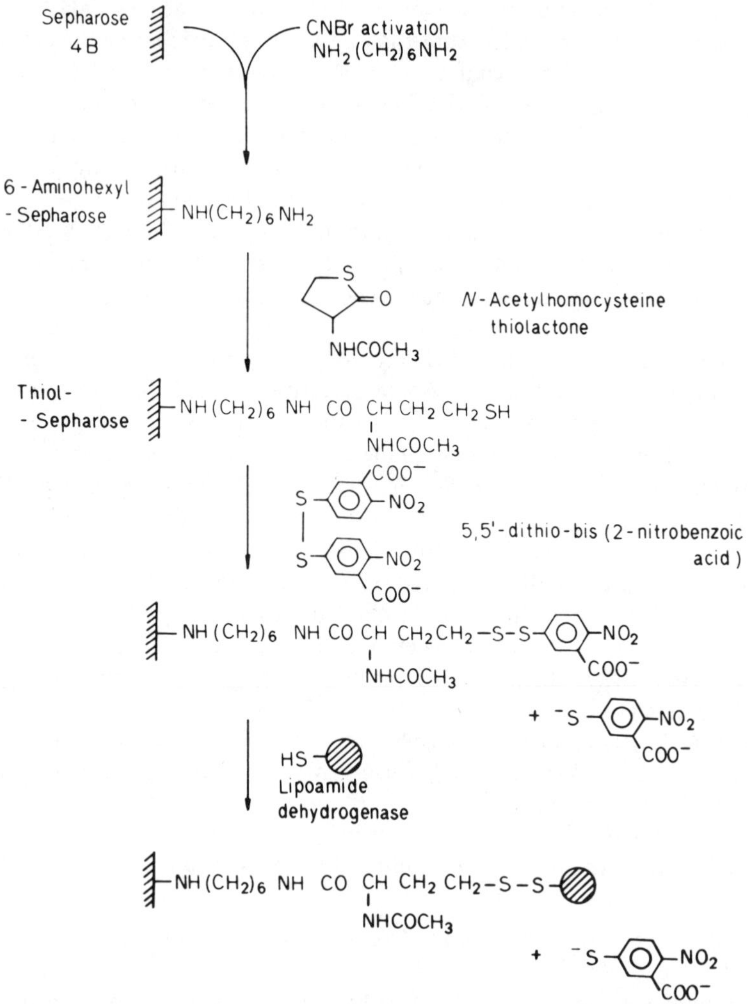

Figure 6.4. Attachment of lipoamide dehydrogenase to thiol-Sepharose by thiol–disulfide interchange. Reprinted with permission from C. R. Lowe (1977), *Eur. J. Biochem.* **76**, 391.

Epichlorohydrin is added to the suspension with stirring at room temperature. These authors report that 50 µmoles of SH per gram of dried agarose will result if 0.75 ml of epichlorohydrin is added. The final thiol content is dependent on the amount of epichlorohydrin used in this step and 4.5 ml of epichlorohydrin will yield a product with 700 µmoles of sulfhydryl group substituents. The epichlorohydrin preparation is mixed at room temperature for 15 minutes, after which the temperature is raised to 60°C and stirring is continued for 2 additional hours. The resulting epoxylated beads are washed with water until neutral, followed with several volumes of 0.5-M sodium phosphate, pH 6.2. Immediately after this, the beads are suction filtered and added to 30 ml of 2-M sodium thiosulfate. The beads are stirred in the thiosulfate for 6 hours at room temperature, followed by washing with several volumes of water and suspension in 60 ml of 0.1-M NaHCO$_3$. Dithiothreitol (8 mg/ml) in 1-mM EDTA is added in *at least* a two-fold excess over the epoxy functions. When 0.75 ml of epichlorohydrin was used in the activation procedure, 4 ml of the dithiothreitol solution was sufficient. After 30 minutes of reaction at room temperature, the beads are removed from the reducing solution by suction filtration and washed successively with 300 ml each of 0.1-M NaHCO$_3$, containing 1-M NaCl and 1-mM EDTA, then 1-mM EDTA, and finally with 0.01-M sodium acetate, pH 4.0, containing 1-mM EDTA. It is important to avoid the passage of air through the beads during washing and to protect the beads from further air oxidation by storing them in deaerated 0.01-M sodium acetate, pH 4.0. Since storage, even under these conditions, results in a reduction of the concentration of free thiol groups by 80% in 1 month (4), reduction of the beads should be performed immediately prior to use. Beads that have been stored for a long period of time can be regenerated to full, or nearly full, capacity by rereducing them with dithiothreitol as described earlier.

To activate the reduced thiopropyl agarose, the beads (which should be reduced if needed to assure a maximal SH content) are washed with water and then 50% acetone–water. The gel is suspended in a 50% acetone–water solution to which 100 mg of 2-pyridine disulfide dissolved in a minimum of acetone–water has been added. The mixture is then stirred for 30 minutes to 1 hour at room temperature and then washed with 50% acetone–water. The beads are finally washed with 1-mM EDTA, pH 7.0. At this point the beads are stable and can be stored for at least 6 months at 40°C with little loss of active thiol content. Activated thiopropyl agarose is superior to activated glutathione agarose or N-acetyl homocysteine agarose in many ways; first, it is hydrophilic, yet uncharged; second, it is much more stable to nucleophilic attack during prolonged storage; and finally, it is more conveniently prepared and does not require cyanogen bromide activation.

Several proteins have been isolated for the first time using thiopropyl agarose. For instance, Ryden and Deutsch (28) have purified the major copper-binding protein from human fetal liver and have identified it as metallothionein. The thiopropyl chromatography step was a major portion of the purification, effecting a five-fold increase in copper content and an approximately 50% yield of copper binding activity (see Table 6.1.).

The method of preparation of copper-binding metallothionein first involved extracting the tissue for 1 hour with 0.3-M ammonium acetate, pH 5.5, containing 1% β-mercaptoethanol. The extract was clarified, applied to a Sephadex G-75

TABLE 6.1. **Preparation of Major Copper-Binding Protein from Human Prenatal Liver**

	Volume, ml	Protein mg	Copper	Copper Content w/w, %	Yield of Copper, %
Extract	211	11,790[a]	1.862	0.0158	100
Sephadex G-75	1395	575[b]	1.716	0.298	92.9
Thiopropyl–					
Sepharose	33	61.2[b]	0.846	1.38	45.4
Sephadex G-50					
Monomer	105	23.4[b]	0.493	2.10	26.5
Dimer	97.5	15.3[b]	0.246	1.61	13.2
Aggregate	76.0	10.4[b]	0.126	1.21	6.8

Source: Reprinted with permission from Ryden and Deutsch (28).

[a] Dry weight after lyophilization.

[b] From amino acid analyses.

column, and then chromatographed using the same buffer without the reducing agent. The copper containing fractions (determined by atomic absorption) were pooled and applied to a small column (2 cm × 7 cm for 500 mg of wet liver as starting material) of activated thiopropyl agarose (15 moles of SH per milliliter of packed gel). The column was washed with (1) 0.3-M ammonium acetate, pH 5.5, (2) 50 ml of the same buffer containing 1-M NaCl, (3) 0.05-M Tris-HCl, pH 8.0, containing 1-M NaCl, and (4) 0.05-M Tris-HCl, pH 8.0. The bound protein was then eluted with 0.05-M Tris-HCl, pH 8.0, containing 1% β-mercaptoethanol. The resulting copper-binding protein is composed of monomeric, dimeric, and polymeric metallothionein, which can be separated further by gel-permeation chromatography.

The resulting metallothionein appeared to be very similar to the zinc- and cadmium-binding proteins isolated from several sources by similar, but more tedious, methods (29–31). Covalent chromatography by thiol–disulfide interchange should be applicable to the purification of these metal-binding proteins, such as ceruloplasm (6) and perhaps to other, as yet undiscovered, metal-binding proteins that often possess several highly reactive and exposed sulfhydryl residues.

Urease has also been purified by using thiopropyl agarose (32). In the covalent chromatography step, a 160-fold purification was effected with a yield of 81%. The three-step procedure (ethanol extraction, covalent chromatography, and gel-permeation chromatography) gave yields of 72% and almost 300-fold purification. The specific activity of this preparation was 2515 units/mg, which is about 50% higher than that obtained by conventional techniques. Similar preparations using activated glutathiane agarose (20) were equally successful but required crystallization of the urease prior to covalent chromatography.

Svenson et al. (7) have also used activated thiopropyl agarose to purify cysteine-containing peptides obtained from proteolytic digestion of proteins. Unlike Egorov et al. (6), these authors reduced and digested the proteins prior to purification and thus effected separation of all cysteine-containing peptides from digestion products of ribonuclease and bovine serum albumin. The peptide mixture was incubated with activated thiopropyl agarose for 6 hours at room temperature. The gel was then washed extensively with suction filtration, packed into a column, and washed with 5 additional volumes of $0.1\text{-}M$ ammonium acetate, pH 8.5.

The bound peptides were eluted from the column with the same buffer containing 50-mM β-mercaptoethanol. A 75% yield of cysteinyl peptides resulted. The authors also attempted the same procedure using activated glutathione-agarose, but when this was used as the matrix, considerable quantities of glutathione were found as a contaminant in the cysteinyl–peptide mixture.

6.1.4. Synthesis and Use of Organomercurial Agarose

Chromatography on immobilized organomercurial compounds has been widely employed to purify such diverse compounds as thiol proteases (33,38), the subunits of lombricine kinase (39), thiol containing histones (40,41), estradiol dehydrogenase (42,43), and even DNA preparations (44). Its wide use may be due to the ease with which organomercurial agar derivatives can be synthesized and to their widespread use as protein modification reagents. The only possible problem in using organomercurial derivatives as covalent chromatography matrices is possible mercurial contamination of the resulting enzyme. Such a problem has never been reported in the literature, possibly because the purified enzymes are normally desalted after organomercurial chromatography by gel filtration.

There are two major methods for synthesizing organomercurial agarose. Either sodium p-chloromercuribenzoate is coupled to aminoethyl agarose by means of a water-soluble carbodiimide such as 1-ethyl-3-(dimethylaminopropyl) carbodiimide (3), or p-aminophenyl mercuri acetate is directly reacted with cyanogen bromide-activated agarose (34,35). Organomercurial glass derivatives have been made in the author's laboratory by the addition of mercury trifluoroacetate to porous glass beads that were previously derivatized with an alkenyl silane such as 7-octenyltrimethoxysilane or tris (7-octenyl)-methoxysilane.

To prepare organomercurial agarose by the carbodiimide method, 25 ml of aminoethyl agarose is washed extensively with water and transferred to 20 ml of

40% (w/v) dimethylformamide in water. Sodium p-chloromercuribenzoate (625 mg) is added with stirring and the pH is adjusted to pH 4.8 by using concentrated HCl. After adjusting the pH, 770 mg of 1-ethyl-3-(dimethylaminopropyl) carbodiimide is slowly added with constant stirring. The pH is maintained at 4.8 for the next 2 hours, followed by stirring the suspension gently for an additional 16 to 24 hours. All operations are performed at room temperature. After the reaction is complete, the beads are cooled to 4°C and washed over a 48-hour period with 10 liters of 0.1-M sodium bicarbonate, pH 8.8. The beads should produce a yellow color in the 2,4,6-trinitrobenzenesulfonic acid test (see page 47). If an orange, rather than yellow color develops, the sodium p-chloromercuribenzoate coupling procedure should be repeated. Failure to repeat the process will result in free amino groups remaining on the agarose, which could impart unwanted ion-exchange characteristics to the chromatographic packing.

The organomercurial agarose prepared in this fashion is stable for at least 1 year of storage. The normal cysteine binding capacity for material prepared in this fashion is 1.2 to 2.5 μmoles per milliliter of settled volumes of support material.

Organomercurial agarose is prepared according to the direct cyanogen bromide attachment method by adding 80 ml of 10% (v/v) aqueous dimethyl sulfoxide at 0°C to 40 ml (settled volume) of cyanogen bromide preactivated agarose (either freshly activated, which is preferable, or commercially available cyanogen bromide activated agarose). To this is added a solution of 6 g of p-aminophenylmercuric acetate in 10 ml of dimethyl sulfoxide. The suspension is stirred for 20 hours at 6°C, warmed to room temperature for 20 minutes, filtered, and then washed by resuspending the beads in 40-ml 20% aqueous dimethylsulfoxide at 37°C for 15 minutes with refiltering. This washing procedure is repeated four times, and then the beads are further washed by packing them into a column and eluting it slowly at room temperature with at least 100 ml of 20% aqueous dimethylsulfoxide over a 3-hour period. The beads are incubated overnight at room temperature in 60 ml of 0.1-M ethylenediamine, pH 8.0, to block any cyanogen bromide-activated functions remaining on the beads. The reacted beads are again washed extensively with water and stored for later use.

The beads prepared in this fashion are as stable as those prepared by the carbodiimide method. By this method, a thiol-binding capacity as high as 8 mM can be achieved. The protein-binding capacities of the organomercurial prepared by either method are approximately equal and usually range from 40 to 50 mg hemoglobin. The exact capacity depends, of course, not only on the nature of the adsorbent, but also on the protein used in each particular case.

Cathepsin B1, a lysozomal thiol protease, has been purified (36) to homogeneity from human liver by using the organomercurial agarose produced by the direct coupling method. Most of the loss in activity occurred with the classical purification techniques used prior to the covalent chromatography step and was especially prevalent during ion-exchange chromatography on carboxymethyl cellulose. The overall yield was 24%, whereas the yield from the organomercurial agarose was 60 to 70%. A by-product of the purification of cathepsin B1 was cathepsin D, a nonthiol lysozomal protease that, since it has no free thiols, passed directly through the organomercurial column (see Table 6.2). In Table 6.2 the yields of protein and enzymic activity are expressed per kilogram of tissue and are mean values based on five separate preparations, each with about 3 kg of tissue. The value for protein in the first three stages is, more precisely, material insoluble in aqueous 80% (v/v) acetone, whereas that for the later stages is the product of volume (liters) and E_{280}. The specific activity of purified cathepsin B1 based on a Lowry determination of protein was exactly twice the value derived from E_{280}.

McDonagh et al. (37) have employed organomercurial agarose prepared by the carbodiimide method and sold commercially as Affi-gel 501 (Bio-Rad, Richmond, Calif.) to purify and study the blood-clotting factor, factor XIII, isolated from plasma and from blood platelets. *Plasma* factor XIII consists of one **a** subunit that contains six free thiol groups and one **b** subunit that lacks a free thiol. *Platelet* factor XIII consists solely of dimeric **a** subunits. The **a** subunit binds to the organomercurial agarose through its free sulfhydryl residues and can be recovered with 80 to 100% yield by elution with dithiothreitol, β-mercaptoethanol, or similar reducing agents. The **b** subunit binds to organomercurial agarose only indirectly through the **a** subunit to which it is attached.

Factor XIII is the clotting factor responsible for catalyzing the cross-linking of fibrin to form the blood clot. Normally, factor XIII exists in the plasma and platelet as an inactive zymogen, but on treatment with thrombin and Ca^{++} ions, it becomes activated through proteolytic cleavage. The proteolysis converts the inactive **a** subunit into an active **a'** subunit. Both **a** and **a'** subunits bind to organomercurial agarose. The **a'** subunit has a large number of surface thiols, whereas the thiols in the inactive **a** subunit are partly buried. Thus the **a'** subunit binds to the organomercurial agarose much more tightly than does the **a** subunit. The **b** subunit is in equilibrium with the **a** or **a'** subunit and can be easily removed by using appropriate buffers in pure form from the entire **ab** complex immobilized on organomercurial agarose. (See Figure 6.5.)

Purification of pure factor XIII or its subunits by this method took no more than 1 day to perform. Time is critically important in preparing the blood clotting factor since, in the crude stages, there is a large amount of proteolytic activity that degrades the enzyme.

In a particularly eloquent experiment, Der Terrossian et al. (39) used organomercurial agarose (prepared also by the direct coupling technique) to separate the two nonidentical subunits of lombricine kinase and also to separate the cysteinyl peptides from a tryptic digest of this enzyme (see Figure 6.6). The

TABLE 6.2. Purification of Human Cathepsins B1 and D[a]

	Cathepsin B1				Cathepsin D		
	Protein, g	Activity Units	Specific Activity, Units/ml/ E_{280} Unit	Yield %	$10^{-3} \times$ Activity, Units	Specific Activity, Units/ml/ E_{280} Unit	Yield %
Homogenate	203	419	0.0021	(100)	113	5.6	(100)
Arquad supernatant	68	309	0.0045	74	78	11	69
Autolyzed extract	50	434	0.0087	104	60	12	53
Acetone (47 to 64%) fraction	1.26	381	0.302	91	55	44	49
CM-cellulose (stepwise elution)	0.54	224	0.415	54	26	48	23
Organomercurial–Sepharose							
Unadsorbed	0.22	5	—	1	23	105	20
Adsorbed	0.076	132	1.74	32	0.3	—	0.3
Sephadex G-75 superfine	0.037	101	2.73	24	—	—	—
CM-cellulose (equilibrium elution)	0.010	27	2.70	6	—	—	—
DEAE-cellulose	0.032	—	—	—	12	375	11

Source: Reprinted with permission from Barrett (36).

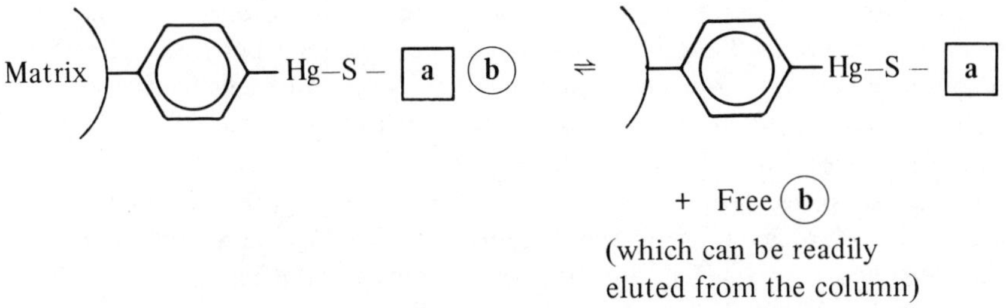

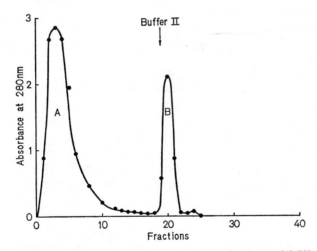

Matrix — ⬡ — Hg–S – [a] (b) ⇌ — ⬡ — Hg–S – [a]

+ Free (b)

(which can be readily
eluted from the column)

Figure 6.5. Purification of the A and B subunits of blood clotting factor XIII.

peptide containing the essential active-site thiol was also separated from
the other cysteinyl peptides.

Lombricine kinase is a dimeric guanidinophosphate kinase that functions in
Lumbricus terrestris muscles in much the same fashion as creatinine kinase does
in mammalian muscle. Despite the fact that it is dimeric, it possesses only one
site for ATP·Mg^{++} binding and only one essential thiol group. This essential thiol
reacts with 2,4-dinitrofluorobenzene at rates much faster than any other thiol or
amine function in the protein. To separate the active site-containing subunit
that possesses the essential thiol, lombricine kinase was treated with 1.75 moles

Figure 6.6. Chromatography of lombricine kinase with free essential–SH group on
Sepharose–mercurial; 70 mg of thiolyzed lombricine kinase in buffer I (10% dimethylsulfoxide, 0.6% butanol, 0.1-M KCl, 10-mM Na$_2$SO$_3$, 0.05-M sodium acetate, 1-mM EDTA,
8-M urea pH 5.5) was applied on a column of 2 ml of Sepharose–mercurial equlibrated
with buffer I. (*A*) Unbound fraction; (*B*) bound fraction eluted with buffer II (0.5-mM
HgCl$_2$ in buffer I). Reprinted with permission from E. Der Terrossian, L.-A. Pradel, R.
Kassab, and G. Desvages (1974), *Eur. J. Biochem.* **45**, 243.

of 2,4-dinitrofluorobenzene per mole of protein. This procedure resulted in the incorporation of 1 mole of 2,4-dinitrofluorobenzene per mole of enzyme and a loss of 97% of the enzymatic activity. Next, the nonessential and less reactive thiols were irreversibly alkylated with excess iodoacetic acid in 8-M urea. This process both alkylates and denatures the protein. After removal of the urea and excess alkylating agent, the 2,4-dinitrophenyl group is removed from the essential thiol by treating the enzyme with 0.1-M β-mercaptoethanol. The incubation was performed under nitrogen to prevent disulfide formation. Next, the excess β-mercaptoethanol was removed by gel filtration on a Sephadex G-25 column. Sodium sulfite was added to the protein (100 mg total) solution. The resulting solution was passed through a 2-ml organomercurial agarose column. The thiol-containing subunit was adsorbed to the column, and the subunit not possessing the essential thiol passed directly through the column and was collected in the initial eluant. The adsorbed catalytic subunit was eluted with the same buffer containing 0.5-mM HgCl$_2$. The eluted protein could be treated further with β-mercaptoethanol to restore the free thiol group. The purified thiol-containing subunit was then subjected to tryptic digestion, and the hydrolysate was again chromatographed in essentially the same fashion to obtain the tryptic peptide that contained the essential thiol.

These authors report that, with the use of classical protein and peptide purification techniques, they spent 1 year of hard work to obtain the essential cysteine-containing peptide with an overall yield, starting from the purified enzyme, of 2.58%. Conversely, a covalent chromatography on organomercurial agarose permitted the purification of the essential cysteine-containing peptide in 10 days with a yield of 41.5%!

Among various other proteins purified by using organomercurial agarose are histone F3, which is the only mammalian histone that contains cysteine (40,41) and 17-β-estradiol dehydrogenase (42,43). This dehydrogenase possesses an active site thiol that can be protected from alkylation with NADP. Thus, to effect specific purification of this enzyme, a crude enzyme solution is incubated with NADP (0.1 mM) and 1.0 mM of sulfhydryl reagent (either N-ethyl maleimide or 5,5'-dithiobisnitrobenzoic acid). Afterward the excess reagent and the NADP is removed by gel filtration and the enzyme is then chromatographed on organomercurial agarose as described previously.

A third type of organomercurial chromatographic media, which has found little use in protein purification but has considerable potential use in separating nucleotides and denatured DNA, is prepared by adding mercuric acetate or mercuric trifluoroacetate to an immobilized alkene (44). The nitrogenous bases of denatured DNA, nucleotides, and nucleosides have a strong affinity for the resulting immobilized alkyl mercury (45–47) and thus are retarded on organomercurial columns. This method of preparing an immobilized organomercurial was first used by Shainoff to produce a covalent chromatography media with the use of cellulose as the matrix for isolation of thiol proteases (48). However, because of the heterogeneity of cellulose, it is rarely used. Gruenwedel et al. (44) modified Shainoff's technique to prepare derivarives of dextrin beads, such as Sephadex G-25, G-50, G-75, G-100 and G-299. First, the beads are reacted with 1-allyloxy 2,3-epoxypropane (allyglycidoxy ether) in concentrated base and sodium borhydride. Mercuric acetate in 10% acetic acid is added to the product of the reaction and incubated at 60°C for 1 hour. The beads are washed extensively and stored in brown bottles away from the light. The beads, which are stable in alkaline solutions but labile when treated with strong acids, have a mercuric ion concentration of 0.2 to 0.4 mmole per gram of dry beads. (See Figure 6.7)

Chromatography on the mercurated dextrin bead is usually performed by applying the nucleotides in a neutral or slightly alkaline buffer (pH < 9) and eluting with a pH gradient. A gradient of increasing ionic strength may also be employed as the eluant.

In the author's laboratory, similar materials have been prepared by silanizing controlled pore glass with 7-octenyltrimethoxysilane or with tris-5-pentenyl methoxysilane. The alkenyl silane glass is then mercuriated in a fashion that is essentially the same as that employed for the dextrin beads. The potential

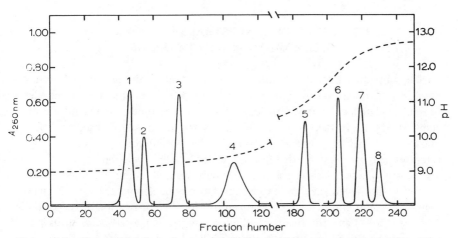

Figure 6.7. Chromatography of mono- and dinucleotides at room temperature on a 0.9-cm × 57-cm column of mercurated Sephadex G-25 using 0.05 Tris buffer plus 0.1-M sodium bicarbonate–carbonate as an eluent. Left-hand ordinate: adsorbence at 260 nm. Right-hand ordinate: pH (– – – – – pH of eluent). Peak 1, Cyd-P-Cyd; peak 2, Cyd-5',P; peak 3, Ado-3',P; peak 4, Ado-P-Ado; peak 5, Guo-2(3')-P; peak 6, dThd-5'P; peak 7, Guo-P-Urd; peak 8, Urd-P-Urd. Flow rate: 60 ml/hour cm². Each fraction contains 3.9 ml. Reprinted with permission from D. W. Gruenwedel, M. G. Heskett, and J. E. Lammert (1975), *Biochim. Biophys. Acta* **402**, 7.

advantage of using controlled pore glass is that a very large pore diameter (up to about 2000 Å) can be obtained through controlled pore glass. With such pore diameters, the larger polynucleotides can occupy the inside of the beads and thus achieve a much higher effective capacity. Additionally, dextrin beads are very fragile and cannot be used in even moderate pressure liquid chromatographic systems, whereas glass is often used in high-pressure liquid chromatography.

An interesting corollary to the use of organomercurials in covalent chromatography of thiol-containing proteins is in the use of immobilized thiols for removing mercury and other heavy-metal ions from solution. A Dutch chemical firm, (49,50), in a patented process, has applied a thiol-containing resin to remove inorganic mercury ions from effluents from their chloro–alkali plant. The process appears to be economically superior to other mercury recovery methods and yields a much higher quality plant discharge. In the author's laboratory, lipoic acid immobilized on very fine porous glass beads has been used as an immobilized reducing agent (51) and employed as a mercury poisoning treatment under certain circumstances (52). If lipoic acid or other thiol beads are given orally, they will chelate most (if not all) inorganic mercury ions that are in the gastrointestinal tract at that time. The mercury is then excreted in the feces with the lipoyl glass beads.

6.2. USE OF QUASISUBSTRATES AS COVALENT CHROMATOGRAPHY PACKINGS

Quasisubstrates are enzyme inhibitors that react at the active site of an enzyme to form an acylated or otherwise modified enzyme that resembles a normal acylated enzyme intermediate but is many orders of magnitude more stable. Many nerve gases are quasisubstrates and are based on diisopropylfluorophosphate, which is a general quasisubstrate and inhibitor for a wide variety of esterases and proteinases that possess an especially nucleophilic serine at the active site. These quasisubstrates are usually reversible and are removed from the active site when they are exposed to a strong nucleophile, such as hydroxylamine. Covalent chromatography procedures for the purification of acetylcholinesterase (53,54) and the five penicillin-binding proteins of *Bacillus subtilis* (55) have been developed on the basis of immobilized quasisubstrates.

Ashani and Wilson (53) coupled the acetylcholinesterase inhibitor 2-aminoethyl-p-nitrophenylmethylphosphate to a carboxylate-containing agarose derivative through a water-soluble carbodiimide linkage as shown in Figure 6.8. When acetylcholinesterase is applied to this covalent chromatography matrix, the active site serine reacts to form a stable enzyme–matrix complex.

Since the only requirement is for the enzyme to possess an "activated" serine

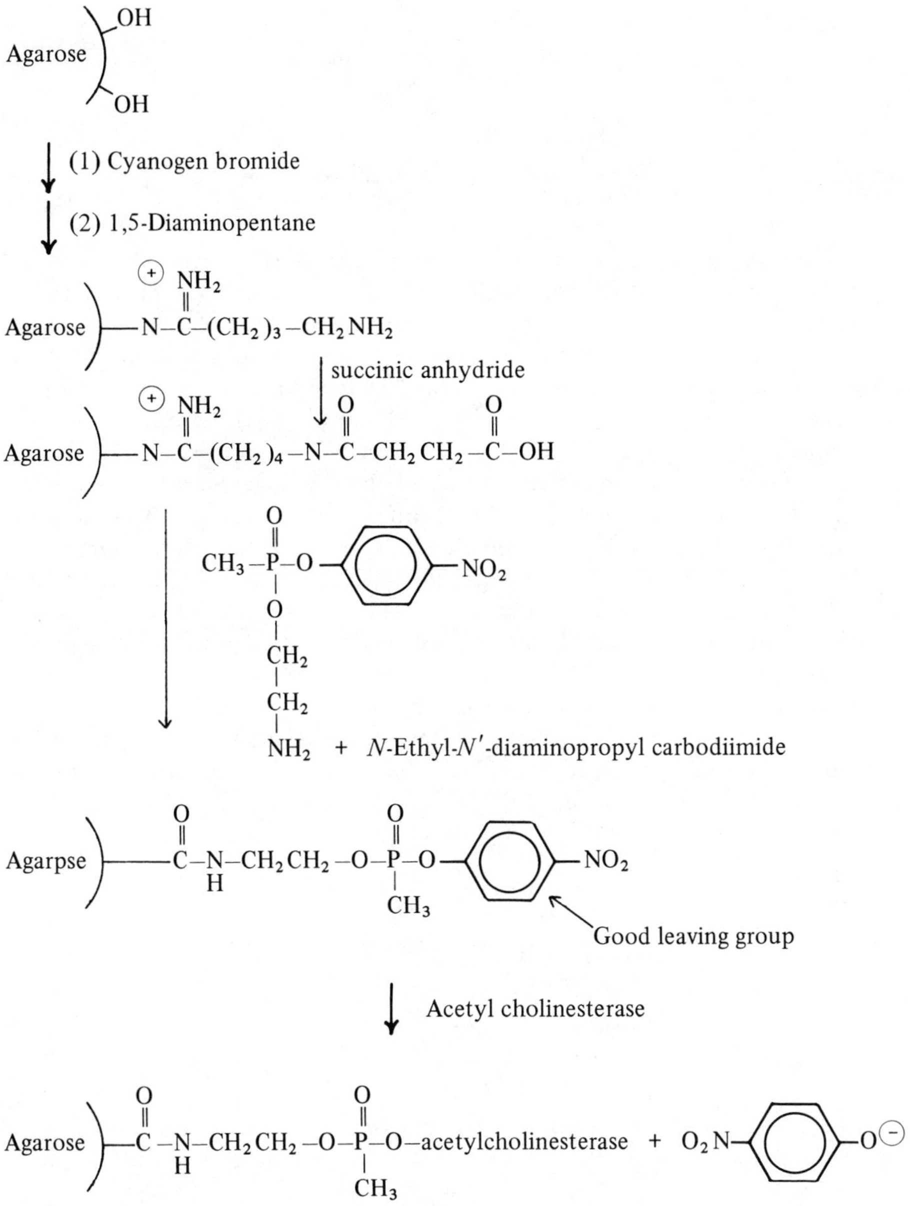

Agarose OH OH

(1) Cyanogen bromide

(2) 1,5-Diaminopentane

Agarose —N–C–(CH$_2$)$_3$–CH$_2$NH$_2$ ($^{+}$NH$_2$)

succinic anhydride

Agarose —N–C–(CH$_2$)$_4$–N–C–CH$_2$CH$_2$–C–OH ($^{+}$NH$_2$)

$$CH_3-\overset{O}{\overset{\|}{P}}-O-\bigcirc-NO_2$$
$$\overset{|}{O}$$
$$\overset{|}{CH_2}$$
$$\overset{|}{CH_2}$$
$$\overset{|}{NH_2} + \textit{N}\text{-Ethyl-}\textit{N}'\text{-diaminopropyl carbodiimide}$$

Agarpse —C–N–CH$_2$CH$_2$–O–P–O–⬡–NO$_2$
H   CH$_3$

Good leaving group

Acetyl cholinesterase

Agarose —C–N–CH$_2$CH$_2$–O–P–O–acetylcholinesterase + O$_2$N–⬡–O$^{(-)}$
H   CH$_3$

Figure 6.8. Synthesis and use of the acetylcholinesterase covalent chromatography medium, α-aminoethyl-p-nitrophenylmethyl phosphate-agarose.

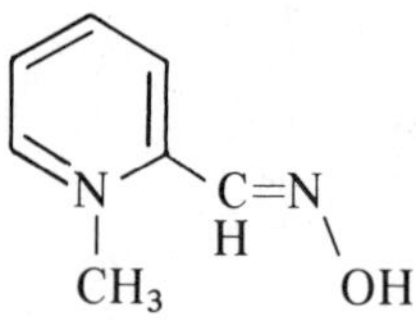

Figure 6.9. N-Methyl pyridinium-aldoxime.

hydroxyl, many serine esterases and proteinases (such as chymotrypsin and trypsin) could potentially be purified on this medium. When the column is treated with a nucleophile, such as hydroxylamine or the acetylcholine active site directed agent, *N*-methyl pyridinium-2-aldoxime, the covalent bond is broken and the esterase can be eluted. (See Figure 6.9.)

The actual mode of operation of the column may well be somewhat more complicated than pictured here. It appears that the enzyme that is first released from the column has the inhibitor covalently bonded to it and that the enzyme-inhibitor complex subsequently reacts with the nucleophile to activate the enzyme. Quite probably, the nucleophile reacts first with the isourea function at the surface of the agarose to release *both* inhibitor and enzyme. (See Figure 6.10.)

Purification of the penicillin-binding proteins from *Bacillus subtilis* membranes were accomplished in a similar manner. A crude protein mixture was obtained by detergent treatment from *Bacillus subtilis* membranes and was applied to a column of immobilized penicillin. The mixture was allowed to remain on the column for 30 minutes, during which the enzyme became acylated by the β-lactam ring of the penicillin.

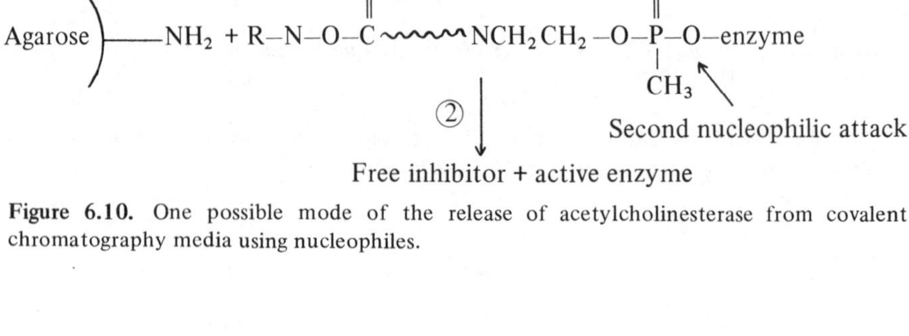

Figure 6.10. One possible mode of the release of acetylcholinesterase from covalent chromatography media using nucleophiles.

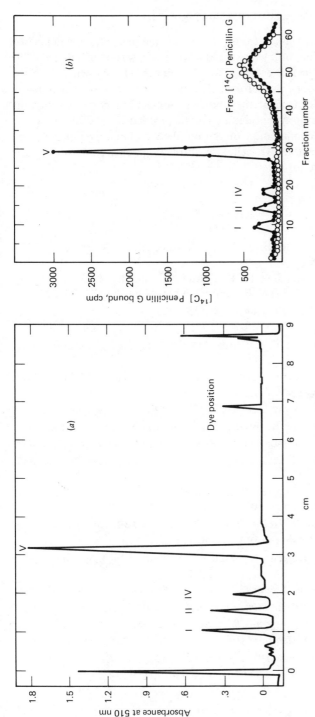

Figure 6.11. Binding of (^{14}C)penicillin G to the penicillin-binding components isolated by affinity chromatography. (*a*) Densitometry of an SDS gel of isolated binding components stained with Coomassie brilliant blue; (*b*) SDS gel of isolated binding components to which (^{14}C)penicillin G had been bound. (^{14}C)penicillin G at a concentration of 170 μg/ml was incubated with the isolated binding components for 10 minutes at 25°C. A 250-fold excess of nonradioactive penicillin G was added, after which the protein was precipitated two times with 80% acetone to remove most of the unbound radioactive penicillin. The samples were electrophoresed on SDS gels in the absence of 2-mercaptoethanol to avoid release of the bound penicillin from the protein. Ammonium persulfate was removed from the gels before application of the samples by prior electrophoresis for 2 hours; ●———●, native penicillinbinding components; 0———0, boiled, control. Reprinted with permission from Blumberg and Strominger (55).

183

Unbound proteins were washed from the column, and one column volume of a neutral buffered solution of hydroxylamine (0.8 M) was then applied. The column was incubated in this solution for 30 minutes, during which time the penicillin–enzyme acyl bond was broken. The freed enzymes were subsequently eluted and characterized. One of the five penicillin-binding compounds of the membrane, D-alanine carboxypeptidase, could be purified about 200 fold with a 45% yield by using a combination of the covalent chromatography step and followed by a subsequent DEAE-cellulose chromatographic separation of the five proteins. (See Figure 6.11.)

References

1. K. Brocklehurst, J. Carlsson, M. P. J. Kierstan, and E. M. Crook (1973), *Biochem. J.* **133**, 573.

2. K. Brocklehurst, J. Carlsson, M. P. J. Kierstan, and E. M. Crook (1974), in *Methods in Enzymology* (W. B. Jakoby and M. Wilchek, eds.) **34**, 531.

3. P. Cuatrecasas (1970), *J. Biol. Chem.* **245**, 3059.

4. R. Axen, H. Drevin, and J. Carlsson (1975), *Acta Chem. Scand. B* **29**, 471.

5. M. Wilchek, T. Oka, and Y. J. Topper (1975), *Proc. Nat. Acad. Sci. (USA)* **72**, 1055.

6. T. A. Egorov, A. Svenson, L. Ryden, and J. Carlsson (1975), *Proc. Nat. Acad. Sci. (USA)* **72**, 3029.

7. A. Svenson, J. Carlsson, and D. Eaker (1977), *FEBS Lett.* **73**, 171.

8. J. P. G. Malthouse and K. Brocklehurst (1976), *Biochem. J.* **159**, 221.

9. M. P. J. Kierstan, J. Carlsson, and K. Brocklehurst (1974), in Ref. 2, p. 532.

10. K. Brocklehurst and M. P. J. Kierstan (1973), *Nature, New Biology* **242**, 167.

11. M. Shipton, T. Stuchburg, K. Brocklehurst, J. A. L. Herbert, and H. Suschitzky (1977), *Biochem. J.* **161**, 627.

12. L. A. E. Sluyterman (1967), *Biochim. Biophys. Acta* **139**, 430.

13. A. N. Glazer and E. L. Smith (1965), *J. Biol. Chem.* **240**, 201.

14. P. T. Englund, T. P. King, L. C. Craig, and A. Walti (1968), *Biochemistry* **7**, 163.

15. B. Friedenson and I. R. Liener (1972), *Arch. Biochem. Biophys.* **149**, 169.

16. L. O. Andersson (1966), *Biochim. Biophys. Acta* **117**, 115.

17. W. L. Hughes Jr. (1947), *J. Am. Chem. Soc.* **69**, 1836.

18. W. L. Hughes and H. M. Dintzis (1964), *J. Biol. Chem.* **239**, 845.

19. J. Carlsson and A. Svenson (1974), *FEBS Lett.* **42**, 183.

20. R. Norris and K. Brocklehurst (1976), *Biochem. J.* **159**, 245.

21. A. Witter and H. Tuppy (1960), *Biochim. Biophys. Acta* **45**, 429.

22. T. P. King and E. M. Spencer (1972), *Arch. Biochem. Biophys.* **153**, 627.

23. C. B. Laurell, E. Thulin, and R. P. Bywater (1977), *Anal. Biochem.* **81**, 336.

24. C. R. Lowe (1977), *Eur. J. Biochem.* **76**, 391.

25. C. R. Lowe (1977), *Eur. J. Biochem.* **76**, 401.

26. C. R. Lowe (1977), *Eur. J. Ciobhem.* **76**, 411.

27. J. Porath and R. Axen (1976), in *Methods in Enzymology* (K. Mosbach, ed.) **44**, 19.

28. L. Ryden and H. F. Deutsch (1978), *J. Biol. Chem.* **253**, 519.

29. J. H. R. Kägi, S. R. Himmelhoch, P. O. Whanger, J. L. Bethune, and B. L. Vallee (1974), *J. Biol. Chem.* **249**, 3537.

30. R. H. O. Bühler and J. H. R. Kägi (1974), *FEBS Lett.* **39**, 229.

31. U. Weser, H. Rupp, F. Donay, F. Linneman, W. Voelter, W. Voetsch, and G. Jung (1973), *Eur. J. Biochem.* **39**, 127.

32. J. Carlsson, I. Olsson, R. Axen, and H. Drevin (1976), *Acta. Chem. Scand. B*, **30**, 180.

33. J. P. Leung, Y. Eshdat, and V. T. Marchesi (1977), *J. Immunol.* **119**, 664.

34. L. A. E. Sluyterman and J. Wijdenes (1970), *Biochim. Biophys. Acta.* **200**, 593.

35. 35. L. A. E. Sluyterman and J. Wijdenes (1974), in *Methods in Enzymology* (W. B. Jakoby and M. Wilchek, eds.) **34**, 544.

36. A. J. Barrett (1973), *Biochem. J.* **131**, 809.

37. J. McDonagh, W. G. Waggoner, E. G. Hamilton, B. Hindenach, and R. P. McDonagh (1976), *Biochim. Biophys. Acta* **446**, 345.

38. A. A. Kortt and T. Y. Liu (1973), *Biochemistry* **12**, 320.

39. E. Der Terrossian, L.-A. Pradel, R. Kassab, and G. Desvages (1974), *Eur. J. Biochem.* **45**, 243.

40. A. Ruiz-Carrillo and V. G. Allfrey (1973), *Arch. Biochem. Biophys.* **154**, 185.

41. A. Ruiz-Carrillo (1974), in *Methods in Enzymology* (W. B. Jakoby and M. Wilchek, eds.) **34**, 547.

42. J. C. Nicolas and J. I. Harris (1973), *FEBS Lett.* **29**, 173.

43. J. C. Nicolas (1974), in *Methods in Enzymology* (W. B. Jakoby and M. Wilchek, eds.) **34**, 552.

44. D. W. Gruenwedel, M. G. Heskett, and J. E. Lammert (1975), *Biochim. Biophys. Acta.* **402**, 7.

45. R. B. Simpson (1964), *J. Am. Chem. Soc.* **86**, 2059.

46. D. W. Gruenwedel and N. Davidson (1966), *J. Molec. Biol.* **21**, 129.

47. D. W. Gruenwedel (1972), *Eur. J. Biochem.* **25**, 544.

48. J. R. Shainoff (1968), *J. Immunol.* **100**, 187.

49. G. J. De Jong and C. J. N. Reker (1974), *J. Chromatogr.* **102**, 443.

50. G. J. De Jong and C. J. N. Reker (1974), First Mercury Congress, Barcelona, Spain.

51. W. H. Scouten and G. Firestone (1976), *Biochim. Biophys. Acta* **453**, 277.

52. W. H. Scouten and J. I. Michelson, unpublished results.

53. Y. Ashani and I. B. Wilson (1972), *Biochim. Biophys. Acta* **276**, 317.

54. H. F. Voss, Y. Ashani, and I. B. Wilson (1974), in *Methods in Enzymology* (W. B. Jakoby and M. Wilchek, eds.) **34**, 581.

55. P. M. Blumberg and J. L. Strominger (1972), *Proc. Nat. Acad. Sci. (USA)* **69**, 3751.

56. P. M. Blumberg and J. L. Strominger (1974), in *Methods in Enzymology* (W. B. Jakoby and M. Wilchek, eds.) **34**, 401.

BIOSELECTIVE ELUTION

Bioselective elution, affinity elution, and *substrate elution* are all terms used to describe the elution of an enzyme from a chromatographic medium by addition of low concentrations of substrate or inhibitor to the elution buffer. This causes little or no change in pH or ionic strength, and thus it is the substrate, often in as low a concentration as 5×10^{-4} *M,* that effects the elution. As has already been indicated, the bioselective elution of enzymes from affinity chromatographic packings is perhaps the best, although not conclusive, evidence that purification is actually occurring because of bioselective interactions, and not through nonspecific binding and elution, such as ion exchange.

Bioselective elution need not be limited only to elution from bioselective adsorbents. Often the addition of a substrate or an inhibitor to a protein adsorbed on an ion-exchange column will effect its elution. Since the introduction of this technique by Pogell (1,2) and Mendicino and Vasarhely (3) for the purification of fructose-1,6-diphosphate, it has been the chief chromatographic tool for the purification of this enzyme obtained from a wide variety of sources (4–22). The technique has been applied, however, to only a fairly small number of other enzymes, which is quite surprising considering the fact that it is much more convenient than affinity chromatography per se. The procedure requires the use of only commercial ion-exchange resins, substrates, and normal laboratory buffers. The purification is usually nearly as good as is found by using affinity chromatography with the possible exception that yields are in general somewhat lower. They are, however, as good as or better than the yields normally obtained by using classical ion-exchange methods.

The basic technique requires the initial determination of enzyme binding to a given ion-exchange column. Next, elution of the column with a salt gradient composed of the application buffer plus added NaCl is performed to determine the ionic conditions needed for elution. After determining the minimum NaCl concentration needed for elution, the bulk of the enzyme is applied to the ion-exchange column and washed with buffer containing 30 to 50% of the minimum NaCl concentration required to elute the enzyme. When no more protein is eluted (as detected by absorbance at 280 nm), the enzyme to be purified is eluted with the same buffer–NaCl concentration, but with the addition of 0.5 to 50 m*M* of substrate or inhibitor. At the properly determined concentration of substrate, the desired enzyme will elute in a sharp peak with a purification factor of at least 10 fold and possibly 200 fold or more. The only major difficulty encountered with this technique is when the enzyme is somewhat more labile in the presence of the substrate or inhibitor than in its absence. In such instances removal of the substrate after elution of the enzyme (either by immediate gel filtration or by desalting with the aid of a membrane protein concentrator) will probably eliminate the problem. Often, however, the presence of substrate will stabilize the enzyme, and thus substrate elution will work to the investigator's advantage.

For example, for the purification of the NAD-specific isocitrate dehydrogenase from yeast, the allosteric effector, citrate, must be present throughout the purification (23). The specific activity of the enzyme obtained in the absence of citrate is less than 10% of that isolated in the presence of citrate. Further, citrate preserves the enzymes kinetic cooperativity with respect to isocitrate. One particularly useful step in the isolation procedure is the bioselective elution of the enzyme from a phosphocellulose column. First, the partially purified enzyme is applied to the phosphocellulose in a pH-7.6 buffer containing 0.5-mM sodium citrate. The column is subsequently washed with the same buffer containing 20-mM sodium citrate until no further protein is eluted. The enzyme is eluted in a sharp peak by applying the same buffer containing 50-mM sodium citrate. A nearly 10-fold purification is effected, which is the largest single purification factor for any step in the preparation. The yield in the bioselective adsorption step is 62%. The overall yield for the enzyme isolation is 26%. The enzyme isolated in this fashion is homogeneous, as judged by electrophoretic and centrifugal analysis. (See Table 7.1.)

Similarly, Illingsworth and Tipton (24) have purified the NADP-dependent isocitrate dehydrogenase from pig liver, using substrate elution from a carboxymethylcellulose column as the final step in obtaining a homogeneous enzyme preparation. Purification factors of 50 could be obtained by eluting the enzyme from the column with 0.2mM sodium isocitrate at pH 6.0. Unfortunately, carboxymethyl–cellulose has a low capacity for the enzyme at this pH, and thus either extremely large columns must be employed, or the enzyme must first be partially purified by more classical techniques. Ammonium sulfate precipitation and DEAE–cellulose chromatography steps were used in this case as preliminary purification procedures.

Penhoet and Rutter (25,26) have purified rabbit muscle aldolase by using substrate elution of the enzyme from phosphocellulose with fructose-1,6-diphosphate. In this case the enzyme is somewhat labile in the presence of the substrate, and it was thus necessary to perform gel filtration immediately after elution. The entire procedure required only two steps, starting with the crude muscle extract, namely, ammonium sulfate precipitation and substrate elution from phosphocellulose. The bioselective elution step effected a 270-fold purification and gave a 63% yield. (See Figure 7.1.)

TABLE 7.1. Purification of Yeast DPN-Specific Isocitrate Dehydrogenase

Fraction	Volume, ml	Protein, mg	Total Activity, Units	Specific Activity, Units/mg	Yield, %
Crude extract	2420	150,040	5420	0.036	(100)
$(NH_4)_2SO_4$ I	1150	53,245	4045	0.076	75
PS supernatant	2260	14,532	4552	0.31	84
$(NH_4)_2SO_4$ II	145	7,308	4329	0.59	80
DEAE-cellulose	1420	1,037	3586	3.46	66
Phosphocellulose	730	77	2224	30.2	41
Concentrated phosphocellulose	16.2		2151		40
Bio-Gel A—1.5mM	381	40.6	1446	35.6	27
Concentrated Bio-Gel A—1.5mM	4.3	39	1424	35.6 (26.5)[a]	26

Source: Reprinted with permission from: L. D. Barnes, G. D. Kuehn, and D. E. Atkinson (1971), *Biochemistry* **10**, 3939. Copyright© American Chemical Society.

[a]Specific activities are based on a bovine serum albumin protein standard, except that the specific activity of 26.5 is based on the isocitrate dehydrogenase protein standard.

189

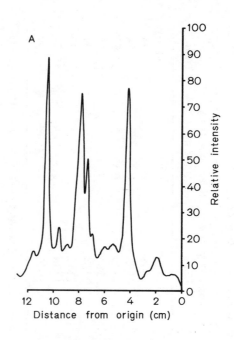

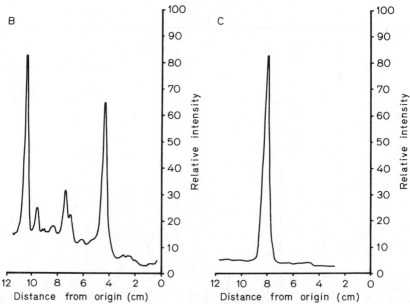

Figure 7.1. Densitograms of SDS gel electrophoresis of tyrosyl-tRNA synthetase. (*A*) CM-50 Sephadex Fraction; (*B*) still bound on CM-40 Sephadex, hence not eluted by tRNA; (*C*) affinity elution fraction. Reprinted with permission from H. G. Faulhammer and F. Cramer (1977), *Eur. J. Biochem.* 75, 561.

Bioselective elution has also been used successfully with another, quite different, class of enzyme, the aminoacyl-tRNA (transfer RNA) synthetases (27, 28). These enzymes are normally difficult to isolate since many aminoacyl tRNA synthetase assays must be performed to identify the elution position of the enzyme from ion-exchange columns. Each assay is both costly and very time consuming. To eliminate the need for so many assays, affinity chromatographic procedures have been widely used, but these, too, are difficult because of the time needed for the chemical synthesis of the bioselective adsorbent. Conversely, bioselective elution requires far fewer assays than classical ion-exchange techniques and does not involve the time-consuming synthesis of an affinity chromatography support.

To purify serine yeast tRNA synthetase, von der Haar (28) has shown that it is necessary to first partially purify the enzymes by ammonium sulfate precipitation and by chromatography on DEAE–cellulose. These procedures effect a 4- to 10-fold purification and partially separate phenyl, alanyl, valyl, seryl, and isoleucyl tRNA synthetases. These partially separated synthetases are then applied to a phosphocellulose column in a $0.03\text{-}M$ potassium phosphate buffer, pH 7.2, that contains $1\text{-}mM$ EDTA, $1\text{-}mM$ dithiothreitol, and 10% glycerol. The column is washed extensively with this buffer, and the enzyme is biospecifically eluted from the column by applying the same buffer containing the appropriate tRNA. The enzyme is subsequently separated from the tRNA by passing the eluant from the first column through a column of DEAE–cellulose previously equilibrated with the same buffer. Both the tRNA and the aminoacylsynthetase are retained on the DEAE–cellulose column. The enzyme is then eluted with $0.13\text{-}M$ potassium phosphate buffer, pH 7.2. After the enzyme has been completely removed, the tRNA is recovered by eluting the column with $1\text{-}M$ NaCl. The tRNA solution is desalted and reused as necessary. Recovery of the tRNA after each use exceeded 90%.

This procedure produces a homogeneous t-RNA synthetase with good yields (see Figure 7.2). Overall purification of phenylalanyl tRNA synthetase was 547 fold and 236 fold for valyl-tRNA synthetase. Less satisfactory results were reported for seryl and isoleucyl tRNA synthetases, and this is quite possibly because they are associated with other proteins that also have an affinity for tRNA. This possibility once again illustrates that "affinity" techniques are *bioselective* and *not biospecific*.

Faulhammer and Cramer (29) have used a similar procedure to purify yeast tyrosyl-tRNA synthetase but employed bioselective elution from carboxymethyl–Sephadex instead of carboxymethylcellulose. They also added the proteinase inhibitor phenylmethylsulfonyl fluoride to all solutions used in the preparation. Two additional steps were added; a pH-4.8 isoelectric precipitation step that removed considerable amounts of undesirable protein and a classical ion-exchange step using carboxymethyl–Sephadex. The ion-exchange

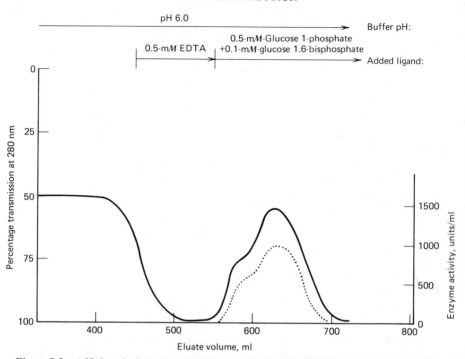

Figure 7.2. Affinity elution of phosphoglucomutase from CM–cellulose column at pH 6.0. The fraction applied was that protein not adsorbed at pH 6.5. . . . . , Phosphoglucomutose; ———, percentage transmission at 280 nm. Reproduced with permission from R. K. Scopes (1977), *Biochem. J.* **161**, 265.

step immediately preceded the use of affinity elution from the same resin. With these modifications they achieved a remarkable 15,000-fold purification. The affinity elution step was essential to obtain a homogeneous enzyme (see Figure 7.2 and Table 7.2).

The use of carboxymethyl–Sephadex may be one of the reasons for their excellent results since the ionic groups in this resin are much more homogeneous than those of the cellulose-based matrix. A second reason for their vastly higher specific activity in the use of the protease inhibitor, phenylmethylsulfonyl fluoride. Without the inhibitor, they found losses of over 50% in total activity and the presence of many low-molecular-weight protein fragments in their final product.

It is probable that affinity elution could be used to bioselectively separate a whole series of enzymes from a single cell extract by merely eluting the enzyme mixture from an ion-exchange column by using a series of substrate-containing eluants. Scopes (30–34) has demonstrated the feasibility of such multiple

TABLE 7.2. Summary of the Isolation of Tyrosyl-tRNA Synthetase from 24-kg Baker's Yeast

Purification Step	Volume ml	Protein Total, mg	Protein Concentration, mg/ml	Total Activity, kU	Specific Activity, U/mg	Yield, %	Protein/RNA, A_{280}/A_{260}
Crude extract	19,300	2,990,000	155	3,500	1	100	0.7
pH-4.8 Supernatant	13,750	653,000	47.5	2,300	3.5	66	0.8
Dialysis after 70% ammonium sulfate precipitation	2,500	432,000	175	1,270	3	36	1.3
DEAE–cellulose	4,000	325,000	81	1,650	5	47	1.5
CM–Sephadex	100	1,460	14.6	1,475	1,000	42	1.8
Affinity elution	18	82.8	4.6	1,243	15,000	35	2.02

Source: Reprinted with permission from Faulhammer and Cramer (29).

193

enzyme purifications by isolating 12 different enzymes from a single rabbit muscle extract and eight from a single chicken muscle extract. In most preparations three single steps were involved in the isolation of each enzyme: (1) ammonium sulfate precipitation, (2) affinity elution from an ion-exchange resin, and (3) gel filtration. The resulting enzymes appear to be homogeneous with specific activities that are at least as high as those reported elsewhere in the literature. In step (1), muscle extracts are prepared by homogenizing fresh tissue in 3 volumes of 0.03-M potassium phosphate buffer, pH 7.0, containing 1-mM EDTA by using a Waring blender at top speed. Cellular debris are removed by centrifugation at 4800 g for 45 minutes. The supernatant is brought to pH 6.0 and fractionated with ammonium sulfate (30). Fraction A precipitates at 0 to 45% saturation, fractions B and C at 45 to 65% saturation, fraction D at 65 to 75% saturation, and fraction E at 75 to 80% saturation. The last fraction, E, was formed by adjusting the pH to 8.0 with ammonium hydroxide. All fractions (except the last) were allowed to stand 30 minutes prior to collection of the precipitate and were performed at 15 to 20°C. For fraction E, the solution was kept over night at 4°C before collecting the precipitate.

The pellets were stored, when necessary, at 4°C as a slurry in the respective final ammonium sulfate concentration. Before use they were dissolved in a small amount of 50-mM Tris-ethanolamine buffer, pH 7.5, and desalted by gel filtration on a column of Sephadex G-25 that was preequilibrated with 0.2-M EDTA and adjusted to pH 8.0 with Tris.

Fraction A contains phosphorylase as a major protein, and fraction E is composed of chiefly glyceraldehyde phosphate dehydrogenase. These enzymes can be readily crystallized from these two fractions. Fractions B and C contain most of the cellular pyruvate kinase, lactate dehydrogenase, aldolase, phosphoglusomutase, phosphoglucose isomerase, and phosphoglycerate mutase, each of which can be separated from the others by bioselective elution on carboxymethylcellulose. Fraction D contains over half of the cellular enolase, triose phosphate isomerase, creatine kinase, and myokinase, which also can be respectively separated by using bioselective adsorption. In each case, the bioselective adsorption step is followed by a gel-permeation chromatography process to yield a purified enzyme.

For example, fractions B and C were adjusted to pH 6.5 with Tris-MES buffer and applied to a carboxymethylcellulose column that was preequilibrated with 0.01-M Tris-MES buffer, pH 6.5. The sample was washed with the same buffer and followed by buffer plus 20-mM KCl. This process eluted a protein peak containing phosphoglucoisomerase, which was then further purified by a separate bioselective elution procedure. Next, the column was eluted with the buffer plus 0.5-mM phosphoenolpyruvate. This eluted pyruvate kinase contaminated with enolase, phosphoglycerate mutase, and phosphoglycerate kinase. These contaminants were later completely removed by gel filtration on Sephadex G-150. Next, the buffer was changed to MOPS buffer, pH 7.2, containing KCl. After a small protein band was eluted, 0.2-mM NADH was added to the eluant to specifically desorb lactate dehydrogenase from the column. When this was completed, the salt concentration in the buffer was increased to 20-mM KCl, and, after a small band of contaminating protein had been eluted, 0.2-mM fructose-1,6-diphosphate was added to bioselectively elute aldolase. The proteins obtained in this fashion were further purified by gel filtration to yield enzymes of high purity. Meanwhile, the protein fraction that had not been adsorbed to the carboxymethylcellulose column at pH 6.5 was treated with 1-M MES to lower the pH 6.0 and reapplied to a carboxycellulose column that was equilibrated at the same pH. The column was then washed with the same buffer containing 0.5-mM EDTA. After washing with this buffer, phosphoglucomutase was bioselectively eluted, using the same buffer containing 0.5-mM glucose-1-phosophate and 0.1-mM glucose-1,6-diphosphate instead of EDTA. As can be seen from Table 7.3 and Figure 7.3, the resulting purification was excellent. A similar scheme was followed to purify the remaining glycolytic proteins from fractions D and E.

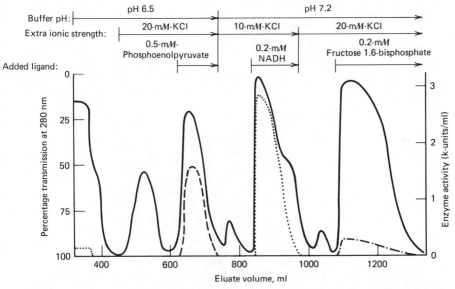

Figure 7.3. Elution diagram for rabbit muscle fraction BC adsorbed to a CM–cellulose column at pH 6.5. The enzymes obtained by affinity elution were pyruvate kinase (– – – –), lactate dehydrogenase (. . . .), and aldolase (– . – .) ———, percentage transmission at 280 nm in 3-mm-path-length cell. Reproduced with permission from R. Scopes (1977), *Biochem. J.*, **161**, 253.

TABLE 7.3. Summary of Purifications[a]

	Pyruvate Kinase	Lactate Dehydrogenase	Aldolase	Phospho-glucomutase
Activity in 1 l of original extract (kilounits)	235	480	42.5	135
Activity in fraction BC (from 1 l) (kilounits)	162	356	32.5	69
Activity not adsorbed at pH 6.5 (kilounits)	0	7	0	69
Activity affinity-eluted (kilounits)	122	228	27.5	56
Specific activity affinity-eluted (units/mg)	260	490	16.5	650
Specific activity after gel filtration (units/mg)	340	600	17.0	750
Highest literature specific activity (units/mg)	340	610	18.0	600
Final yield of enzyme from 1 l (mg)	320	380	1350	68

Source: Reprinted with permission from Scopes (31).

[a]The "highest literature specific activities" are extrapolated to assay conditions used in the present experiments.

7.1. THEORY OF BIOSELECTIVE ELUTION

One possible reason for the relatively small number of investigations performed by use of bioselective elution could be that no coherent theory of bioselective elution has been developed. Indeed, bioselective elution most probably possesses a complex and multifaceted basis of action, and no single theory will explain *all* the as yet discovered elution patterns. This has also been shown to be true for the other bioselective techniques currently being used, such as affinity chromatography, immunoadsorption, and lectin chromatography. Nevertheless, single predictive models that work much of the time have been developed for these techniques.

In a similar fashion, Scopes (3) has demonstrated that *most* of the known examples of bioselective elution can be explained if we consider that a negatively charged ligand can bind to an enzyme and make it more negative. The resulting enzyme–ligand complex is less able to be bound to a negatively charged ion exchange and is thus eluted. Conversely, a protein with a slight negative charge can be adsorbed to a positively charged ion-exchange resin. If the protein subsequently binds a positively charged ligand, the enzyme–ligand complex will become positively charged or, at least, have its negative charge sufficiently decreased to desorb it from the ion-exchange resin. (See Figure 7.4.)

There is strong evidence that this is the mode of action of bioselective elution of most of the glycolytic proteins. As shown by Scopes (3), the binding of the appropriate eluting ligand changes the charge of the protein that is calculated as negative charge introduced per 100,000 daltons (molecular weight units) from 3 to over 14 charged functions per 100,000 daltons (see Table 7.4).

The ability of a given ligand to elute an enzyme would thus appear to be dependant solely on two factors: (1) the dissociation constant for the enzyme–substrate complex and (2) the number of charges introduced by the ligands that bind to the enzyme. In most cases this seems to be true, as shown by the fact that bioselective elution of many enzymes can be most readily applied at pH values or ionic strengths where the enzyme is weakly adsorbed to the ion-exchange resin. (See Figure 7.5.)

It would seem that if this were the only factor involved, the type of ion-exchange resin used would be rather inconsequential. Scopes suggests that phosphocellulose is usually less satisfactory for bioselective elution than carboxymethyl cellulose because it is a more variable and heterogeneous material. Nonetheless, phosphocellulose is the best adsorbent for phsophoglucose isomerase and phosphoglycerate mutase. Possibly this is because phosphocellulose has a strong resemblance to the enzyme substrates. In such cases, apparent bioselective elution could be affinity chromatography per se. This is especially true if the enzyme active site recognizes some portion of the matrix. This

TABLE 7.4. Some Buffer Systems and Ligands for Affinity Elution of Rabbit Muscle Glycolytic Enzymes[a]

Enzyme	$(NH_4)_2SO_4$ Fraction	Adsorbent	Buffer, pH for Absorption	Buffer, pH for Preelution Wash and Elution	Ligands in Buffer to Cause Elution, Concentration, (mM)	Charge Introduced per 100,000 Daltons	Approximate Specific Activity (at 25°C) of Eluted Enzyme
Phosphorylase	A	CM-cellulose	MES, 6.0	MES, 6.5	AMP, 0.2+Glc-6-P, 0.2	4	18
Phosphoglucomutase	BC*	Phosphocellulose	MES, 6.0	MES, 6.5	Glc-1-P, 0.2+Glc-1, 6-P₂, 0.05	3 or 6	750
Phosphoglucose isomerase	BC	Phosphocellulose	MOPS, 7.2	MOPS, 7.2	Glc-6-P, 0.5 or Fru-6-P, 0.5	3.5	750
Phosphofructokinase	—	CM-cellulose	MOPS, 6.9	MOPS, 7.2+ 75mM-KCl	Citrate, 1.0	3.5	140
Aldolase	BC,B	CM-cellulose	MOPS, 7.2	Tricine, 8.0+ 20mM-KCl	Fru-1,6-P₂, 0.2	10	18
Triose phosphate isomerase	CD*, D*, E*	CM-cellulose	Picolinate, 5.5	MES, 5.8	Grn-P, 0.5	7.5	4000
Glycerol phosphate dehydrogenase	BC*	CM-cellulose	MES, 6.0	MES, 6.5	NADH, 0.2	5	220

Glyceraldehyde phosphate dehydrogenase	E	CM-cellulose	MOPS, 7.2	Tricine, 8.0	NAD^+, 0.2+Gra-3-P, 0.2+P_1, 1.0	14	150
Phosphoglycerate kinase	CD,D	CM-cellulose	MES, 6.5	Tricine, 8.0	Gri-3-P, 0.5	6	600
Phosphoglycerate mutase	BC	Phosphocellulose	MES, 6.5	MES, 7.2	Gri-2,3-P_2, 0.2	16	1000
Enolase	CD,C	CM-cellulose	MES, 6.5[b]	MOPS, 7.0[b]	Gri-2-P or Prv-P, 0.5	7.5	50
Pyruvate kinase	BC	CM-cellulose	MES, 6.5	MOPS, 7.2	Prv-P, 0.5	5	300
Lactate dehydrogenase	BC	CM-cellulose	MOPS, 7.2	Tricine, 8.0	NADH, 0.2	5	450
Myokinase	CD	CM-cellulose	MES, 6.0	MES, 6.5	ADP, 0.5	14+	800
Creatine kinase	CD,D	CM-cellulose	MES, 6.0	MES, 6.2	ATP, 0.5	7	100

Source: Reprinted with permission from Scopes (30).

[a]The best $(NH_4)_2SO_4$ fraction to use for a particular enzyme is indicated. In the asterisked (*) examples, a heat pretreatment for 5 minutes at pH5.5, 50°C before desalting removed proteins (mainly creatine kinase) that might be adsorbed and denatured on the column, and so interfere with the running of the column. The preelution wash buffer always included extra EDTA at the concentration indicated in the next column for ligands, to represent the "dummy substrate." CBN recommended abbreviations have been used: Glc, glucose; Fru, fructose; Grn, dihydroxyacetone; Gra, glyceraldehyde; Gri, glycerate; Prv, enolpyruvate.

[b]Magnesium acetate (1.0 m*M*) included in buffers to stabilize enzyme.

199

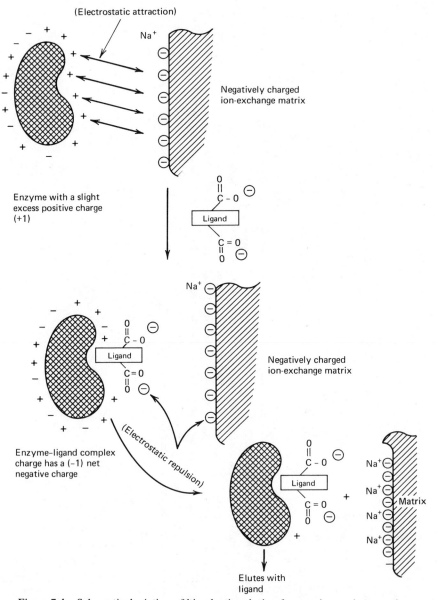

Figure 7.4. Schematic depiction of bioselective elution from an ion exchange resin.

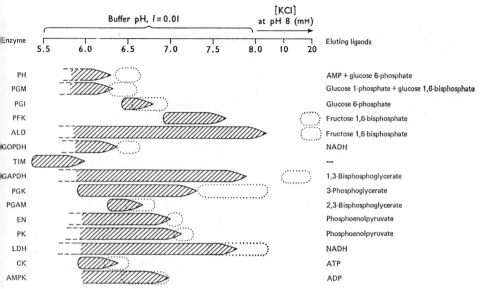

Figure 7.5. Adsorption and affinity elution of rabbit muscle glycolytic enzymes on CM–cellulose at I-0.01, 4°C. The hatched pH ranges indicate suitable conditions for adsorption of the enzyme on a column of CM–cellulose. In examples where the lower end of the range is closed, the enzyme is unstable at pH values below this, and so although adsorbed, may be denatured. In the other examples with open ends of ranges, the enzyme may be stable at lower pH values, but relevant experiments have not been carried out. In general, few enzymes can be recovered if adsorption is at or below pH 5.5. The pH areas enclosed by dotted lines indicate the approximate range at which the enxyme can be eluted by using between 0.2 and 0.5 mM of the ligand indicated on the right-hand side of the figure. Abbreviations for enzymes: PH, phosphorylase; PGM, phosphoglucomutase; PGI, phosphoglucose isomerase; PFK, phosphofructokinase; ALD, aldolase; GOPDH, glycerol phosphate dehydrogenase; TIM, triose phosphate isomerase. GAPDH, glyceraldehyde phosphate dehydrogenase; PGK, phosphoglycerate kinase. PGAM, phosphoglycerate mutase; EN, enolase; PK, pyruvate kinase; LDH, lactate dehydrogenase; CK, creatine kinase; AMPK, AMP kinase. Reproduced with permission from R. Scopes, (1977), *Biochem. J.* **161**, 265.

could be very likely for enzymes which have carboxylic acids as substrates (see Figure 7.6).

Likewise, conformational changes caused by binding a substrate or inhibitor could cause substrate elution. Enolase, for example, is monomeric in the absence of Mg^{++} but dimeric in its presence. If monomeric enolase is adsorbed to carboxymethylcellulose in the absence of Mg^{++}, the enzyme can be eluted, in poor yield, by elution with Mg^{++}. This occurs in spite of the fact that Mg^{++} should increase the positive charge and the enzyme Mg^{++} complex should have an increased ionic attraction for the negatively charged ion-exchange resin.

Nonetheless, in most instances it is basically an ionic phenomenon related

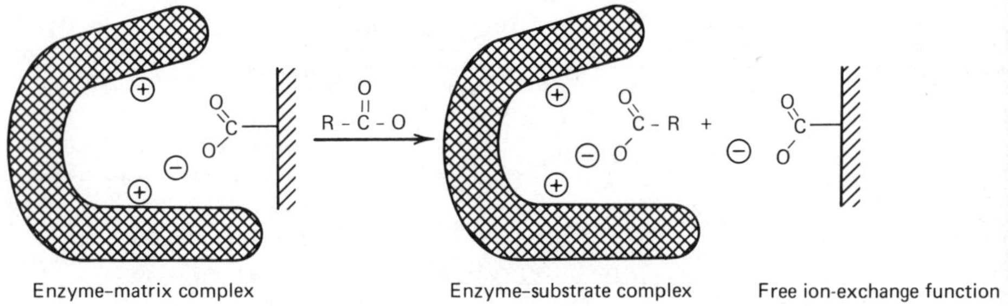

Figure 7.6. Ion-exchange resin functioning as an affinity chromatography medium.

only to the charges introduced in binding an ionic substrate. Thus very good elution should occur when multiple-charged substrates are utilized. For example, pancreatic ribonuclease (36) is eluted in a sharp peak from phosphocellulose by using RNA in the presence or absence of either Mg^{++} or EDTA. Conversely, DNA, which is also highly charged but binds ribonuclease only very weakly, bioselectively elutes the enzyme in a very broad, diffuse band. Partially digested RNA also elutes the enzyme in a broad band, probably in part because of its reduced charge, whereas RNA, totally digested by treatment with alkalai, does not bioselectively elute the enzyme.

7.2. BIOSELECTIVE ELUTION FROM HYDROPHOBIC AMPHILYTES

Many of the spacer arms used for attaching bioselective adsorbent ligands to matrices (especially agarose) contain hydrophobic regions and possess amphilyte character and often exist in zwitterionic form. Yon (37) has prepared such immobilized hydrophobic amphilytes, which he terms *imphilytes,* as ion-exchange and hydrophobic resins for the purification of various proteins (see page 256). (See Figure 7.7.)

Attempts to utilize such resins for enzyme purification by bioselective elution have shown that some enzymes plus their substrates bind more tightly to the imphilyte resin whereas others, including numerous dehydrogenases are desorbed. Of seven enzymes tested, only alkaline phosphotase was completely unaffected by the addition of substrate (37).

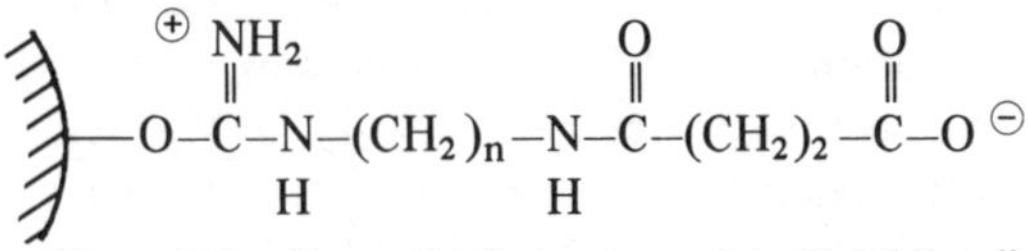

Figure 7.7. Generalized structure of an "imphilyte" resin.

Since many affinity chromatography packings are also imphilytes, elution of an enzyme from these adsorbents does not necessarily *prove* that the enzymes were actually adsorbed onto the affinity matrix. The same effect could be explained by bioselective desorption of the enzyme that was held only by the small portion of imphilyte ion-exchange functions found in the adsorbent. Obviously, care must be taken to demonstrate that adsorption or desorption of an enzyme is actually due to bioselective factors and even bioselective elution of an enzyme from an affinity chromatography system does not alone prove this conclusively.

7.3. ANALYTICAL BIOSELECTIVE ELUTION— A MICROASSAY FOR 3'5'-CYCLIC AMP

Perhaps one of the most exciting possibilities for the application of bioselective elution is in the determination of microamounts of substrates present in biological fluids. The technique is basically a reverse of bioselective elution in that the substrate is adsorbed onto an ion-exchange column, which is subsequently eluted with the appropriate enzyme. In principle, the concentration of the eluted substrate could be determined directly. In practice, to increase the sensitivity of the assay, an indirect approach utilizing radioactive substrates is employed.

This method was first developed by Sinha and Colman (38) specifically for the determination of cyclic-AMP (cAMP) levels in plasma, although the principle would seem to be applicable to a wide variety of other substances. To assay for cAMP, the cAMP was extracted by adding 2 ml of 100% ethanol to a 0.5-ml sample of plasma. If tissue samples were employed, 1 mg of tissue was suspended in 210 ml of 80% ethanol and then sonicated for 5 minutes at 0°C. Following sonication, the protein-bound cAMP was released by heating the mixture to 60°C for 5 minutes, cooling it, and centrifuging it to remove the precipitated protein. The supernatant was used for the subsequent cAMP assay. Extraction of cAMP into ethanol was a necessary prerequisite to the actual analytical procedure since protein and/or large amounts of other ionic materials could limit the assay sensitivity. One milliliter of the ethanol extract was added to 1 ml of 20-mM sodium acetate, pH 4.0, mixed with a predetermined amount of radioactive cAMP and then applied to a QAE cellulose ion-exchange column (1.0-mole volume) equilibrated with 10-mM sodium acetate, pH 4.0. The column was washed with 4 ml of the same buffer. Subsequently, 20 μg of protein kinase plus 28 μg of protein kinase inhibitor in the buffer was passed onto the column. The column with the enzyme was incubated at room temperature for 3 to 4 hours, and the cAMP complex was eluted with 1.5 ml of the same buffer. The radioactivity of the eluate was determined by using a scintillation counter.

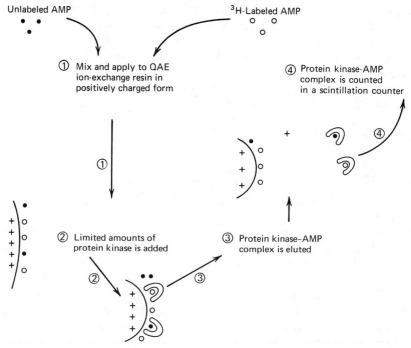

Figure 7.8. Analytical bioselective elution as the basis for a microassay for AMP.

Since any unlabeled cAMP from the tissue will compete with the added radio-active cAMP for the limited number of protein kinase cAMP binding sites, the lower the radioactivity eluted from the column, the greater the tissue cAMP concentration. The actual cAMP concentration can be determined by using a standard curve prepared from known concentrations of unlabeled cAMP. (See Figure 7.8.)

Classical methods of cAMP determination, such as radioimmunoassay, requires prior enrichment of the cAMP by lyophilization. Such methods are not always linear and are quite time consuming. Conversely, analytical bioselective elution is rapid, requires no lyophilization or other laborious preparation, and follows a linear standard curve over a wide concentration of cAMP. (See Figure 7.9.)

References

1. B. M. Pogell (1962), *Biochem. Biophys. Res. Commun.* **7**, 225.
2. B. M. Pogell (1966), in *Methods in Enzymology* (W. A. Wood, ed.) **9**, 9.

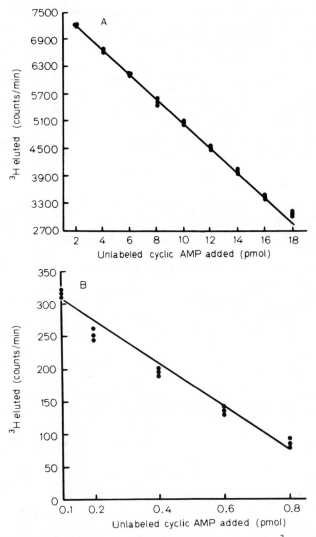

Figure 7.9. Standard curves of competitive displacement of cyclic (^{3}H)-AMP by cyclic AMP, produced by the elution of the radioactive nucleotide bound to protein kinase. (*A*) Cyclic (^{3}H)AMP [2.0 pmole: 36,000 cpm (counts per minute)] adsorbed in the QAE cellulose column was incubated with 20 μg of protein kinase for 3 hours at 23°C with different amounts of unlabeled cyclic AMP added as indicated. Elution of the radioactive cyclic AMP bound to protein kinase was done as described in the text. (*B*) Identical experiments were done with 0.1 pmole of cyclic (^{3}H)-AMP (1600 cpm) incubated with protein kinease and subpicomolar quantities of the cyclic nucleotide. Reproduced with permission from A. K. Sinha and R. W. Colman, *Eur. J. Biochem.* **73**, 367 (1977).

3. J. Mendicino and F. Vasarhely (1963), *J. Biol. Chem.* **240**, 3528.

4. O. M. Rosen (1966), *Arch. Biochem. Biophys.* **114**, 31.

5. S. Pontremoli (1966), in *Methods in Enzymology* (W. A. Wood, ed.) **9**, 625.

6. O. M. Rosen, S. M. Rosen, and B. L. Horecker (1966), in *Methods in Enzymology* (W. A. Wood, ed.) **9**, 632.

7. O. M. Rosen, S. M. Rosen, and B. L. Horecker (1965), *Arch. Biochem. Biophys.* **112**, 411.

8. S. Pontremoli, S. Traniello, B. Luppis, and W. A. Wood (1965), *J. Biol. Chem.* **240**, 3459.

9. B. M. Pogell and M. G. Sarngadharan (1971), in *Methods in Enzymology* (W. B. Jakoby, ed.) **22**, 379.

10. J. Fernando, M. Enser, S. Pontremoli, and B. L. Horecker (1968), *Arch. Biochem. Biophys.* **126**, 599.

11. H. Carminatti, E. Rozengurt, and L. Jiminez de Asua (1969), *FEBS Lett.* **4**, 307.

12. S. Pontremoli and S. Traniello (1975), in *Methods in Enzymology* (W. A. Wood, ed.) **42**, 347.

13. S. Pontremoli and E. Melloni (1975), in *Methods in Enzymology* (W. A. Wood, ed.) **42**, 354.

14. O. M. Rosen (1975), in *Methods in Enzymology* (W. A. Wood, ed.) **42**, 360.

15. A. M. Geller and W. L. Byrne (1975), in *Methods in Enzymology* (W. A. Wood, ed.) **42**, 363.

16. J. Mendicino and H. Abou-Issa (1974), in *Immobilized Biochemicals and Affinity Chromatography* (R. B. Dunlap, ed.), Plenum, New York, pp. 85–97.

17. J. Mendicino, H. S. Prihar, and F. M. Salama (1968), *J. Biol. Chem.* **243**, 2710.

18. N. Kratowich and J. Mendicino (1970), *J. Biol. Chem.* **245**, 2483.

19. W. J. Black, A. Van Tol, J. Fernando, and B. L. Horecker (1972), *Arch. Biochem. Biophys.* **151**, 576.

20. F. Marcus (1967), *Arch. Biochem. Biophys.* **122**, 393.

21. S. Traniello, S. Pontremoli, Y. Trashima, and B. L. Horecker (1971), *Arch. Biochem. Biophys.* **146**, 161.

22. C. L. Sia and B. L. Horecker (1972), *Arch. Biochem. Biophys.* **149**, 222.

23. L. D. Barnes, G. D. Kuehn, and D. E. Atkinson (1971), *Biochemistry* **10**, 3939.

24. J. A. Illingsworth and K. F. Tipton (1970), *Biochem. J.* **118**, 253.

25. E. E. Penhoet and W. J. Rutter (1975), in *Methods in Enzymology* (W. A. Wood, ed.) **42**, 241.

26. E. E. Penhoet, M. Kochman, and W. J. Rutter (1969), *Biochemistry* **8**, 4391.

27. F. Von der Haar (1973), *Eur. J. Biochem.* **34**, 84.

28. F. Von der Haar (1974), in *Methods in Enyzmology* (W. B. Jakoby and M. Wilchek, eds.) **34**, 163.

29. H. G. Faulhammer and F. Cramer (1977), *Eur. J. Biochem.* **75**, 561.

30. R. K. Scopes (1977), *Biochem. J.* **161**, 253.

31. R. K. Scopes (1977), *Biochem. J.* **161**, 265.

32. R. K. Scopes and T. Fifis (1975), *Proc. Aust. Biochem. Soc.* **8**, 17.

33. A. A. Stewart and R. K. Scopes (1975), *Proc. Aust. Biochem. Soc.* **8**, 15.

34. A. Chappel, R. K. Scopes, and R. S. Holmes (1976), *FEBS Lett.* **64**, 59.

35. R. K. Scopes (1969), *Biochem. J.* **113**, 551.

36. J. Eley (1969), *Biochemistry* **8**, 1502.

37. R. J. Yon (1976), *Biochem. J.* **161**, 233.

38. A. K. Sinha and R. W. Colman (1977), *Eur. J. Biochem.* **73**, 367.

DNA–RNA AFFINITY CHROMATOGRAPHY

There are many proteins that have a DNA- or RNA-related function and thus recognize and bind immobilized DNA and RNA. These proteins are of immense interest to both the molecular biologist and the biochemist. Among them are several DNA polymerases, the RNA polymerases, DNA regulatory proteins, RNA-dependent DNA polymerase, repressor proteins, ribosomal proteins, tRNA synthetases, and viral DNA- and RNA-binding proteins. Moreover, various types of RNA and DNA also bind selectively to immobilized RNA and/or DNA. In fact, immobilized DNA and RNA have such a wide variety of uses in purifying biomolecules involved in polynucleotide functions that they are probably the most extensively investigated and widely used of all the general types of affinity chromatographic packings currently in use.

The most widely used form of an immobilized nucleic acid is DNA immobilized on cellulose that is prepared by either codrying cellulose and DNA (1–3) or by irradiating a mixture of DNA and cellulose with ultraviolet light (4,5). The former is more easily synthesized and has no ultraviolet-light-induced changes in the DNA, whereas the latter produces DNA–cellulose, which is much more tightly bound and thus does not "leak" DNA from the cellulose matrix (see page 10). Deoxyribonucleic acid–cellulose is widely used for the purification of DNA- and RNA-binding proteins.

Oligomers of poly(dT) bound to cellulose have been widely used to purify messenger RNA (mRNA). Messenger RNA contains large regions of poly(A) that are homologous to the poly(dT) and thus hybridizes to poly(dT)-cellulose. All RNAs other than m-RNA have no poly(A) regions and thus do not bind to oligo (dT) cellulose (6,7). Poly (U) has also been immobilized by using agarose (8) and has been employed in the same fashion for purification of mRNA. [The poly(A) regions of RNA are equally homologous to either poly(U) or poly(dT).] Described in the following section, in some detail, is the use of these and other forms of affinity chromatography involving nucleic acids.

8.1. SYNTHESIS AND USE OF DNA–CELLULOSE CHROMATOGRAPHY

Deoxyribonucleic acid–cellulose performance depends heavily on the quality and the type of cellulose used for the synthesis. Almost any type of cellulose will bind or trap DNA and thereby yield a DNA–cellulose suitable for obtaining DNA-binding proteins. However, to minimize nonspecific binding of non-DNA-related proteins and thus achieve maximum purification, the cellulose used must be of high purity. Since cellulose is chiefly prepared from wood, cotton, and other woody plant material, it is often contaminated with the polyphenolic component of wood, lignin, which is in large measure responsible for the strength of wood. Such polyphenols bind to many proteins (9,10) and must be removed to obtain a product capable of maximal purification of DNA-binding proteins.

Alberts and Herrick (1) recommend the use of a very highly purified grade of cellulose (e.g., Munktel 410, Bio-Rad, Richmond, Calif.) further treated to remove any impurities. First, the cellulose is washed several times with boiling ethanol, followed successively by brief washes with 0.1-M NaOH, 0.001-M tri-sodium EDTA, 0.01-M, and finally with distilled water. Less purified cellulose starting material may additionally require the removal of sulfate groups that could function as ion-exchange centers and thereby cause nonspecific adsorption. This is accomplished by incubating cellulose with 50-mM NaOH containing 2mg/ml NaBH$_4$ at 60°C for 1 hour (11). Extensive washing and reduction with borohydride is required only when the presence of small amounts of contaminates could alter the interpretation of the results of an experiment or interfere with the purification. Often large batches of DNA–cellulose can be prepared from any "reagent-grade" cellulose and used successfully for the initial step in purifying large amounts of DNA-binding proteins. The author has found this to be a useful first step when followed by one or more classical purification steps and finally by a second DNA–cellulose chromatography step on a much smaller DNA–cellulose column prepared from washed and treated cellulose.

To prepare DNA–cellulose by the physical trapping technique, a solution of from 1 to 3 mg/ml of DNA in 10-mM Tris-HCl, pH 7.4, containing 1-mM EDTA (or any similar buffer) is slowly added to the cleaned and dried cellulose until a thick paste results. (Usually 1 g of cellulose will require 3 to 4 ml of DNA solution.) The paste is stirred for several minutes to establish uniformity and then spread over the bottom of a petri dish or a crystallizing dish. The cellulose layer should be as thin as possible (in no case over about a quarter of an inch thick) to speed the drying. The paste is dried and then broken into pieces, ground with a mortar, and placed in a lyophilizer or vacuum desiccator for 16 to 48 hours to assure complete drying. The dry powder is suspended in 20 volumes of appropriate buffer (e.g., 10-mM Tris HCl, pH 7.4) containing 1-mM EDTA and 0.15-M NaCl and placed for 16 to 24 hours at 4°C. The DNA–cellulose is then collected by suction filtration and washed twice by suspending the DNA–cellulose in 20 volumes of the same buffer at 4°C and immediately collecting it by suction filtration. The washed DNA–cellulose is resuspended as a slurry in a convenient quantity of the wash buffer and stored frozen. Immediately prior to use it should be washed again by suction filtration with 20 volumes of wash buffer (1).

To determine the amount of DNA bound, an aliquot is heated to 100°C for 20 minutes in wash buffer to release the DNA from the cellulose. The DNA released is separated from the cellulose by suction filtration, and the amount of DNA is determined by the absorbance of the filtrate at 260 nm or by chemical methods (12). This procedure usually results in a DNA content of 0.5 to 1.5 mg of DNA per packed multiliter of ecllulose. The drying step is the most crucial part of the process, and no DNA is bound without drying. Slow air drying is considerably more efficient than lyophilization. Higher DNA concentrations may be obtained, or lesser concentrated DNA solutions may be employed, if repeated cycles of codrying the DNA and cellulose are employed.

This is probably the most convenient method to use in preparing DNA–cellulose, but the DNA is bound reversibly and is slowly desorbed from the cellulose. At equilibrium very little adsorbed DNA remains. However, since the half-life of the DNA–cellulose in a neutral buffer is 400 hours at 37°C and several orders of magnitude longer at 4°C, any experiment done at 4°C will result in no apparent loss in the DNA-binding capacity of the column. The small amount of DNA released will contaminate the proteins purified by this method and may cause considerable difficulty in some experiments. Therefore, the investigator should be aware of its presence in his final DNA–cellulose eluate. Normally these small amounts of DNA will not cause significant difficulty, but if any problem arises, DNA can be readily removed from the eluate by subsequent chromatography on DEAE- cellulose.

If covalent binding of DNA to cellulose is desired, a number of methods have been developed to irreversibly fix the DNA to cellulose. Probably the first method of binding DNA to cellulose in aqueous media was the ultraviolet irradiation method of Litman (4). In this procedure the DNA is physically trapped in the cellulose matrix and then cross-linked by irradiation to fix the DNA. Washed cellulose (0.75 g) is mixed with an aqueous solution of DNA (e.g., 6 ml of 2 mg/ml calf thymus DNA in $0.01\text{-}M$ NaCl) and a smooth paste prepared. This slurry is air-dried overnight as in the previous method, and the dry DNA–cellulose is suspended in 20 ml of absolute ethanol. The alcohol suspension is placed under a low-pressure mercury lamp held 10 cm from the surface and irradiated for 15 minutes at about 100,000 ergs/min. The DNA suspension is slowly stirred with a small magnetic stir bar throughout the operation to ensure that the DNA is irradiated in a homogenous fashion. The alcohol suspension is then suction filtered and the DNA–cellulose cake is washed three times by suspending it in 50 ml of $1\text{-}mM$ NaCl for 10 minutes followed by suction filtration. After the final DNA–cellulose filter cake is washed, it is spread out to air-dry overnight. About 90% of the DNA is bound by this method. The irradiation step is essential for the preparation and only about 10% of the DNA is bound in its absence. The amount of DNA bound is determined directly by the diphenyl-amine method by using a preweighed amount of DNA–cellulose (12). No DNA leakage occurs with this form of DNA–cellulose, and it can be stored for several years at room temperature as a dry powder. Many proteins recognize the ultra-violet-irradiated DNA, including a few whose function may be specifically related to repair of radiation-damaged DNA (13). Similar ultraviolet irradiation methods have been used to immobilize viral RNA (14), and ribosomal RNA to cellulose (15) and to immobilize RNA, DNA, and polynucleotides to vinyl, nylon, and fiberglass supports (5,16).

Gilham (10,19) and Richwood (18) have immobilized DNA to cellulose and other polysaccharide materials by way of their terminal phosphate functions, uisng a water-soluble carbodiimide as the coupling agent. To wash the cellulose before coupling, Gilham (19) suspended standard-grade chromatographic cellulose in 1% methanolic HCl for 3 days at room temperature followed by extensively washing the cellulose with water on a suction filter with subsequent air-drying. The DNA or other polynucleotide is dissolved in water, and the pH of the solution is adjusted to pH 6.0 with dilute hydrochloric acid or dilute sodium hydroxide, as needed. This solution, containing 2.5 to 250 μmoles of DNA/ml, is added to one-quarter volume of 0.2-M MES buffer, pH 6.0, to which water-soluble carbodiimide is added (final concentration of 100 mg/ml). [Either N-cyclohexyl-N'-β-(γ-methylmorpholinium)ethyl carbodiimide p-toluene sulfonate or 1-ethyl-3(3-dimethylamine propyl)-carbodiimide hydrochloride would appear to be suitable water-soluble coupling agents.] The buffer–coupling agent mixture is added to the clean and dry cellulose powder to make a thick paste. The paste is spread out in a petri dish bottom and placed in the laboratory at room temperature to dry slowly (relative humidity 30 to 40%). At 2 hours, 5 hours, 8 hours, and 14 hours after the paste was prepared the petri dish is placed in a glass tank lined with wet filter paper. The paste is maintained in the water-saturated atmosphere of this tank for 1 hour and afterward is returned to the laboratory atmosphere. After 24 hours of reaction time, the largely dry cellulose powder is washed extensively by agitation in a large volume of 0.05-M potassium phosphate buffer, pH 7.0, followed by collection using suction filtration. This procedure is repeated three to five times, followed by a final wash with water. The DNA–cellulose is then collected by suction filtration and air-dried. Alternately, the cellulose may be suspended in 0.1% ammonium hydroxide between the first and second postassium phosphate buffer washes. This latter procedure removes small amounts of carbodiimide that may be present as substituents on the uracil, thymine, or guanine bases.

Deoxyribonucleic acid–cellulose has been used in numerous studies of DNA-polymerase, including its purification from a wide variety of sources. (No single purification process of DNA-polymerase can be considered to be representative of the use of DNA-polymerases, in general, because of the wide variety in sources, physical and enzymatic properties, or even in the physiological functions of these enzymes.) For example, Gardner and Kado (20) used DNA-cellulose chromatography to purify DNA polymerase obtained by the physical trapping technique from normal and gall tumor tissues of *Vinca rosea* (periwinkle). Starting with 850 g of the plant cells from tissue culture, they isolated about 100 μg of highly purified polymerase (see Table 8.1). The low final protein concentration created an instability problem with the enzyme, losing 50 to 80% of its activity in less than a week at -20°C. Even so, a final purification of over 2000 fold was obtained. Greater purification could undoubtedly be attained if substantially larger quantities of tissue were initially employed. The difficulties inherent with the physically trapped DNA-cellulose became apparent during this purification procedure. As seen in Figure 8.1, substantial quantities of DNA were eluted from the column when the salt conventration was increased from 0.15-M KCl to 0.64-M KCl.

To purify the enzyme, tissue cultures were harvested and kept at -70°C until use. The cell-free homogenate was prepared by homogenizing 200 to 400 g of

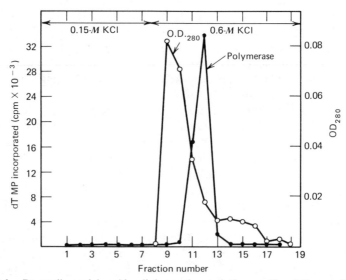

Figure 8.1. Deoxyribonucleic acid–cellulose column elution profile of *V. rosea* DNA polymerase activity. A 2.5-cm × 8-cm column of DNA–cellulose was equilibrated with TME-glycerol buffer containing 100 μg/ml of bovine serum albumin and 0.15-M KCl. Enzyme from phosphocellulose column (0.62 mg of protein) was adsorbed on the column, and after a 500-ml wash with TME-glycerol buffer wihout bovine serum albumin, the KCl concentration was changed to 0.6 M. Fractions (5 ml) were collected and assayed with denatured DNA (system A). Most of the initial absorbence at 280 nM was due to DNA leakage (i.e., high 260/280 ratio) from the cellulose when the KCl concentration was raised to 0.6 M. Reproduced with permission from Gardner and Kado (20).

TABLE 8.1. Purification of *Vinca rosea* 6S Polymerase from Sterile Tissue Cultures

Fraction	Total Protein, mg	Activity[a], Units	Specific Activity, Units/mg	Purification, Fold	Yield, %
I. 40 000 g crude supernatant[b]	1237	74.2	0.06[c]		100
II. DEAE–cellulose	94	58.2	0.62	10.3	78.4
III. Phosphocellulose	2.37	62.5	26.4	440	84.2
IV. DNA–cellulose	0.102	13.6	133	2217	18.3[d]

Source: Reprinted with permission from Gardner and Kado (20).

[a] Assays were carried out with poly [d(A-T)$_n$] (1.0 OD$_{260}$ unit/ml) as template. Specific activity and concentration of dTTP were 1272 cpm/mole and 5 μM, respectively. One unit of enzyme activity is defined as the amount of enzyme that catalyzes the incorporation of 1 nmole of dTTP into product in 30 min.

[b] Enzyme extraction from 850 g of *V. rosea* tissue.

[c] Incorporation was not perfectly linear with crude supernatants indicating the presence of possible inhibitors.

[d] Dilute solutions of this enzyme fraction lost activity rapidly at –20°C in 20% glycerol solutions (>50% in 7 days).

213

tissue in an equal volume of TME buffer (50-mM Tris-HCl, pH 7.85, 10-mM β-mercaptoethanol, and 1-mM EDTA). After homogenization, the extract was passed through miracloth and cheesecloth and then brought to 0.8 in KCl and 20% in glycerol. The extract was stirred for 8 hours at 4°C and then centrifuged to remove cell debris and other unwanted solids.

The extract was dialyzed against TME, and the dialyzed extract was applied to a DEAE–cellulose column (2.5cm × 15cm). The column was washed with 400 ml of TME and developed with a 400-ml linear gradient of 0- to 0.4-M KCl in TME. The DNA-polymerase-containing peak was determined by enzyme assay, pooled, and dialyzed against TME plus 0.05-M KCl and 20% glycerol. The resulting enzyme preparation was applied to a 10-cm × 2.5-cm phosphocellulose column. The column was washed with buffer containing 0.05-M KCl and the enzyme subsequently was eluted with a 400-ml linear gradient from 0.05-M KCl to 0.8-M KCl. The fractions containing DNA polymerase activity were pooled, dialyzed against TME plus 0.15-M KCl, and applied to a DNA–cellulose column. The column was washed with buffer until the OD_{280} of the wash was less than 0.01, at which point the enzyme was eluted with 0.6-M KCl.

The resulting plant DNA polymerase was immunologically distinct from the 6 to 8 S mammalian DNA polymerases but was similar in its molecular size, sensitivity to certain inhibitors (N-ethylmaleimide, heparin, ethanol, etc.) pH optimum, template preference, and requirements for Mg^{++} and Fe^{++} ions. The gall tumor enzyme obtained was identical to that isolated from normal tissue.

Although this may not be a "model" for all DNA polymerase preparations, it does illustrate a potential purification method utilizing DNA–cellulose. The use of DNA–cellulose chromatography as a final step in a purification sequence is a fairly common process since the initial steps in the purification, particularly DEAE–cellulose chromatography, removes the endogenous RNA and DNA that might competitively elute DNA polymerase from the column. It is essential that all endogenous polynucleotides be removed from the crude enzyme preparation prior to DNA–cellulose chromatography.

Although DEAE–cellulose chromatography is a good method for removing endogenous DNA from crude cell extracts, several other methods have been more widely used. For example DNA can be removed by partitioning between two aqueous polymer phases, such as polyethylene glycol and polydextran (21), or by precipitating the nucleic acids by using a high-ionic-strength polyethylene glycol solution (1,2). At least 90% of the endogenous DNA can be precipitated by adding NaCl to a final concentration of 0.6-M and polyethylene glycol 6000 at a final concentration of 10% to a crude cell extract, followed by centrifugation at 10,000 g for 15 minutes to remove the precipitated DNA (see Figure 8.2).

Alternatively, DNA-free cell extracts can be prepared by treating the crude extract with pancreatic DNAase and Mg^{++} in appropriate concentrations. Sen and Todaro (22) in studying DNA-binding proteins from mouse cells, first

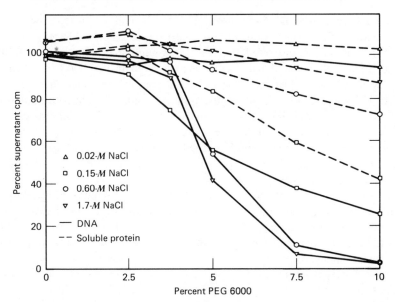

Figure 8.2. Precipitation of DNA as a function of polyethylene glycol and salt concentrations. *Escherichia coli* B were infected with T4 bacteriophage and [³H] thymide, and [¹⁴C] leucine were used to label T4 DNA and early proteins, respectively. An extract was prepared as described for PEG precipitation, except that the concentration of NaCl was 50 mM and the extract was clarified by centrifugation for 1 hour at 38,000 rpm in a Spinco 40 rotor. To 100 μl of supernatant, 200 μl of 1.5 × concentrated PEG–NaCl solution was added. After 15 minutes at 0°C a visible precipitate was pelleted at 10,000 g for 15 minutes. Samples of the supernatants were applied to GF/A glass filters (Whatman) and rinsed with cold 5% trichloroacetic acid, followed by cold ethanol and warm diethylether rinses [D. Pettijohn (1967), *Eur. J. Biochem.* **3**, 25]. The samples were counted by standard scintillation techniques, and the data were corrected for overlaps. Treatment of a 10-fold dilution of the extract at 0.60-M NaCl (not shown) gave less severe loss of protein (12% vs. 35% at 10% PEG) and nearly equivalent precipitation of DNA. High-molecular-weight DNAs precipitate at lower PEG concentrations than does the sonicated DNA shown here; thus a cleaner separation of DNA and protein should be obtainable after gentle lysis procedures. Reprinted with permission from B. Alberts and G. Herrick (1971), *Methods in Enzymology* (L. Grossman and K. Moldave, eds.), **21D**, 198.

isolated the nucleus by gentle homogenization and differential centrifugation. They resuspended the isolated nucleii in a hypotonic buffer, pH 7.4, containing Ca^{++} and Mg^{++}, and the suspension was sonicated briefly. After centrifuging to remove the debris, the soluble nuclear material solution was adjusted to 10-mM MgCl$_2$, and DNAase I (100 μg/ml) and ribonuclease T1 (50 μg/ml) were added. The nucleic acids were digested for 1 hour at 37°C, and the digestion process was terminated by adding EDTA to a final concentration of 25 mM. The extract was dialyzed versus 10-mM Tris, pH 7.4, containing 10-mM EDTA, 1-mM β-mercaptoethanol, and 5% glycerol (v/v). The dialyzed extract was applied to a column of denatured DNA–cellulose. The DNA binding proteins were then eluted with buffers of increasing ionic strength.

This technique demonstrated that the DNA-binding proteins obtained from mouse tumor cells created with either mouse sarcoma virus or some chemical carcinogen were no different from DNA-binding cells isolated from normal mouse cells. On the other hand, mouse tumors created by transformation using SV-40 virus or other chemical carcinogens did reveal more DNA-binding proteins than were found in normal mouse cells. These extra DNA-binding proteins were identical to proteins previously identified as T (for tumor)-antigen proteins. These proteins appear in many types of tumor and may be useful in identifying or classifying tumors (22).

Removal of endogenous DNA in bacterial cell extracts requires far less rigorous conditions than needed in animal cell extracts. Alberts (1) reports that treatment of bacterial cells with 20-μg/ml of pancreatic DNAase I for 15 minutes at 20°C or 1 hour at 10°C is sufficient to digest over 98% of the endogenous DNA. Digestion of mammalian DNA using pancreatic DNAase I is possible, but high salt concentrations are required to dissociate the complexes between DNA and proteins (e.g., histones, which are present in mammalian nuclei). Typically, digestion of mammalian DNA was accomplished by using 100 μg of DNAse I for 1 hour at 20°C in the presence of 0.6-M NaCl. In either case, 10-mM MgCl$_2$ is required since DNAase I is active only in the presence of Mg^{+2} or Ca^{+2} ion. The reaction is effectively quenched on removal of the ions usually caused by adding excess EDTA, which chelates the divalent cations.

The presence of histones in mammalian extracts can also be a problem since these proteins, as well as some other proteins, bind DNA in a "nonspecific" fashion. They can be removed from a DNA–cellulose column by washing the column with a solution of a polyanionic material such as dextran sulfate (22), which competes with the DNA for any nonspecifically bound highly cationic material (such as the histones.)

Deoxyribonucleic acid–cellulose can be used for purifying many proteins other than the DNA polymerases) that bind to DNA. For example, it has been used widely to purify the DNA-dependent RNA polymerases (2, 23–37), nucleases (28,29), and the various untwisting or melting proteins found in bacteriophages (2,3, 30–35), bacteria (36–39), and plant (40–42) and mammalian cells (43). Numerous proteins that stimulate DNA polymerase or replication activity have also been isolated by this method (44–47).

8.2. ALTERNATIVE METHODS OF DNA–CELLULOSE SYNTHESIS

Although the physical trapping method of Alberts (1–3) and the physical trapping-ultraviolet-cross-linked methods of Litman (4) are the major methods

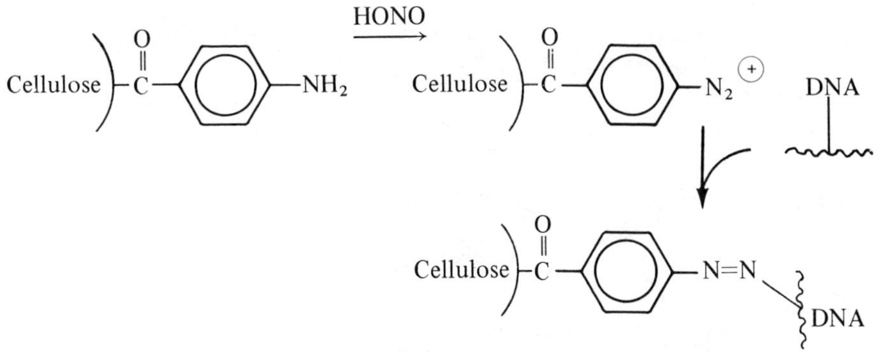

Figure 8.3. DNA coupling to *p*-aminobenzoyl cellulose by diazotization.

for DNA–cellulose preparation, several alternative methods have been employed. Besides the carbodiimide method of Gilham (see page 221), various types of single-stranded DNA, for example, denatured DNA or native DNA with single-stranded regions, can be coupled directly to CNBr-activated cellulose (48). Alternatively, an aryl amine can be coupled to cellulose; diazotized; and added to denatured, single-stranded, or partially single stranded DNA (49,50). The coupling occurs through the bases, predominantly to the dG, dA, and dC moieties with a lesser degree of coupling to dT residues (51,52).

8.3. POLY(dT) AND OLIGO(dT)–CELLULOSE AS mRNA SELECTIVE AGENTS

Messenger RNA was first proposed by Jacob and Monod in 1961 (53) as an intermediate in protein synthesis. It is the template that directs the synthesis of proteins and thus determines what protein is to be synthesized. Although the existence of mRNA has been well documented for many years (54), only recently have effective methods for isolating and purifying mRNA been developed. Many of these techniques have been based on the discovery of enzymes that add a sequence of poly(A) to the 3′-terminus of mRNA or its precursors (55,57) while they are still within the nucleus. The length of the poly(A) sequence varies, with values of about 50 to 200 nucleotides bieng reported. A 3′-terminal poly(A) sequence has been found on all types of mRNA studied to date, with the single exception of the histone mRNA (58,59).

The synthesis of mRNA seems to be initiated by the transcription of DNA into heterologous nuclear RNA (hnRNA), which probably then undergoes several modifications, the last of which is the addition of poly(A) to some apparently specific, but as yet undetermined, sequence at the 3′-terminus.

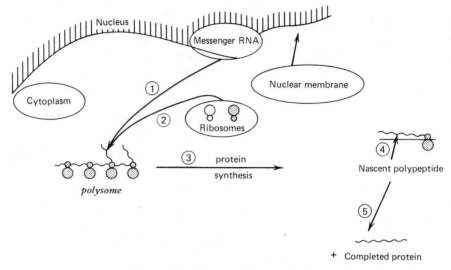

Figure 8.4. Messenger-RNA synthesis.

Several cellulose-bound polymers of thymidine or uridine have been used to purify mRNA. Thymidine or uridine polymers attached to a small number of noncellulosic support materials have also been employed. The most widely used is oligo(dT)–cellulose, which is produced by polymerizing thymidine-5'-phosphate, using dicyclohexylcarbodiimide as the condensing agent. Undoubtedly one important factor in the widespread use of this material is its commercial availability. An adsorbent prepared from poly(U) bound to cyanogen bromide-activated agarose is also commercially available.

To prepare oligo(dT) cellulose (6,19,60), 2 mmoles of thymidine-5'-phosphate (as the pyridinium salt) in dry pyridine (3 ml) is added to a two-fold excess of dicyclohexylcarbodiimide. A few glass beads are added, and the thick gum that forms is shaken for 5 days at room temperature. Then 5 g of washed cellulose (see page 209) dried at 100°C for 10 hours is added to the mixture along with 50 ml of dry pyridine and two additional millimoles of dicyclohexylcarbodiimide. This mixture is shaken for an additional 5 days at room temperature, after which the cellulose is collected by suction filtration and washed with pyridine. The washed oligo(dT) cellulose is suspended in 50% aqueous pyridine for 10 hours, collected by centrifugation, and then washed further with copious quantities of warm ethanol to remove all of the dicyclohexylurea that accumulates during the synthesis. Finally, the product is washed by way of suction filtration with distilled water until the absorbance of the washing at 267 nm is essentially zero. About 60% of the thymidine is incorporated into the matrix by this method. Other polynucleotide derivatives of cellulose can readily be prepared in this fashion. Jovin and Kornberg (61) have used oligonucleotide celluloses to purify DNA polymerase, RNA polymerase, and deoxynucleotidyl-transferase. Oligo(dT) cellulose has also been used to purify DNA ligase (62).

A similar oligo(dT) cellulose preparation can be made in aqueous media by way of water soluble carbodiimides (10,17). The coupling of thymidine-5'-phosphate is nearly identical (ca. 60%) to that prepared by the nonaqueous method described previously. The aqueous method is essentially the same as that described for the attachment of DNA to cellulose by way of carbodiimide-mediated coupling (see page 211); however, thymidine-5'-phosphate in MES buffer, pH 6.0, is used instead of aqueous DNA solutions.

Aviv and Leder (63) were the first investigators to use oligo(T) cellulose prepared by this method to purify mRNA, although Nakazato and Edmonds (55) and Sheldon et al. (16) had earlier employed poly(dT) cellulose and poly(U) cellulose that was prepared by much simpler synthetic routes to purify poly(A)-rich RNA (presumably mRNA). Aviv and Leder (63) isolated the poly(A)-rich RNA fraction from rabbit reticulocyte polysome. First, the RNA was extracted from the isolated polysomes by phenol chloroform-isoamyl alcohol extraction. The polysomal RNA was precipitated with ethanol and dissolved in 0.1-M Tris-HCl, pH 7.5, containing 0.5-M KCl. The polysomal RNA solution (100 A_{260} units) was applied to a 2-ml column of oligo(dT) cellulose containing about 0.5 g (dry weight) of the cellulose adsorbent. The column was washed extensively with the same Tris-KCl application buffer to remove all the RNA, presumably ribosomal RNA (rRNA) that was not adsorbed. The poly(A)-rich RNA was eluted with two buffers of decreased ionic strength; (1) 0.01-M Tris-HCl, pH 7.5, containing 0.1-M KCl and (2) 0.01-M Tris-HCl, pH 7.5. The eluted material was made to 2% in potassium acetate, adjusted to pH 5.5, and RNA was precipitated by adding 2 volumes of ethanol. The oligo(dT) cellulose was regenerated by washing with 0.1-M KOH and reequilibrated with the application buffer. The entire procedure can be scaled up appreciably (at least 20 fold). The globulin mRNA isolated in this fashion was capable of coding for the *de novo* synthesis of rabbit globin by use of an ascites tumor cell protein synthesis system.

Crawford and Wells (64) extend this process one step further. Using chicken globin mRNA (isolated by the same method as a substrate for RNA-dependent DNA polymerase), they were able to prepare ^{32}P-labeled complementary DNA (cDNA) of the globin mRNA. By hybridizing this ^{32}P-labeled cDNA to newly synthesized ^{3}H-labeled RNA, production of a high-molecular-weight RNA precursor to globin mRNA was illustrated. They also demonstrated that this precursor (1) is newly synthesized, (2) has a higher molecular weight (145 S) than steady-state globin mRNA (105 S), (3) is the precursor to globin mRNA, and (4) has a very short half-life ($\geqslant$ 3 to 4 minutes). Further, the globin mRNA-145 precursor can be adsorbed onto oligo(dT) cellulose, thereby demonstrating that it contains a poly(A) sequence. (See Figure 8.5.)

Venetianer and Leder (65) have employed oligo(dT) cellulose as the basis for

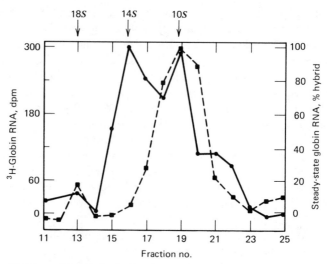

Figure 8.5. Chicken erythroblasts (0.25 ml packed cells) were pulse labeled with 4 mCi of [^{3}H] uridine in 4 ml of Eagles medium for 5 minutes. Ribonucleic acid was isolated from the washed cells and centrifuged on 10 to 40% sucrose gradients at 38,000 rpm for 16 hours at 4°C. Fractions (0.4 ml) were collected dropwise from the bottom of the gradient and the 18S and 28S ribosomal RNA was located by measuring the optical density of each fraction. Fractions that contained RNA sedimenting between 5S and 22S were ethanol precipitated and hybridized separately to 0.5 µg of chicken globin cDNA in a 70-µl hybridization volume. Control hybridizations contained no chicken globin cDNA. Labeled RNA sequences hybridizing to globin cDNA were detected by RNAse assay. Steady-state globin sequences were located by RNA excess hybridization and S1 nuclease assay. ^{3}H-labeled globin RNA (● – ●); steady-state globin RNA (■ – ■). Reproduced with permission from R. J. Crawford and J. R. E. Wells (1978), *Biochemistry* **17**, 1591. Copyright © American Chemical Society, 1978.

synthesis of cellulose that contains covalently bound DNA (cDNA) that is complimentary to globin mRNA. This DNA–cellulose, in turn, was employed as an affinity chromatography medium for the isolation of additional mRNA. To prepare cDNA-cellulose, they first isolated highly purified mouse globin mRNA (66). This mRNA (10 µg/ml) was added to oligo(dT) cellulose (30 mg/ml) in a RNA-dependent DNA polymerase synthetase mixture. The reaction mixture was stirred vigorously for 150 minutes at 37°C. The initial step of the reaction was the noncovalent binding or annealing of the globin mRNA to the thymidine bases of the oligo(dT) cellulose. Next, the RNA-dependent DNA polymerase, using the oligo(dT) as a primer copied the globin mRNA, creating a cDNA strand that was covalently bound to the oligo(dT), which, in turn, was coupled to the cellulose. The globin mRNA, on the other hand, was noncovalently bound to its complimentary strand. To terminate the reaction, the cellulose mixture was poured into a water-jacketed column and washed with 0.01-M KOH. This step

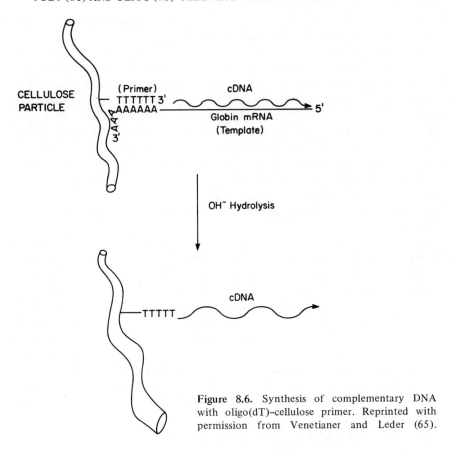

Figure 8.6. Synthesis of complementary DNA with oligo(dT)–cellulose primer. Reprinted with permission from Venetianer and Leder (65).

dissociates and removes the mRNA bound to the cDNA and ultimately hydrolyzes the RNA to mononucleotides (RNA, but not DNA, is very readily hydrolyzed under such alkaline conditions). Next, the excess KOH was removed by washing the column with distilled water (see Figure 8.6).

To purify mouse globin mRNA using the cDNA cellulose, a partially purified mRNA sample was dissolved in a buffer containing 10-mM Tris·HCl, pH 7.5, 0.3-M NaCl, and 0.1M EDTA and passed through the column with a water-jacketed temperature at 72°C and at a flow rate of 2 ml/hour or less. The column was washed with the application buffer (25 ml) at the same temperature, and then the temperature was lowered to 40°C and the column was further washed with 10-mM Tris·HCl, pH 7.5. Finally, the jacket temperature was raised to 70°C and the purified mRNA was eluted as a sharp peak, with 10-mM Tris·HCl, pH 7.5. When polysomal RNA was used as the source of mRNA, 98% of the total RNA passed unadsorbed through the column, whereas only 2% of the RNA (identified as essentially pure mouse globin mRNA) was eluted as a sharp peak at 70°C. The overall yield of the initial mRNA present in the crude polysomal RNA applied to the column was 67%.

Levy and Aviv (67) have used cDNA cellulose obtained through this procedure to devise a method to quantify the amount of labeled globin mRNA produced under various conditions, even if the globin mRNA was an extremely small percentage of the total labeled RNA. This procedure was accomplished by adding a small amount of labeled RNA to globin cDNA cellulose in a deforming buffer. After washing extensively with the same buffer, only globin mRNA remains bound to the cDNA cellulose and can then be quantified by using standard curves prepared from known amounts of previously purified globin mRNA. Globin cDNA cellulose could also be employed to purify globin mRNA from a highly heterogenous polysomal RNA sample isolated from spleen (68). It has also been used to isolate *its* complimentary sequence from sheared nuclear DNA (69).

Woo et al. (70,71) have utilized the same technique to purify ovalbumin mRNA and the coding strand of nuclear DNA for ovalbumin (72). An improvement in the technique was accomplished by modifying the RNA-dependent DNA polymerase reaction step by including actinomycin D to arrest double-stranded DNA synthesis. Rhoades and Hellman (73) have attempted to optimize the conditions needed for cDNA cellulose synthesis and use. They achieved yields of 98% of the mRNA with a purification of actor of over 40 fold by chromatography of ovalbumin polysomal RNA on ovalbumin cDNA cellulose.

In all cases one of the major problems involved in mRNA purification is the presence of RNAase in almost every tissue. It is a highly active enzyme and tends to adsorb to glassware in an active state. Hence all glassware used in RNA isolation must be autoclaved to inactivate potential RNAase. To further inactivate RNAase, Jost et al. (84) added diethylpyrocarbonate to all buffers used in the purification of vittellogenin mRNA isolated from chicken liver. The result was a three- to seven-fold improvement in the recovery of mRNA.

Recently, oligo(dT) cellulose has been used to isolate the mRNA induced by illuminating barley plants grown in the dark (75). This light-induced mRNA was shown to yield predominantly a polypeptide of M_r 29,500. This is 4000 amu (atomic mass units) greater than the chlorophyll a/b protein. This is the major protein added to chloroplast membranes during illumination and is the major protein synthesized by the cells after they are subjected to growth in the light. The M_r 29,500 protein was immunologically cross-reactive with the chlorophyll a/b protein and peptide mapping of both chlorophyll a/b, and the polypeptide yielded peptide patterns that were essentially identical. Apparently this polypeptide is a precursor of the chlorphyll a/b protein. Interestingly, the chlorophyll a/b protein does not appear on the illumination of chlorophyll b minus mutants, but the mRNA coding for the M_r 29,500 polypeptide does appear in these mutants after dark-grown plants have been illuminated. It would appear that chlorophyll b synthesis is not the signal for the production of the mRNA but is needed to convert the M_r 29,500 polypeptide into the chlorophyll a/b protein.

Oligo(dT)$_{12-18}$cellulose is also a valuable medium for isolation of reverse transcriptase (RNA-dependent DNA polymerase) (76). Reverse transcriptase has a higher affinity for poly(dT)$_{12-18}$ as compared to its affinity for larger polymers, for example, poly(dT)$_{>100}$. Cellular DNA polymerases conversely prefer the larger polymer. Samples of various tissue culture cells or purified viral particles, such as Rauscher murine leukemia virus (MuLv), were sonicated in isotonic saline and then diluted with a detergent-containing buffer and then incubated at 37°C for 30 minutes. The disrupted cells were centrifuged to remove debris and dialyzed versus a manganese-containing buffer since Mn^{++} ions are needed for the tight binding of the enzyme to oligo(dT). Oligo(dT)$_{12-18}$ (PL Biochemicals, Milwaukee) was coupled to cellulose by using dicyclohexylcardobiimide (6,19). The sample was applied to a 2-ml column of oligo(dT)$_{12-18}$ cellulose equilibrated with the same buffer. The enzyme was eluted with a linear gradient of KCl in the absence of manganese. As seen in Figure 8.7, a substantial purification of the enzyme (ca. 100 fold) is obtained by the use of this procedure.

8.4. SYNTHESIS AND APPLICATION OF POLY(U) CELLULOSE AND POLY(U) AGAROSE TO mRNA PURIFICATION

Probably the easiest to synthesize of all "affinity hybridization" media is poly(U) cellulose as prepared by Sheldon et al. (16,77). This is an adaptation of Litman's ultraviolet-irradiation method of covalently coupling DNA to cellulose (4). Washed cellulose (1 g) is added to a solution of commercial poly(U) (8 mg/ml) in distilled water. The cellulose and poly(U) solutions are thoroughly mixed, and the paste is lyophilized and then resuspended in 10 ml of 95% ethanol. The suspension is poured into a petri dish and treated by irradiation for 15 minutes with a 30-W germicidal lamp that is held 20 cm from the surface. The irradiation causes cross-links to form between the various poly(U) chains and also possibly causes covalent bond formation between the poly(U) and the cellulose matrix. The irradiated poly(U) cellulose is washed well with distilled water and subsequently resuspended and stored in 0.01-M Tris·HCl pH 7.5 containing 9.1-M NaCl. Although only about 5% (0.4 mg) of the initial poly(U) is coupled to the cellulose, the binding capacity of poly(A) to a column of poly(U) cellulose is very good. A 0.8-cm × 20-cm column of this material will retain at least 1.5 mg of poly(A).

Poly(U) was also coupled (16) to glass fiber filters by the same method. Fiberglass filters (2.4-cm diameter) were supported on a rubber framework, and

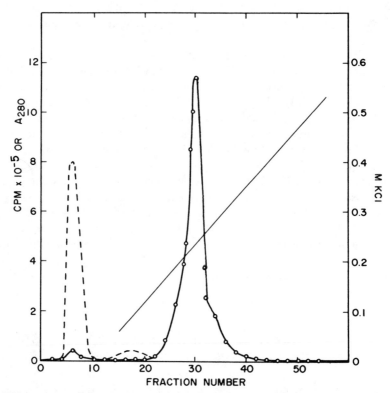

Figure 8.7. Chromatography of MuLv DNA polymerase on $(dT)_{12-18}$ cellulose and 2 ml (2 mg of protein) of dialyzed extract (S-100) of MuLv was applied to the column (0.9 cm × 12 cm) and eluted. Aliquots (0.01 ml) of the 2.0-ml fractions were assayed in 0.05-ml reaction mixtures for 60 minutes at 37°C: (– – –), A_{280}; (0), cpm (^{3}H)TTP incorporated with the poly(rA) $(dT)_{12-18}$ assay; (———) M KCl. Reproduced with permission from Gerwin and Milstein (76).

0.15 ml of a solution of poly(U) (1 mg/ml) in distilled water was applied to each filter. The filters were air-dried at 37°C and then irradiated on each side for 2.5 minutes under the same conditions employed in the preparation of poly(U) cellulose. The filters were washed with water and dried. About 0.1 mg of poly(U) was bound to each filter under these conditions. Substantially more poly(U) could be bound if the initial concentration was increased. These poly(U) filters appear to be very useful for the rapid assay of the poly(A) or mRNA content in multiple samples.

Poly (U) coupled to CNBr-activated agarose is also very useful in the isolation of mRNA (78–83). While poly(U) agarose can be purchased ready-made its synthesis is quite simple. Agarose (4%) beads (40 g wet weight) are washed and suspended in 60 ml of ice water. Cyanogen bromide (9 g) dissolved in 135 ml of water is added and the pH adjusted to 11.5 and maintained with the addition of concentrated KOH as necessary. After 20 minutes the agarose is collected by suction filtration and washed with 600 ml of ice-cold 0.1-M potassium phosphate buffer, pH 8.0. The washed agarose is resuspended in 40 ml of the same buffer, which contains 20 mg of poly(U). The mixture is incubated at 4°C for 18 hours and then washed with 200 ml of 90% formamide in 10-mM potassium phosphate (pH 7.5)–10-mM EDTA–0.2% Sarkosyl, followed by another 200-ml wash with 25% formamide in 50-mM Tris·HCl, pH 7.5, 0.7 M NaCl–10-mM EDTA.

Application and development of poly(U) cellulose columns can be performed essentially as described for oligo(dT) cellulose by using a water-jacketed column and changing the temperature to effect elution (80) (see page 221). Alternatively, buffers containing various concentrations of deforming agents (which break the hydrogen bonds between complimentary bases) can be employed (78,79,81). Thermal elution appears to produce somewhat better results, possibly because the deforming agents used often contain undesirable impurities.

Thyroglobulin mRNA was isolated from a crude RNA extract from the membrane-bound polysomal fraction isolated from horse thyroid cells (80). The crude RNA fraction was dissolved in 10-mM Tris·HCl (pH 7.5)–300-mM NaCl–1-mM EDTA–0.2% SDS and then passed through the column at room temperature. The mRNA fraction was then eluted with the same buffer minus NaCl using a stepwise increase in column temperature. The RNA that eluted at 25°C, 34°C and 50°C was sedimented in a linear sucrose gradient (see Figure 8.8) and then ultracentrifuged. Each peak in the ultracentrifugation pattern was tested for its ability to stimulate *Xenopus* oocytes to form material that was precipitable by antibodies to thyroglobulin. Two mRNA fractions, 16-S and 33-S, were able to stimulate oocyte synthesis of such a product. The authors attribute their observation to the possible occurrence of two different sized subunits of thyroglobulin. The size of the subunits of thyroglobulin (which itself is a large, multisubunit protein of 600,000 molecular weight) have not yet been established. These results could be consistent with the existence of different-sized subunits or, as is the case with the chlorophyll a/b protein, with different sized mRNA and/or polypeptide precursors to the physiological thyroglobulin subunit.

Recently the mRNA of the procollagen molecule was isolated and purified from the frontal bones (calvaria) of chick embryos using successive chromatography on both oligo(dT) cellulose and poly(U) agarose (81). The oligo(dT) cellulose step removed substantially all but traces of rRNA and possessed a higher capacity for mRNA than did poly(U) agarose. Conversely, poly(U) agarose had a greater specificity for mRNA and thus could remove even traces of rRNA. The yield of mRNA after both chromatographic procedures exceeded 70%.

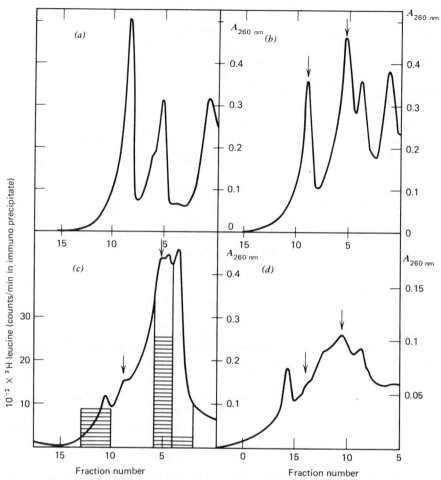

Figure 8.8. Sedimentation pattern of tRNA(A) and RNA retained on a poly(U)–Sepharose column. Total RNA extracted from a 27,000g pellet was dissolved in hybridization buffer and filtered through a 3-ml poly(U)–Sepharose column. After extensive washing of the column with hybridization buffer, retained RNA was eluted by lowering the ionic strength and by stepwise increase of the temperature. The RNA was precipitated with ethanol, dissolved in 10-mM Tris-HCl, (pH 7.5)–10-mM NaCl–1-mM EDTA and layered onto 5-ml 5 to 30% linear sucrose gradients in the same buffer. Centrifugation was for 120 minutes at 60,000 rpm in an SW65 T1 rotor. 1% aliquot tRNA (*a*): RNA eluted at 25°C (*b*), at 35°C (*c*), and at 50°C (*d*); RNAs from regions *a*, *b*, *c*, and *d* of a gradient like that of panel *c* were recovered and injected separately in a sample of 10 oocytes. The results of immunoprecipitations performed in duplicate in 100-μl aliquots of oocyte extracts are shown (shaded area). The radioactivity of a precipitate obtained from buffer-injected oocytes has been subtracted from each value (200 cpm). The arrows point to 18-5 and 28-*S* tRNA. Reproduced with permission from G. Vassart, H. Brocas, P. Nokin, and J. E. Dumont (1973), *Biochim. Biophys. Acta* **324**, 575.

The purified mRNA was subjected to sucrose gradient ultracentrifugation, and the various A_{260} peak fractions were assayed for their ability to induce the synthesis of collagenase-sensitive protein (presumably procollagen) using the cell free protein synthesis system from wheat germ. The results indicated that procollagen mRNA has a sedimentation constant of $26S$ and has a very long lifetime as compared to other mRNAs present in chicken embryo calvoria.

8.5. DNA ATTACHED TO MATRICES OTHER THAN CELLULOSE

Although it would appear that cellulose is generally a poor matrix for affinity chromatography, DNA–cellulose is more often employed than any alternative form of immobilized DNA. Nonetheless, several very attractive alternatives exist. Bendich and Bolton (84) have physically entrapped various types of DNA within an agarose matrix by dissolving DNA in a hot aqueous solution of agarose that sets quickly on cooling. Although single-stranded DNA is most commonly used in this process, many DNAs (including circular DNA) can also be entrapped (85). Highly purified agarose powder is suspended in distilled water (8 g/100 ml) and is dissolved by either boiling or autoclaving briefly. An equal volume of denatured DNA (ca. 1 mg/ml) in standard citrate-saline (SSC; 0.15-M NaCl–0.015-M sodium citrate) is added to the hot agarose, and the mixture is capped and shaken vigorously by hand for about 30 seconds. The mixture is poured onto a cold beaker that is placed in an ice bath. The agar gels quickly and is cut into small (2-cm^2) pieces, which are pressed through a 30-mesh stainless steel screen held in the bottom of a glass syringe with the tip broken off. The irregular DNA particles are washed by soaking them overnight in water and then collected by suction filtration. The DNA–agarose so prepared is employed in essentially the same fashion as DNA–cellulose. Unfortunately, the flow rate of this material is quite slow because of the heterogeneity of the particle size and shape. The author has found that the flow rate can be considerably improved if the hot DNA-containing agarose solution is shaken briefly and poured into a 10-fold or greater volume of rapidly stirred mineral oil. The flask employed is best if it has been heated and shaped to form three or four substantial indentations. These surface irregularities offer considerable help in keeping the two phases separate. After the agarose has formed nearly spherical beads, the stirring is stopped, and the beads are collected by gravity filtration (see also page 65). The beads are subsequently washed and used like the particles obtained by the cutting procedure. If faster flow rates are needed, the beads produced can be sized by sieving. Schaller et al. (85) have isolated *E. coli* DNA polymerase I, DNA polymerase II, RNA polymerase, exonuclease III, and T_4 polynucleotide kinase by this method. Purification factors as high as 200 fold were obtained with yields varying from 75 to 100% .

Both native and denatured DNA have been entrapped in an agarose–acrylamide mixture that was created by a polymerization process that never exceeds 50°C and thus does not denature double-stranded DNA (86). Again the maximal flow rate is about 12 ml/hour but could be considerably increased if spherical beads were made by using the mineral oil emulsion method. To prepare DNA–acrylamide-agarose, DNA (~60 mg) is dissolved in 100 ml of 0.05-M Tris, pH 7.8, containing 9.7 g of acrylamide and 0.3 g of N,N'-methylenebisacrylamide. The mixture is cooled to 15°C and 0.8 ml of TEMED (N,N,N',N'tetramethlyethylenediamine), a polymerization accelerator, is added. Afterward the arcylamide–DNA solution is poured into an equal volume of 1% agarose solution at 50°C and the mixture shaken rapidly, and 0.15 g of ammonium persulfate dissolved in a minimal volume of water is added with continued shaking. The mixture is then allowed to set for 30 minutes, cut into 1-cm^2 cubes, and made into particles by pushing the cubes through a stainless steel mesh essentially as described for DNA–agarose. Deoxyribonucleic acid–acrylamide–agarose produced by this method has been successfully employed to purify DNA polymerase from $E.$ $coli$ cell free extracts (88) and used to isolate and study the nonhistone DNA-binding proteins from rat liver chromatin (87). This technique would seem to be very useful because the DNA does not "leak" from the acrylamide matrix and because the DNA is essentially unaltered by its entrapment. Both techniques can and have been employed in studying hybridization of nucleic acids (84, 88–92). (See Figure 8.9.)

Besides physical entrapment, DNA has been covalently attached to agarose by use of chemical coupling methods. Poonian et al. (93) and Weissbach and Poonian (94) covalently coupled single-stranded DNA to agarose by cyanogen bromide activation. Apparently the cyanogen bromide-activated agarose reacts with the purine and pyrimidine bases of the nucleic acids. Double-stranded DNA in which the bases are buried on the inside of the helix binds very inefficiently to cyanogen bromide-activated agarose. The cyanogen bromide coupling procedure is quite easy. Agarose is cyanogen bromide activated if any of the classical methods (see pages 44 and 50) are used, or commercial cyanogen bromide activated agarose is swollen according to the directions. The CNBr–agarose is washed by suction filtration with ice-cold 0.05-M potassium phosphate buffer. The interstitial buffer is removed by filtration, and the gel is added to an equal volume of a solution of single-stranded DNA [RNA can also be employed (95)] in a pH-8.0 buffer. The mixture is stirred 16 to 48 hours and then the gel is collected by suction filtration and washed with the same buffer until the A_{260} was essentially zero. Further washing with a higher ionic strength buffer can be added to ensure complete removal of unbound DNA. Ribonucleic acid (95,96) can be bound to agarose using the same method, although a pH of 6.0 is preferred over pH 8.0 when RNA is being coupled.

Deoxyribonucleic acid–agarose prepared by cyanogen bromide activation has

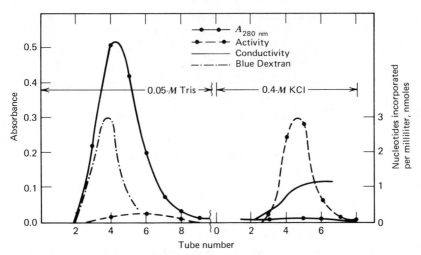

Figure 8.9. Deoxyribonucleic acid–acrylamide gel chromatography; 5.2 mg of DNA polymerase dissolved in 0.3 ml of 0.05-M Tris buffer, pH 7.8, was applied to a 1.5-cm × 15-cm column. The column was washed with 90 ml of the same buffer (30 fractions, 3 ml/tube), and then the polymerase was eluted with 0.4-M KCl. The solid line on the left indicates absorbence at 280 mn and the dotted line, 650 mn for Blue Dextran. The dashed line on the right represents polymerase activity; some activity appears on the left. The solid sigmoid line on the right is the conductivity (arbitrary units). Polymerase activity was assayed by adding 0.1 ml from each tube to 0.4 ml of a mixture containing 3 μg native calf thymus DNA, 3 mmole of each of the four deoxynucleoside-5^1-triphosphates, and 1.6 μmole each of $MgCl_2$ and mercaptoethanol, in 0.05-M Tris, pH 7.8. The mixture was incubated for 20 min at 37°C, and then chilled in ice; 0.1 ml of 0.07% bovine serum albumin was added, then 1 ml of 10% trichloroacetic acid. The mixture was filtered through 11A Millipore filters, dried on planchets, and counted. Reproduced with permission from L. F. Cavalieri and E. Carroll (1970), *Proc. Nat. Acad. Sci. (USA)* **67**, 807.

been used to isolate Hela cell DNA polymerase (93), bovine pancreatic DNase, and hog spleen DNase (97). In the purification of Hela cell DNA polymerase, DNA-agarose proved to be 50% more efficient (e.g., more enzyme bound per mole of ligand) than was DNA–cellulose. The DNA–agarose beads, since they were prepared from commercially available spherical beads with defined sizes, had a much faster flow rate than did DNA cellulose. Even so, only a three-fold purification of DNA polymerase was reported.

Other types of DNA-containing bioselective adsorbents include DNA immobilized on controlled pore glass (98,99) using either the carbodiimide procedure or by fixation with glutaraldehyde. The former has been successfully employed to purify DNA polymerase from *E. coli* and was capable of separating several RNA polymerases from the same material. In addition, DNA–glass is suitable for use in high-pressure chromatography systems.

Recently a new coupling method has been reported in which carboxyl groups are introduced into single-stranded DNA by treatment with a limited quantity of 4-diazobenzoic acid. The carboxyl groups introduced in this fashion were coupled to aminoalkyl agarose by way of a water-soluble carbodiimide (51). The product was employed to purify the RNA that was complementary to the immobilized DNA. As yet, no enzyme purification has been attempted using this type of DNA–agarose.

8.6. IMMOBILIZATION OF RNA

Ribonucleic acid has been immobilized to cellulose, agarose, and similar matrices by most of the methods used to immobilize DNA (95,96,100), as well as several additional methods that take advantage of the vicinal diol present or potentially present in RNA (19, 101–108).

Probably the simplest method to couple RNA to cellulose is by ultraviolet cross-linking of physically adsorbed RNA (100). A solution of RNA in 5-mM NaCl is added to enough methylated cellulose to make a thick paste (usually about 1 g of methylated cellulose requires 4 to 5 ml of RNA solution). The mixture is air-dried and then is resuspended in absolute ethanol. The ethanolic-DNA cellulose is irradiated with a low-pressure mercury lamp (30 mesh at 15 cm). The cellulose is maintained in suspension by stirring during the irradiation, after which it is collected by suction filtration and washed with several liters of 5-mM NaCl.

Ribonucleic acid can also be easily attached to aminocellulose or aminoalkyl-agarose by first oxidizing the free RNA diol and then adding the resulting alde-hydro–RNA to the amine matrix to form a Schiff base, which is finally reduced with NaBH$_4$ to irreversibly couple the RNA to the matrix. Alternatively, the aldehydro-RNA may be simply added to a hydrazine or a hydrazide-containing matrix to form a stable hydrazone bond. (See Figure 8.10.)

In a typical preparation sodium periodate (10 mg) is added to RNA (10 mg dissolved in about 1.5 ml of ice-cold distilled water). The reaction is continued (with stirring) in an ice bath for 30 minutes, after which the reaction is terminated by adding an equal volume of 4-M NaCl–0.2-M Tricine buffer, pH 8.0, and 0.5 g of precycled and washed (2-M NaCl–0.2-M Tricine, pH 8.2) aminoethyl cellulose. The Schiff base is formed between the oxidized RNA and the amino-ethyl cellulose by stirring the mixture for 6 hours, after which sodium boro-hydride (50 mg) in 2 ml of 2-M NaCl–0.1-M Tricine, pH 8.2, is added to reduce the Schiff base to a secondary amine. Finally, the RNA–cellulose is thoroughly washed with the same buffer. Ribonucleic acid–agarose or RNA–acrylamide can be synthesized by the same method, but with the substitution of aminoalkyl-agarose or aminoalkyl–acrylamide for aminoethyl cellulose.

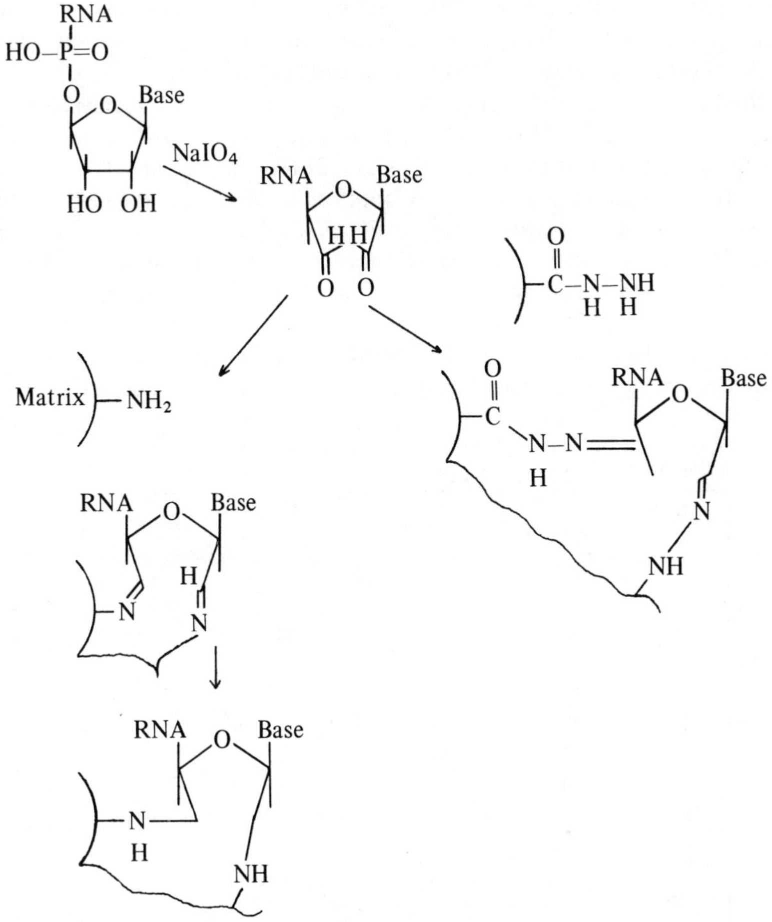

Figure 8.10. Immobilization of RNA via peroxide oxidation. The resulting aldehyde group, which is only formed on the 3′ terminus, can be either attached to a hydrazide containing matrix in a single step or be attached to an amine containing matrix by a two step process of imine formation followed by reduction.

Alternatively, the periodate-oxidized RNA can be coupled to hydrazide-agarose (105,106,109) or acrylamide (110) by mixing the agarose-hydrazide with 0.1-M sodium acetate, pH 5.0, and then mixing this with agarose-hydrazide (see page 45) suspended in the same buffer. The pH should be adjusted to 5.0 (if necessary) and the suspension stirred for 3 to 4 hours at 4°C, after which NaCl is added to a final concentration of 1 M. Stirring is continued for 30 minutes, and then the RNA–agarose is collected by suction filtration and washed with 0.2-M NaCl, followed by a final wash with distilled water. Several variations of this procedure have been suggested (106,110,111), and the various methods should be carefully considered before proceeding further.

Burrell and Horowitz (112,113) have used ribosomal RNA covalently coupled to agarose by way of a hydrazide bond as a method to purify and study ribosomal protein obtained from *E. coli*. Three ribosomal probeins, L-18, L-25, and L-5 were adsorbed on *E. coli* 5-S RNA-agarose but were not adsorbed on tRNA-agarose or 30-S-RNA–agarose. Many of the 50-S ribosomal RNAs were adsorbed on tRNA agarose, whereas protein $S3$ was the major protein of the 30-S ribosomal subunit that had an affinity for tRNA–agarose. (See Figures 8.11 and 8.12.)

Joyce and Knowles (114) have used the periodate oxidation–hydrazide method to isolate specific tRNAs rapidly and in a highly purified state. First, the

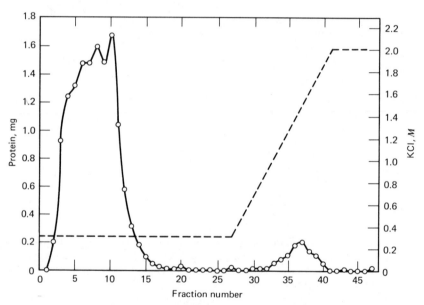

Figure 8.11. Affinity chromatography of *E. coli* 50-S ribosomal proteins on 5-S-RNA–agarose: 15 mg 50-S ribosomal proteins in 16-ml binding buffer (0.005-M potassium phosphate, pH 7.4: 0.3 M KCl; 0.02 M MgCl$_2$; 0.006-M 2-mercaptoethanol) was chromatographed on a 1.0-cm × 4.5-cm 5-S-RNA-agarose column, containing 11 mg RNA; 2.0-ml fractions were collected. The column was washed with binding buffer to remove unbound proteins. Bound proteins were eluted with a linear gradient of KCl (0.3 to 2.0 M) and EDTA (0 to 0.005 M). (– – – –) KCl concentration. Reproduced with permission from H. R. Burrell and J. Horowitz (1977), *Eur. J. Biochem.* **75**, 533.

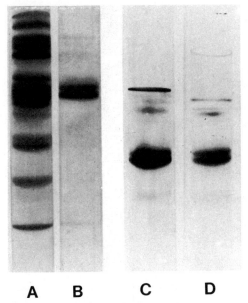

A   B   C   D

Figure 8.12. Electrophoresis of 50-S ribosomal proteins bound to 5-S-RNA coupled to agarose. (A) Total 50-S ribosomal proteins. (B) The 50-S ribosomal proteins bound to a 5-S-RNA–agarose column and eluted with high-salt-content EDTA buffer. A batch proce-dure for examining the interaction of proteins with immobilized RNA was used in experi-ments (C) and (D). (C) The 50-S ribosomal proteins bound to native 5-S-RNA (1.9 mg) coupled to agarose. (D) The 50-S ribosomal proteins bound to denatured 5-S RNA (2.1 mg) coupled to agarose. 5-S Ribonucleic acid was denatured by heating at 60°C in magnesium-free buffer. Each gel contained all the protein recovered in the high-salt-content EDTA wash. Reproduced with permission from H. R. Burrell and J. Horowitz (1977), *Eur. J. Bio-chem.* **75**, 53.

tRNA desired (e.g., valyl-tRNA) is charged with the appropriate amino acid using the cognate aminoacyl tRNA synthetase. The uncharged tRNAs are oxi-dized as described previously and passed into a column of hydrazide–agarose. If any excess oxidized RNA passes through the column, it is reduced with boro-hydride, followed by deacylation of the valyl-t-RNA. To prepare a specific ad-sorbent for valyl-tRNA synthetase, the purified tRNA is periodate oxidized and applied to hydrazide–agarose as described previously. Such specific tRNA columns can be used to obtain highly purified tRNA synthetase. Other, but similar, methods for purification of tTNA synthetase by affinity chromatog-raphy on immobilized tRNA have also been employed to yield homogeneous or nearly homogeneous enzymes (115–117). (See Figures 8.13 and 8.14.)

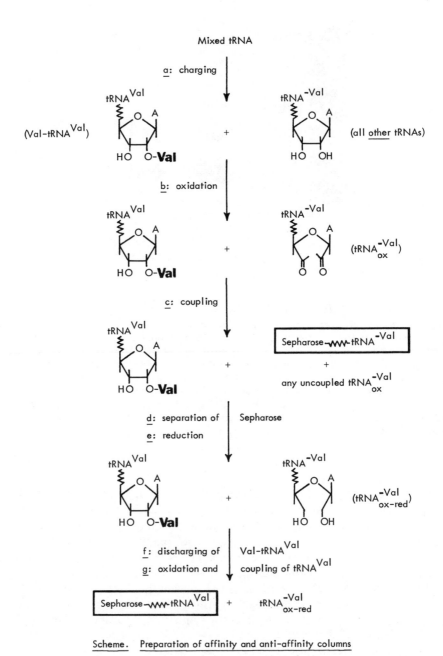

Scheme. Preparation of affinity and anti-affinity columns

Figure 8.13. Preparation of affinity and antiaffinity columns. Reproduced with permission from C. M. Joyce and J. R. Knowles (1974), *Biochem. Biophys. Res. Commun.* **60**, 1278.

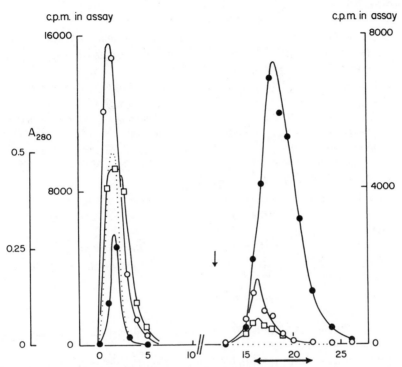

Figure 8.14. Elution profiles of tRNA columns. All columns were run using 0.05-M acetate buffer, pH 5.5, containing glycerol (10% v/v), MgCl$_2$ (10 mM), EDTA (0.1 mM), and 2-mercaptoethanol (20 mM). The enzyme sample was a partially purified mixture containing Val-, Leu-, Met-, and Tyr-tRNA synthetases from *B. stearothermophilus* (kind gift of Drs. K. Sargeant and A. Atkinson, Microbiological Research Establishment, Porton Down, Salisbury, Wiltshire, UK). After extensive washing to remove nonbound protein, the bound protein was eluted with a linear gradient (at ↓) of KCl (0 to 1 M). Fractions were assayed for Val-RS (– ● –), Leu-RS (– 0 –), and Tyr-RS (– □ –); A_{280} nm (. . . . . .). Fractions marked ←→ were pooled and concentrated. Reproduced with permission from C. M. Joyce and J. R. Knowles (1974), *Biochem. Biophys. Res. Commun.* **60**, 1278.

References

1. B. Alberts and G. Herrick (1971), in *Methods in Enzymology* (L. Grossman and K. Moldave, eds.) **21D**, 198.
2. B. M. Alberts, F. J. Amodie, M. Jenkins, E. D. Gutman, and F. J. Ferris (1968), *Cold Spring Harbor Symp. Quant. Biol.* **33**, 289.

3. B. M. Alberts (1970), *Fed. Proc., Fed. Am. Soc. Exp. Biol.,* **29**, 1154.

4. R. M. Litman (1968), *J. Biol. Chem.* **243**, *6222.*

5. R. J. Britten (1963), *Science* **142**, 963.

6. P. T. Gilham (1964), *J. Am. Chem. Soc.* **86**, 4982.

7. M. Edmonds and M. G. Caramela (1969), *J. Biol. Chem.* **244**, 1314.

8. A. J. Bendiah and E. T. Bolton (1968), in *Methods in Enzymology* (L. Grossman and K. Moldave, eds.) **12B**, 635.

9. J. DeLarco and G. Guroft (1973), *Biochem. Biophys. Res. Commun.* **50**, 486.

10. P. T. Gilman (1974), in *Immobilized Biochemicals and Affinity Chromatography* (R. B. Dunlap, ed.), Plenum, New York, pp. 173–185.

11. J. Porath and S. Hjerten (1962), in *Methods of Biochemical Analysis* (D. Glick, ed.), Vol. 9, Wiley-Interscience, New York, p. 193.

12. K. Burton (1956), *Biochem. J.* **62**, 315.

13. R. S. Feldberg and L. Grossman (1976), *Biochemistry* **15**, 2402.

14. N. V. Fedoroff and N. D. Zinder (1971), *Proc. Nat. Acad. Sci. (USA),* **68**, 1838.

15. I. Smith, H. Smith, and S. Pitko (1972), *Anal. Biochem.* **48**, 27.

16. R. Sheldon, C. Jarale, and J. Kates (1972), *Proc. Nat. Acad. Sci. (USA),* **69**, 417.

17. P. T. Gilham (1968), *Biochemistry* **7**, 2809.

18. D. Richwood (1972), *Biochim. Biophys. Acta* **269**, 47.

19. P. T. Gilham (1971), in *Methods in Enzymology* (L. Grossman and K. Moldave, eds.) **21D**, 191.

20. J. M. Gardner and C. I. Kado (1976), *Biochemistry,* **15**, 688.

21. B. M. Alberts (1967), in *Methods in Enzymology* (L. Grossman and K. Moldave, eds.) **12A**, 566.

22. A. Sen and G. J. Todaro (1978), *Proc. Nat. Acad. Sci. (USA)* **75**, 1647.

23. R. R. Burgess, A. A. Travers, J. J. Dunn, and E. K. F. Bautz (1969), *Nature (Lond.),* **221**, 43.

24. E. K. F. Bautz and J. J. Dunn (1972), in *Procedures in Nucleic Acid Research,* Vol. 2, (G. L. Cantoni and D. R. Davies, eds.), Harper, New York.

25. T. J. Guilfoyle and J. J. Jendrisak (1978), *Biochemistry* **17**, 1860.

26. T. M. Wandzilak and R. W. Benson (1978), *Biochemistry* **17**, 426.

27. D. Uyemura and I. R. Lehman (1976), *J. Biol. Chem.* **251**, 4078.

28. J. E. Naber, A. M. J. Schepman, and A. Rorsch (1966), *Biochim, Biophys. Acta* **114**, 326.

29. J. C. Kaplan, S. R. Kushner, and L. Grossman (1969), *Proc. Nat. Acad. Sci. (USA)* **63**, 144.

30. B. M. Alberts and L. Frey (1970), *Nature (Lond.)* **227**, 1313.

31. B. M. Alberts, L. Frey, and H. Delius (1972), *J. Molec. Biol.* **68**, 139.

32. R. C. Reuben and M. L. Gefter (1974), *J. Biol. Chem.* **249**, 3843.

33. J. A. Huberman, A. Kornberg, and B. M. Alberts (1971), *J. Molec. Biol.* **62**, 39.

34. W. M. Huang and J. M. Buchanan (1974), *Proc. Nat. Acad. Sci. (USA)* **71**, 2226.

35. R. B. Carroll, K. E. Neet, and D. A. Goldthwait (1972), *Proc. Nat. Acad. Sci. (USA)*, **69**, 2741.

36. R. E. Depew, L. F. Liu, and J. C. Wang (1978), *J. Biol. Chem.* **253**, 511.

37. N. Sigal, H. Delius, T. Kornberg, M. L. Gefter, and B. Alberts (1972), *Proc. Nat. Acad. Sci. (USA)*, **69**, 3537.

38. W. E. Masker and H. Eberle (1971), *Proc. Nat. Acad. Sci. (USA)*, **68**, 2549.

39. M. Abdel-Monem and H. Hoffmann-Berling (1976), *Eur. J. Biochem.*, **65**, 431.

40. Y. Hotta and H. Stern (1971), *Nature (New Biol.)* **234**, 83.

41. Y. Hotta and H. Stern (1978), *Biochemistry* **17**, 1872.

42. J. Mather and Y. Hotta (1977), *Exp. Cell. Res.* **109**, 181.

43. G. Herrick and B. Alberts (1976), *J. Biol. Chem.* **251**, 2124.

44. D. W. Mosbaugh, D. M. Stalker, G. S. Probst, and R. R. Meyer (1977), *Biochemistry* **16**, 1512.

45. D. M. Stalker, D. W. Mosbaugh, and R. R. Meyer (1975), *J. Cell. Biol.* **67**, 416a.

46. G. S. Probst, D. M. Stalker, D. W. Mosbaugh, and R. R. Meyer (1975), *Proc. Nat. Acad. Sci. (USA)* **72**, 1171.

47. M. Melchali and A. M. De Recondo (1978), *Biochem. Biophys. Res. Commun.*, **82**, 255.

48. D. J. Arndt-Jovin, T. M. Jovin, W. Bühr, A.-M. Frischauf, and M. Marquardt (1975), *Eur. J. Biochem.* **54**, 411.

49. L. E. M. Miles and C. N. Hales (1968), *Biochem. J.*, **108**, 611.

50. B. E. Noyes and G. R. Stark (1975), *Cell*, **5**, 301.

51. H. W. Dickerman, T. J. Ryan, A. I. Bass, and N. K. Chatterjee (1978), *Arch. Biochem. Biophys.* **186**, 218.

52. E. N. Moudrianakis and M. Beer (1965), *Proc. Nat. Acad. Sci. (USA)*, **53**, 564.

53. F. Jacob and J. Monod (1961), *J. Molec. Biol.* **3**, 318.

54. G. Brawerman (1974), *Annu. Rev. Biochem.* **43**, 621.

55. H. Nakazato and M. Edmonds (1972), *J. Biol. Chem.* **247**, 3365.

56. R. T. Mans and T. J. Walter (1971), *Biochim. Biophys. Acta* **247**, 113.

57. M. L. Giron and J. Huppert (1972), *Biochim. Biophys. Acta* **287**, 438.

58. M. Adesnik and J. E. Darnell (1972), *J. Molec. Biol.* **67**, 397.

59. J. R. Greenberg and R. P. Perry (1972), *J. Molec. Biol.*, **72**, 91.

60. P. T. Gilham and W. E. Robinson (1964), *J. Am. Chem. Soc.* **86**, 4985.

61. T. M. Jovin and A. Kornberg (1968), *J. Biol. Chem.* **243**, 250.

62. N. R. Cozzarelli, N. E. Melechen, T. M. Jovin, and A. Kornberg (1967), *Biochem. Biophys. Res. Commun.* **28**, 578.

63. H. Aviv and P. Leder (1972), *Proc. Nat. Acad. Sci. (USA)*, **69**, 1408.

64. R. J. Carwford and J. R. E. Wells (1978), *Biochemistry* **17**, 1591.

65. P. Venetianer and P. Leder (1974), *Proc. Nat Acad. Sci. (USA)*, **71**, 3892.

66. W. Prensky, D. M. Steffensen, and W. L. Hughes (1973), *Proc. Nat. Acad. Sci. (USA)*, **70**, 1860.

67. S. Levy and H. Aviv (1976), *Biochemistry* **15**, 1844.

68. T. G. Wood and J. B. Lingrel (1977), *J. Biol. Chem.* **252**, 457.

69. J. N. Anderson and R. T. Schmike (1976), *Cell*, **7**, 331.

70. S. L. C. Woo, J. M. Monahan, and B. W. O'Malley (1977), *J. Biol. Chem.* **252**, 457.

71. S. L. C. Woo, R. G. Smith, A. R. Means, and B. W. O'Malley (1976), *J. Biol. Chem.* **251**, 3868.

72. S. L. C. Woo, T. Chandra, A. R. Means, and B. W. O'Malley (1977), *Biochemistry* **16**, 5670.

73. R. E. Rhoades and G. M. Hellman (1978), *J. Biol. Chem.* **253**, 1687.

74. J.-P. Jost, G. Pehling, S. Panyim, and T. Ohno (1978), *Biochim. Biophys. Acta* **517**, 338.

75. K. Apel and K. Kloppstech (1978), *Eur. J. Biochem.* **85**, 581.

76. B. I. Gerwin and J. B. Milstein (1972), *Proc. Nat. Acad. Sci. (USA)*, **69**, 2599.

77. J. Kates (1970), *Cold Spring Harbor Symp. Quant. Biol.* **35**, 743.

78. U. Lindberg and T. Persson (1972), *Eur. J. Biochem.* **31**, 246.

79. U. Lindberg and T. Persson (1974), in *Methods in Enzymology* (W. B. Jakoby and M. Wilchek, eds.) **34**, 496.

80. G. Vassart, H. Brocas, P. Nokin, and J. E. Dumont (1973), *Biochim. Biophys. Acta*, **324**, 575.

81. D. Breitkreutz, L. Diaz De Leon, L. Paglia, M. Zeichner, J. Wilczek, and R. Stern (1978), *Biochim. Biophys. Acta* **517**, 349.

82. U. Lindberg, T. Persson, and L. Philpson (1972), *J. Virol.* **10**, 909.

83. J. N. Ihle, K.-L. Lee, and F. T. Kenney (1974), *J. Biol. Chem.* **249**, 38.

84. A. J. Bendich and E. T. Bolton (1968), in *Methods in Enzymology* (L. Grossman and K. Moldave, eds.) **12B**, 635.

85. H. Schaller, C. Nüsslein, F. J. Bonhoeffer, C. Kurz, and I. Neitzschmann (1972), *Eur. J. Biochem.* **26**, 474.

86. L. F. Cavalieri and E. Carroll (1970), *Proc. Nat. Acad. Sci. (USA)* **67**, 807.

87. K. Wakabayashi, S. Wang, G. Hord, and L. S. Hnilica (1973), *FEBS. Lett.,* **32**, 46.

88. E. T. Bolton and B. J. McCarthy (1962), *Proc. Nat. Acad. Sci. (USA),* **48**, 1390.

89. B. J. McCarthy and E. T. Bolton (1963), *Proc. Nat. Acad. Sci. (USA),* **50**, 156.

90. E. T. Bolton and B. J. McCarthy (1964), *J. Molec. Biol.* **8**, 201.

91. B. J. McCarthy and B. H. Hoyer (1964), *Proc. Nat. Acad. Sci. (USA),* **52**, 915.

92. D. B. Cowie and A. D. Hershey (1965), *Proc. Nat. Acad. Sci. (USA),* **53**, 57.

93. M. S. Poonian, A. J. Schlabach, and A. Weissbach (1971), *Biochemistry* **10**, 424.

94. A. Weissbach and M. S. Poonian (1974), in *Methods in Enzymology* (W. B. Jakoby and M. Wilchek, eds.) **34**, 463.

95. J. Kempf, N. Pfleger, and J. M. Egly (1978), *J. Chromatogr.* **147**, 195.

96. A. F. Wagner, R. L. Bagianesi, and T. Y. Shen (1971), *Biol. Chem. Biophys. Res. Commun.* **45**, 184.

97. J. C. Schabort (1972), *J. Chromatogr.* **73**, 253.

98. W. H. Scouten (1974), *Am. Lab.* **6**, 23.

99. L. Jervis and N. M. Pettit (1974), *J. Chromatogr.* **97**, 33.

100. L. Smith, H. Smith, and S. Pifko (1972), *Anal. Biochem.* **48**, 27.

101. D. L. Robberson and N. Davidson (1972), *Biochemistry,* **11**, 533.

102. J. Petre, A. Bollen, P. Nokin, and H. Grosjean (1972), *Biochimica,* **54**, 823.

103. D. G. Knorre, S. D. Misina, and L. C. Sandachtchiev (1964), *Izvest. Sib. Otd. Akad. Nauk SSR, Ser. Khim. Nauk* **11**, 134.

104. O. D. Nelidova and L. L. Kiselev (1968), *Molec. Biol. (USSR),* **2**, 47.

105. M. Wilchek and R. Lamed (1974), in *Methods in Enzymology* (W. K. Jakoby and M. Wilchek, eds.) **34**, 475.

106. R. Barker, I. P. Trayer, and R. L. Hill (1974), in *Methods in Enzymology* (W. B. Jakoby and M. Wilchek, eds.) **34**, 479.

107. J. C. Lee and P. T. Gilham (1966), *J. Am. Chem. Soc.* **88**, 5685.

108. J. C. Lee, H. L. Weith, and P. T. Gilham (1970), *Biochemistry* **9**, 113.

109. M. Wilchek and T. Miron (1974), in *Methods in Enzymology* (W. B. Jakoby and M. Wilchek, eds.) **34**, 72.

110. J. K. Inman (1974), in *Methods in Enzymology* (W. K. Jakoby and M. Wilchek, eds.) **34**, 30.

111. R. Lamed, Y. Levin, and M. Wilchek (1973), *Biochim. Biophys. Acta* **304**, 231.

112. H. Burrell and J. Horowitz (1975), *FEBS Lett.* **49**, 306.

113. H. R. Burrell and J. Horowitz (1977), *Eur. J. Biochem.* **75**, 533.

114. C. M. Joyce and J. R. Knowles (1974), *Biochem. Biophys. Res. Commun.* **60**, 1278.

115. P. Remy, C. Birmele, and J.-P. Ebel (1972), *FEBS Lett.* **27**, 134.

116. J. J. Befort, N. Befort, G. Petrissant, P. Remy, and J.-P Ebel (1974), *Biochemie* **56**, 625.

117. S. Bartkowiak and J. Pawelkiewicz (1972), *Biochim. Biophys. Acta* **272**, 137.

HYDROPHOBIC CHROMATOGRAPHY

Hydrophobic chromatography is a natural outgrowth from affinity chromatography. At first, though, hydrophobic interactions were merely one of the problems in affinity chromatography. Hofstee (1-4) and Shaltiel (5-10) observed that many proteins were adsorbed to the hydrophobic portions (e.g., the long spacer arms), of many bioselective adsorbents. The resulting nonspecific adsorption did, in fact, interfere with obtaining good results in some affinity purifications but could also be used to obtain purification of other proteins. Shaltiel, in particular, recognized the potential value of hydrophobic chromatography and developed it in a series of investigations (5-10) and it has been used widely since that time (11-15).

Hydrophobic chromatography can be divided into two different categories; (1) "true" hydrophobic chromatography and (2) "detergent" or amphiphilic chromatography. The first category contains those separations that depend only on the presence of an immobilized hydrophobic residue, and the second, those purification methods that depend on the presence of *both* a hydrophobic residue and a hydrophilic (usually ionic) residue immobilized on the matrix. A third category, those for which the need for an ionic residue has not been determined, is by far the largest group since the difference between hydrophobic and detergent chromatography has only recently been understood (16,17).

9.1. HYDROPHOBIC INTERACTIONS WITH PROTEINS

To understand how hydrophobic groups interact with protein, it is useful to first review the general nature of hydrophobic and hydrophilic interactions in proteins. Although there is a great variation between proteins, a general picture of an enzyme is a somewhat globular macromolecule with an internal hydrophobic core, formed to minimize the entropy of the system, surrounded by a hydrophilic outer sphere. Internally there may be some hydrophilic residues, and conversely, there may be regions of hydrophobicity on the outside surface. In addition, there are crevices of various sizes reaching into the hydrophobic core. The entire molecule is changing shape, albeit slightly, due to thermal movement of portions of the molecule in what we often term the "breathing" of the molecule. Thus the shape is dynamic and in many respects resembles the earth's

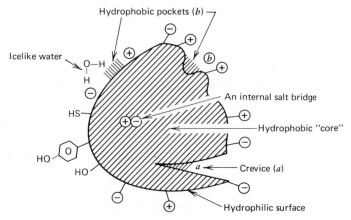

Figure 9.1. Stylized protein structure.

own shape, complete with earthquakes and eruptions that are rapidly occurring. (See Figure 9.1.)

This dynamic protein molecule would probably favor binding to hydrophobic materials if an equilibrium can be reached and often proteins bind irreversibly to such totally hydrophobic materials as polystyrene (18). The protein is also irreversibly denatured by this procedure, actually turning inside out to minimize the entropy of the system. (See Figure 9.2.)

The adsorption of proteins in this fashion, however, is often slow, and such a thermodynamically stable enzyme–matrix complex is never reached because of the high-energy barrier present in the process whereby the outer protein surface unfolds. The presence of a crevice (d in Figure 9.1) would cause such unfolding to proceed more rapidly, whereas the absence of hydrophobicity on the protein surface might cause the reaching of such a thermodynamically favored state, as shown in Figure 9.2, to take an almost infinitely long time. If it was not for this kinetically slow process, most if not all proteins would be adsorbed to any hydrophobic matrix.

Useful hydrophobic chromatography media do not have a totally hydrophobic surface but instead have hydrophobic *regions* separated by various distances from each other by a hydrophilic surface, as for example, epoxypropyl-agarose (Figure 9.3). If a protein with no hydrophobicity on the surface (P_I) is applied to the epoxypropyl agarose, no binding should occur while proteins (P_{II}) with hydrophobic regions will bind. However, since hydrophobic interactions are very weak, such proteins should bind very weakly or even be only slightly retarded by chromatography on such materials. Since hydrophobic bonding in general is increased upon heating or increasing the ionic strength, binding to a hydrophobic chromatography media would be increased by performing the chromatography at higher ionic strength.

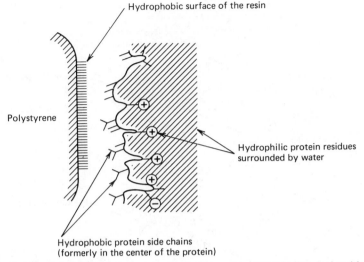

Figure 9.2. Binding of a protein to a hydrophobic matrix. The binding depicted here is so strong as to invert the protein.

However, proteins with a hydrophobic crevice surrounded by *negatively* charged residues will not bind tightly to epoxypropyl agarose, but may bind to cyanogen bromide-coupled aminopropane–agarose, since both ionic and hydrophobic forces are involved in this adsorption process. (See Figure 9.4.)

In this case we will get a composite effect. In the presence of high salt concentration the protein will bind tightly as a result of the strong hydrophobic inter-

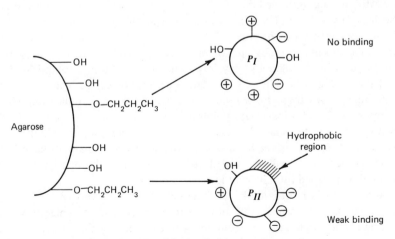

Figure 9.3. "True" hydrophobic chromatography.

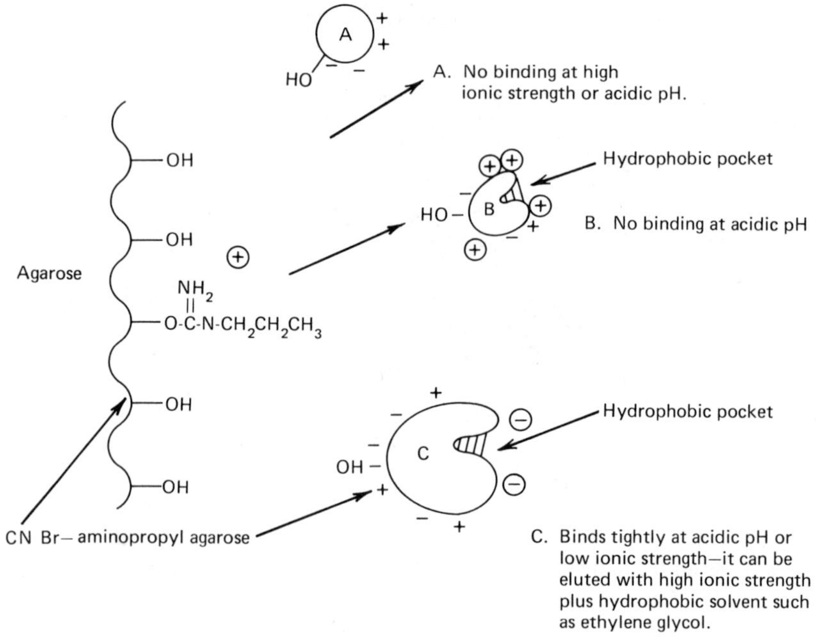

Figure 9.4. Amphiphile chromatography.

action, but at a low ionic strength, where the hydrophobic interaction is weak, the adsorption will be due to the ionic attraction between the protein and the charged matrix. This is one example of amphiphilic or "detergent" chromatography. Elution of the protein will be difficult, except at very alkaline pH values (where the protein may well be unstable) or by elution with an aqueous solution of both organic solvent plus a high salt concentration. (See Figure 9.5.)

These contrasting descriptions of "true" hydrophobic chromatography and amphiphilic chromatography demonstrate the general nature of the process. Hydrophobic chromatography may in practice be much more complicated than the elementary considerations show. Quite easily, however, it can be seen that amphiphilic chromatography will be of much more general application whereas true hydrophobic chromatography will be much more selective in use. To date only a very few proteins have been established as binding via purely hydrophobic interactions (16,17), although most certainly many proteins with large hydrophobic surfaces or crevices, needed for hydrophobic chromatography, exist and could be purified by this technique.

A third type of "hydrophobic" chromatography is "charge transfer" or "π–π" chromatography, in which the alkylhydrophobic ligand is replaced with an aromatic ligand, preferably one containing an electron withdrawing substitutent.

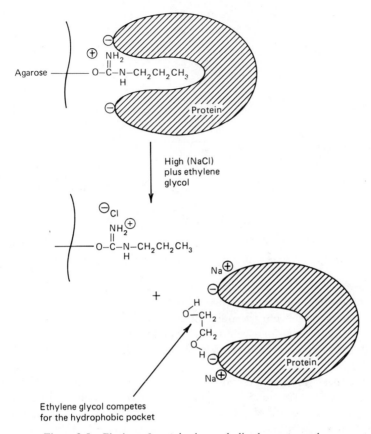

Ethylene glycol competes
for the hydrophobic pocket

Figure 9.5.   Elution of proteins in amphylic chromatography

Porath and Caldwell (19) have synthesized a dinitrophenyl-agarose particularly suited for this purpose, but the commonly employed purification of many bio-molecules on phenylagarose (20), benzoylated cellulose and agarose (21,22), and nitrophenyl glass (23) probably employ some aspects of π-π or charge-transfer interactions. (See Figure 9.6.)

9.2. NOMENCLATURE IN HYDROPHOBIC CHROMATOGRAPHY

There have been many names given to the process of hydrophobic chroma-tography, most of which were related to the specific way in which hydrophobic chromatography was employed. For example, Yon and Simmonds (24) con-trolled the degree of hydrophobicity of their resins by using systems in which

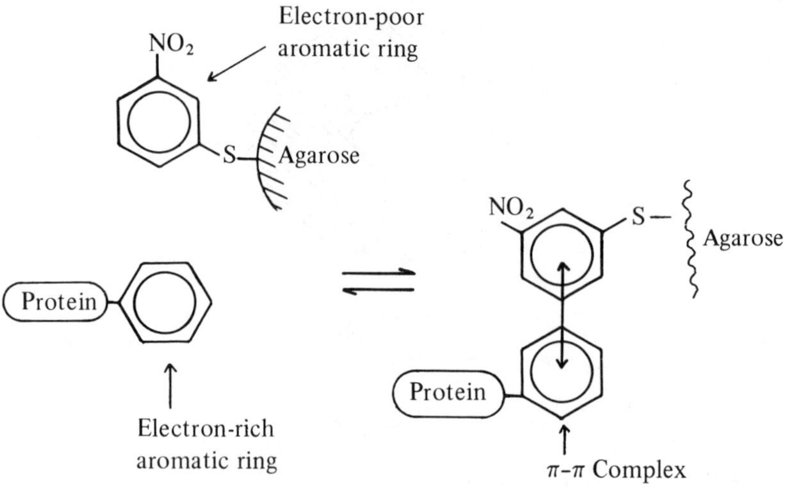

Figure 9.6. Protein–matrix interactions in charge transfer chromatography.

both the protein and the hydrophobic resin had a negative charge; thus hydrophobic interactions were the only binding forces between protein and matrix. Since the ionic repulsions were modulating the hydrophobic interactions, they termed this "repulsion-controlled hydrophobic chromatography" (20). "Phosphate"-induced chromatography (25) and "imphilyte chromatography" (26) have also been used to describe various special applications of hydrophobic chromatography. The term "hydrophobic chromatography," although it does not distinguish between these various applications or the presence or absence of additional factors (e.g., charge transfer), seems to be the best general description of the process and is used here to describe any type of chromatography in which immobilized hydrophobic groups, with or without additional interactions, are essential to the chromatographic process.

9.3. HISTORY OF HYDROPHOBIC CHROMATOGRAPHY

The first use of hydrophobic chromatography has been attributed (21) to the Swedish biochemist, Tiselius (27), who as early as 1948 noted that some proteins could be adsorbed unto cellulose and silica at high salt concentrations, especially at concentrations slightly below their normal "salting-out" range. He termed this "adsorption by salting out," although it was actually due to the slight hydrophobic nature of even such hydrophillic material as cellulose.

Probably the first practical observations of hydrophobic chromatography were noted by Shaltiel (5–10) and Hofstee (1–4) as forms of nonspecific adsorption in affinity chromatography. Fortunately, they were also aware of the possibility of harnessing this undesirable interference in bioselective adsorption as a new and powerful method of protein purification.

Shaltiel et al., in studying the various enzymes of glycogen metabolism, attempted to attach glycogen particles to agarose by way of spacer arms of four and eight carbons in length. To their surprise, the glycogen coupled by way of the eight-carbon spacer retained glycogen phosphorylase, whereas the one containing the four-carbon spacer did not even retard the enzymes (7). Using radioactive glycogen, they showed that an identical amount of glycogen had been bound to both columns. (See Figure 9.7.) This suggested that the glycogen was only slightly, if at all, involved in the retention of the enzyme, but rather, as shown in Figure 9.7, the glycogen phosphorylase interacted chiefly with the hydrophobic coating on the agarose surface. If they prepared the same derivative without any glycogen present, essentially the same results were obtained. Both ω-aminoalkyl agaroses and alkyl agaroses (see Figure 9.8) were effective adsorbents. Chromatography of crude muscle extracts on butyl-agarose yielded a purification of 100 fold in one step (9,10).

Shaltiel continued to develop this technique, purifying glycogen synthetase (5), histone binding protein (6), and glutamine synthetase (4) in rapid succession. For each of these purifications an aminoalkane or a diaminoalkane was coupled to cyanogen bromide activated agarose, which, as we know now, produces a cationic, amphiphylic, adsorbent. The influence of the ionic portion of the adsorbent can be seen from the fact that the enzymes could be eluted by using gradients of increasing salt concentration (see Figure 9.8), whereas hydrophobic interactions are reinforced by increased ionic strength. The hydrophobic involvement is also clear since matrixes with larger number of methylene groups were able to bind many enzymes with increased strength.

At about the same time, Hofstee had noted hydrophobic interactions as a source of interference in affinity chromatography that he, too, harnessed as a useful tool for protein purification (1–4). He also used hydrophobic alkyl and aryl amines attached to cyanogen bromide-activated agarose and eluted the enzymes by using salt gradients. In addition, ethylene glycol was added to the eluant, thus effecting *both* hydrophobic adsorption and hydrophobic elution, aided by ion interaction. Ovalbumin, serum albumin, α-chymotrypsin, γ-globulin, and β-lactoglobulin were chromatographed and separated on 4-phenyl butylamine and alkyl amines of various chain lengths (3,4). Using n-octylamine agarose, Hofstee immobilized xanthine oxidase, lactate dehydrogenase, DNAse I, and alkaline phosphatase, each in its active form. The adsorption of the enzymes to octylamine–agarose was irreversible, but, if n-octyl amine were replaced by n-butyl amine, the enzymes could be readily removed by using high-ionic strengths, for example, 1-M NaCl. Similar immobilization by hydrophobic adsorption has been used for a number of other enzymes (28,29).

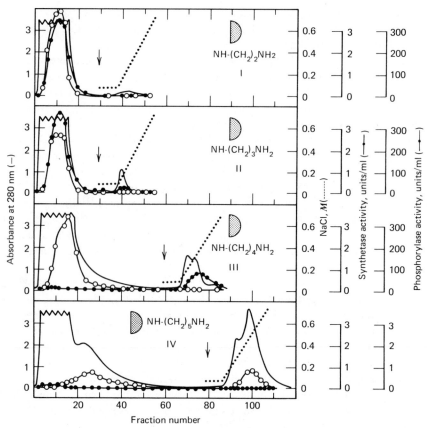

Figure 9.7. Preferential adsorption of glycogen synthetase and glycogen phosphorylase on ω-aminoalkyl–agaroses that vary in the length of their hydrocarbon chains. First, 10 ml of muscle extract (protein concentration: 26 mg/ml) was applied on each of four ω-aminoalkyl–agarose columns (8 × 0.9 cm) equilibrated at 22°C with 50-mM β-glycero-phosphate–50-mM 2-mercaptoethanol–1-mM EDTA (pH 7). Unadsorbed protein was washed off until the absorbence at 280 nm (– – –) dropped below 0.05; then a linear NaCl gradient (in the same buffer) was applied, and 1.7-ml fractions were collected and their synthetase (● – ●), as well as their phosphorylase (0 – 0), activies were monitored. The concentration of NaCl in the various fractions (. . . .) was determined by conductivity measurements, by use of a calibration curve. Reprinted with permission from Shaltiel and Er-el (5).

Hofstee also demonstrated that the cyanogen bromide-coupled alkyl amines yielded hydrophobic chromatography materials that were extremely inhomogeneous (30). As shown in Figure 9.9, if purified ovalbumin is chromatographed on *n*-butylkanine-agarose at a saturating level, the resulting chromatograph showed many protein peaks. If the same protein were chromatographed at much lower concentrations, there was only one protein peak. Since this protein

Abbreviation	Structure
Seph-C_2-NH_2	NH-CH_2-CH_2-NH_2
Seph-C_3-NH_2	NH-CH_2-CH_2-CH_2-NH_2
Seph-C_4-NH_2	NH-CH_2-CH_2-CH_2-CH_2-NH_2
Seph-C_5-NH_2	NH-CH_2-CH_2-CH_2-CH_2-CH_2-NH_2
Seph-C_6-NH_2	NH-CH_2-CH_2-CH_2-CH_2-$CH_2$$CH_2$-$NH_2$
Seph-C_8-NH_2	NH-CH_2-CH_2-CH_2-CH_2-CH_2-CH_2-CH_2-CH_2-NH_2

Figure 9.8. Structure of ω-aminoalkyl–agaroses. Reproduced with permission from Shaltiel and Er-el (5).

preparation was shown to be homogeneous by several independent methods, the only reasonable interpretation is that the column packing itself is highly heterogeneous.

This problem and many similar ones can be related to the presence of the positively charged isourea function. To minimize this, Hjerten (13,31,32) and Läas (21) coupled alkyl and aryl groups to agarose, using nonionic coupling procedures. Conversely, Wilchek (16,17, 33-36) lowered the pK of cyanogen bromide-activated alkyl amines by acylating the product or by replacing the alkyl amine with an alkyl hydrazide. Using these materials, they demonstrated that for many proteins the positively charged isourea was an essential part of the chromatographic medium. This demonstrated, for the first time, that "detergent" amphiphilic chromatography functioned in a fashion distinctly different from that of "pure" hydrophobic chromatography. Nonetheless, both are forms of hydrophobic chromatography since they each depend on hydrophobic interactions.

9.4. PROBLEMS ENCOUNTERED IN HYDROPHOBIC CHROMATOGRAPHY

Fortunately, the problems that occur in the application of hydrophobic chromatography are relatively few. Irreversible binding, usually due to the choice of a much too hydrophobic chromatography medium, is probably the

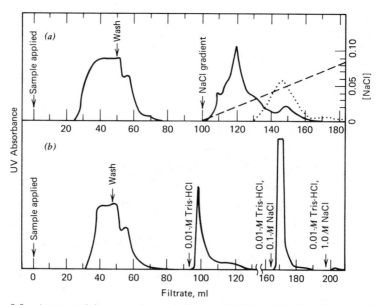

Figure 9.9. Apparent inhomogeneity and "irreversibility" of the binding of ovalbumin by n-butylamine-substituted Sepharose 4B. The protein, at a concentration of 1 mg/ml in 0.001-M Tris-HCl, pH 8, was applied to a 2-ml cooled ($5°C$) column of the adsorbent until the latter was saturated, that is, until the light absorbence ($A_{280 nm}$, solid curves) of the filtrate reached a constant value. The column then was washed with the buffer until no further protein was released. The loading, washing, and subsequent elution was done continuously by means of a peristaltic pump, and the adsorbence of the filtrate was continuously monitored. (a) The loaded and washed column was eluted with the aid of an NaCl gradient in the buffer. The dotted curve was obtained when only 2 mg of the protein (instead of a saturating amount) was applied to the column and the eluate was monitored at 225 instead of 280 nm. (b) Elution was carried out by a stepwise increase of the ionic strength. Reproduced with permission from B. H. J. Hofstee, *J. Macromol. Sci.* **10A**, 111 (1976), Courtesy Marcel Dekker, Inc.

most serious problem. If the chromatographic medium is carefully chosen through the use of small-trial separations, this problem should rarely, if ever, occur. A more common but less serious problem is the small loss of specific activity that results from the detergent effect of the adsorption–desorption process (16,21). In such cases employing noncharged hydrophobic chromatography media and/or less hydrophobic medium may result in a good purification. Unfortunately, it would appear likely that in a very few cases hydrophobic chromatography may never yield preparations with as high a specific activity as those purified by bioselective adsorption or by classical techniques, simply because of this usually minor degree of denaturation. The problems of hydrophobic chromatography are thus certainly no more significant than the problems encountered in classical procedures. Indeed, the common ion-exchange medium, DEAE–cellulose, may function as a weakly amphiphilic medium (7,30,37) in addition to its electrostatic effect.

9.5. EXAMPLES OF HYDROPHOBIC CHROMATOGRAPHY

9.5.1. Plant Aminoacyl-tRNA Synthetases

Attempts to purify aminoacyl-tRNA synthetases have been very difficult. Classical techniques, in general, have failed to yield reasonable purification schemes, although purification by way of bioselective adsorption has been reasonably successful (38–43) (see also pages 230 through 235). Jakubowski and Pawelkiewicz (44) were able to obtain an excellent purification of several aminoacyl-tRNA synthetases from yellow lupine seeds by using aminohexyl-agarose. As shown in Figure 9.10 and Table 9.1, aminohexyl agarose, but not aminobutyl–agarose, was capable of separating the synthetases from the bulk of the seed protein. In theory, successive columns of these two hydrophobic matrices should effect considerable purification and a separation of these enzymes. As shown in Table 9.1, a procedure that couples ammonium sulfate precipitation and hydrophobic chromatography gave considerable purification for many of the aminoacyl synthetases (e.g., 370 fold for the phenylalanyl enzyme), but was not particularly useful for the leucyl or tyrosyl enzyme.

TABLE 9.1. Purification of Aminoacyl-tRNA Synthetases
of Yellow Lupin Seeds

| Step | Specific Activities, nmole of Amino Acid/mg of Protein/10 min | | | | | |
	Trp	Phe	Ser	Val	Leu	Tyr
Crude extract	0.43	0.07	0.53	2.6	1.1	1.4
Ammonium sulfate fraction						
35 to 50%	0.65	0.2	–	7	0.5	–
50 to 70%	–	–	2	–	–	2
Aminohexyl–Sepharose	9	26	18.5	106	2	2

Source: Reprinted with permission from H. Jakubowski and J. Pawelkiewicz (1973), *FEBS Lett.* **34**, 150.

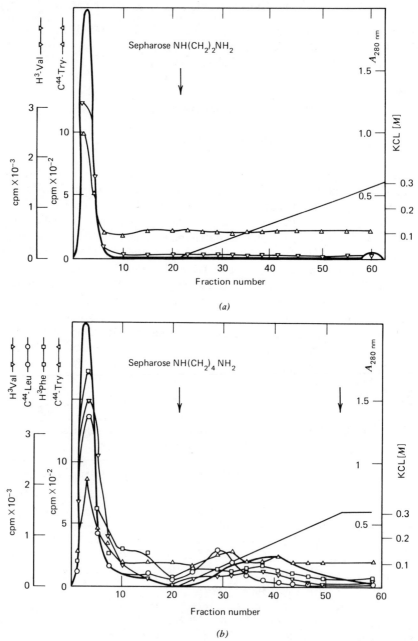

Figure 9.10. Chromatography of aminoacyl-tRNA synthetases on modified Sepharoses containing (*a*) ω-aminoethyl-, (*b*) ω-aminobutyl-, and (*c*) ω-aminohexyl groups; 60 mg of protein 30 to 50% ammonium sulfate saturation fraction in 3 ml of buffer *b* was applied on each of three modified Sepharoses. Reproduced with permission from H. Jakubowski and J. Pawelkiewicz (1973), *FEBS Lett.* **34**, 150.

252

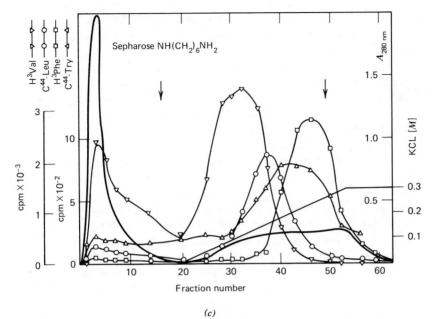

(c)

Figure 9.10. (Continued)

Note that the enzymes were eluted from this column with a linear salt gradient. This chromatographic medium would be expected to behave as an ion-exchange medium due to the two charged on its surface. However, both the aminobutyl and aminohexyl materials could be expected to have basically the same ion-exchange properties. Actually, with the use of aminobutylagarose or an aminohexyl agarose, deaminated by treatment with nitrous acid, separation as good as that shown in Figure 9.10 was obtained. These results establish firmly that hydrophobic residues are needed for this purification, although it is not clearly established whether "pure" hydrophobic factors or an amphiphilic chromatographic process was responsible for the results. The fact that the proteins eluted with increasing ionic strength strongly favors the latter.

9.5.2. Yeast α-Isopropylmalate Isomerase

Bigelis and Umbarger (45) employed hydrophobic chromatography with the use of valyl agarose and leucyl agarose in the presence of high ionic strengths to purify this very labile enzyme. α-Isopropylmalate (IPM) isomerase is stabilized by both high ionic strength and by glycerol. High ionic strength increases hydrophobic interactions between protein and alkyl amino acid-substituted agarose, whereas glycerol decreases the hydrophobic bonding. Using a combination of these two mutually antagonistic effects of these stabilizers, Bigelis and Umbarger (45) were able to purify the enzyme in such a fashion that it was always in the presence of one or both of these stabilizing agents, glycerol or high $(NH_4)_2SO_4$ concentration. (See Table 9.2.)

The resulting purification, consisting of an ammonium sulfate precipitation with subsequent chromatography on leucyl and valyl agarose, resulted in a purification of 200 fold with a yield of almost 30%. Figure 9.11 shows the elution of the enzyme from a substantial amount of protein contamination, followed by the subsequent elution of fumarase that was copurified in this system. Electrophoresis of the resulting protein showed a single protein band that was eluted from the polyacrylamide gel and shown to contain all the α-isopropylmalate isomerase.

Note here that elution of the enzyme was effected by using decreasing ionic strength or an increasing glycerol concentration or both (Figure 9.12), implying that hydrophobic interactions were more important than the ionic factors, a fact that might be attributed to the almost zero net charge on valyl or leucyl agarose prepared by the cyanogen bromide method. (See Figure 9.13.)

Hydrophobic chromatography in the presence of high-ionic-strength buffers has been of general use in several such enzyme preparations (4,46–48) and may

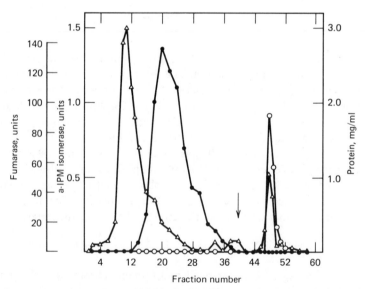

Figure 9.11. Chromatography of a 50 to 65% $(NH_4)_2SO_4$ fraction obtained from yeast strain S288cα on more highly substituted valine-Sepharose. The eluting buffer was potassium phosphate containing 1.24 M $(NH_4)_2SO_4$. The arrow indicates the point at which the buffer was changed to one consisting of potassium phosphate, 30% glycerol. Column dimensions were 2 cm × 11.5 cm △ – △, protein; ● – ●, α-IPM isomerase; 0 – 0, fumarase. Reproduced with permission from Bigelis and Umbarger (14).

TABLE 9.2. Purification of α-IPM Isomerase

Procedure	Volume, ml	Total Protein, mg	Specific Activity, units/mg	Total Activity, units	Yield, %	Purification, fold
Crude extract	136.0	2285	0.031	70.84	100.0	1.0
0 to 50% (NH₄)₂SO₄ supernatant	149.0	1013	0.068	68.88	97.2	2.2
50 to 65% (NH₄)₂SO₄ fraction	4.0	160	0.355	56.80	80.2	11.5
Valine–Sepharose	65.0	40	1.02	40.51	57.2	32.8
Leucine–Sepharose concentrate	1.0	3.4	6.17	20.98	29.6	199.0

Source: Reprinted with permission from Bigelis and Umbarger (14).

255

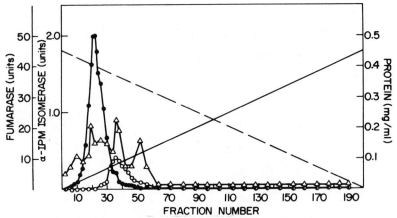

Figure 9.12. Chromatography of fractions, obtained from the valine–Sepharose step, on leucine Sepharose. Protein was adsorbed in the potassium phosphate 1.24-M $(NH_4)_2SO_4$ buffer and eluted with a buffer of increasing glycerol (——) and decreasing $(NH_4)_2SO_4$ (– – –) concentrations: $\triangle$ – $\triangle$, protein, ● – ●, α-IPM isomerase; 0 – – – 0, fumarase. Reproduced with permission from Bigelis and Umbarger (14).

well find further application in isolating labile enzymes since high salt concentrations, with the use of lyotropic salts such as $(NH_4)_2SO_4$, stabilizes many otherwise labile proteins.

9.5.3. Purification of Detergent-Solubilized Membrane Proteins

Hydrophobic chromatography would appear to be a very promising method for the purification of membrane proteins that, in general, have large hydrophobic regions. Simmonds and Yon (49,50) employed an amphiphilic resin for the purification of the sodium dodecyl sulfate (SDS)-solubilized proteins from human erythrocyte membrane with excellent results. They used N-(3-carboxypropionyl)aminalkyl–agarose (CPAA–agarose) prepared from cyanogen bromide-activated agarose to yield a zwitterionic structure very similar to the valyl and leucyl agarose prepared by Bigelis and Umbarger (45). The length of the hydro-

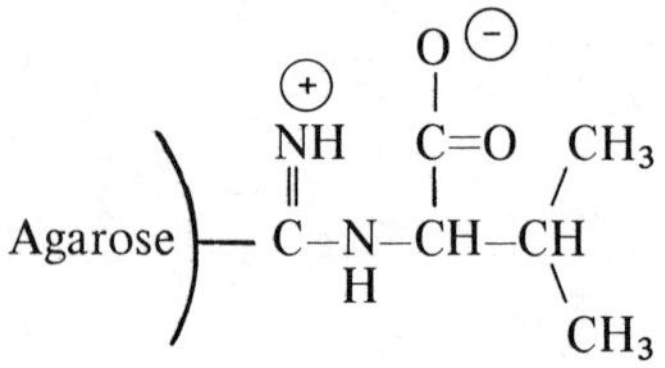

$$
\text{Agarose}
\begin{array}{c}
\qquad\qquad\quad\overset{\ominus}{O} \\
\qquad\qquad\overset{\oplus}{\quad}\;\; | \\
\qquad NH \quad C=O \quad CH_3 \\
\qquad \| \qquad | \qquad / \\
C-N-CH-CH \\
\qquad H \qquad\quad \backslash \\
\qquad\qquad\qquad\quad CH_3
\end{array}
$$

Figure 9.13. Zwitterionic form of valyl-agarose.

phobic portion of the chromatographic medium determines the usefulness of the chromatographic medium.

Solubilized membrane proteins are best separated on N-(3-carboxypropyl) aminodocyl–agarose or its aminohexyl and aminoacetyle analogues. Proteins were eluted by changing the pH, which, in turn, changes the hydrophobicity of the CPAA–agarose matrix. Glycophorin was purified in 85% or higher purity with yields of 89 to 100% of the applied glycophorin by using this technique. (See Figure 9.14.)

Liljas et al. (51) have also used hydrophobic chromatography to isolate glyco-phorin by using a two-step (hydroxylapatite and phenyl–agarose chromatog-raphy) procedure. First, membrane proteins solubilized with Tween 20 were applied to a hydroxylappatite column in a 0.005-M potassium phosphate, pH 6.8 buffer. The column was washed with the same buffer, and then the glycophorin was eluted with 0.07-M potassium phosphate, pH 6.8. This eluate was made up to 0.3-M potassium phosphate, pH 6.8, and applied to a pentyl–agarose column that was prepared by the epoxide coupling procedure. The column was then

TABLE 9.3. Titratable Groups in CPAD–Sepharose and Homologous Adsorbents

The majority of active groups in these adsorbents are assumed to have the general structure

$$\text{---O--C--NH--[CH}_2]_n\text{--NH--CO--[CH}_2]_2\text{--CO}_2\text{--}$$
$$\overset{\parallel}{\underset{+}{\text{NH}_2}}$$

where the carbohydrate matrix is indicated by hatching and the charge state at pH 7 is shown. Carboxyl groups are taken as the sum of all groups dissociating between pH 3.0 and 6.0. Cationic (isourea and other) groups are taken as the sum of all groups dissociating between pH 7.5 and 10.5. (See Table 9.3.)

	Titratable Groups μmole/ml of Settled Gel	
n	Carboxyl	Cationic
4	5.6	7.3
6	6.6	9.0
8	5.0	7.1
10	6.0	7.1
12	6.3	9.1

Source: Reprinted with permission from Simmonds and Yon (49).

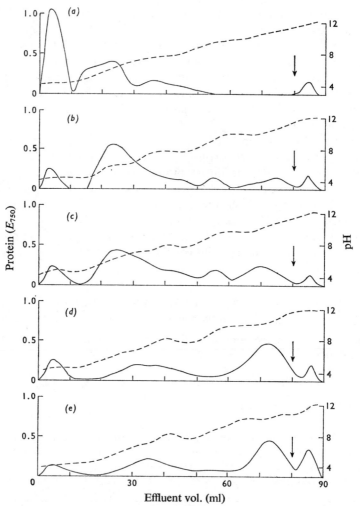

Figure 9.14. Elution of solubilized membrane proteins by increasing pH: comparison of homologous adsorbents. Identical columns (bed volume 3-ml diameter 0.8 cm) of the adsobent materials listed in Table 9.3 were equilibrated with buffer containing 0.1% (w/v) SDS. The hydrocarbon chain length (C_4) of the adsorbent groups was C_4 (*a*), C_6 (*b*), C_8 (*c*), C_{10} (*d*), C_{12} (*e*). A sample of solubilized membranes containing 4 mg of protein was applied to each column. The columns were then developed with 20 ml of buffer 1, followed by successive 10-ml portions of buffers of increasing pH and finally (where indicated by arrow) with 10 ml of buffer containing 7% (v/v) butan-1-ol. All buffers contained 0.1% (w/v) sodium dodecyl sulfate. Effluent fractions were assayed for protein (- - -) and pH (– – – –). The flow rate was approximately 1 ml/min, and the temperature was 22°C. Reproduced with permission from Simmonds and Yon (49).

developed with a lower-ionic-strength buffer (0.005-M potassium phosphate, pH 6.8), and the glyophorin eluted as a sharp peak. The purity was about the same as the protein isolated by Simmonds and Yon (50), but gave a much higher yield. These two isolations of the same protein using gels with either "pure" hydrophobic or "detergent" chromatography characteristics is a good illustration of the variety of ways in which hydrophobic interactions can be used in protein purification.

9.5.4. High-Pressure Hydrophobic Chromatography of Protein

Hydrophobic chromatography, or "reversed-phase" chromatography, as it is termed by the analytical chemist, can be applied to the separation of proteins and polypeptides by using high-pressure, high-performance liquid chromatographic (HPCL) systems. Mönch and Dehnen (52,53), Frei et al. (54), Hancock et al. (55), and Radhakrishnan et al. (56) have recently adapted commercial programmed liquid chromatographic systems for separation of these biomolecules. The best results have been seen when hydrophilic (e.g., phosphate) or hydrophobic (e.g., an alkyl sulfonic acid) ion-pairing reagents are added to the eluant. Excellent separation of as little as 10 μg of protein is seen, and the area under each peak is proportional to the amount of protein applied from 10 to 50 μg. This technique is normally employed as an analytical method, although it could be scaled up to preparative levels for the isolation of small amounts of very pure protein.

The best results were obtained with octadecylsilane-coated silica particles (Nucleosil 10 C18, Macherey and Nagel, Düren, CFR) at 6000 psi and 4°C. Elution of the proteins was performed by using a linear gradient of 0.05-M KH_2PO_4 (pH 2.0)-2 methoxyethanol (95:5) and isoproponol:2 methoxy-

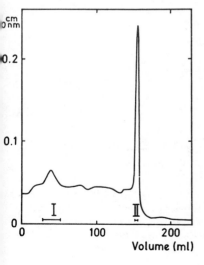

Figure 9.15. Pentyl-Sepharose chromatography of fraction II from hydroxyapatite chromatography. Sample: 25 ml ($A_{280}^{1cm} = 0.1$) in 0.3-M potassium phosphate buffer (pH 6.8) of the concentration: 0.3 M and 0.005 M. The high absorption in the beginning of the chromatogram is due to adsorbing material in the 0.3-M buffer. Reproduced with permission from L. Liljas, P. Lundahl, and S. Hjerten (1976), *Biochim. Biophys. Acta* **426**, 526.

methanol (95:5) adjusted to pH 2.0 with H_3PO_4. Unfortunately, very few proteins remain in active form under these conditions.

It would seem that development of HPLC systems for proteins under non-denaturing conditions would offer significant advantage. For example, eluants could be assayed for enzyme activity and so on. The zwitterion or "imphylyte" system of Yon and Simmonds (49,50) would seem worth attempting since elution can be done with the use of a purely aqueous solvent and a pH gradient in the physiological range. At any rate, HPLC will be a very valuable analytical, and possible preparative, tool if it can be adapted to less drastic elution conditions. (See Figures 9.16 and 9.17.)

9.6. HYDROPHOBIC-ACTIVE SITE CONTAINING PROTEINS

Hydrophobic interactions are, normally, undesirable if an investigator wishes to purify an enzyme with the use of affinity chromatography. The one major exception occurs when the active site itself is hydrophobic. In such cases it is not always possible to separate hydrophobic and "bioselective" interactions, and they may, in fact, both occur in a complementary fashion. Thus in the purification of lactate dehydrogenase with the use of affinity chromatography, hydrophobic interactions that appear only as an interference in the process

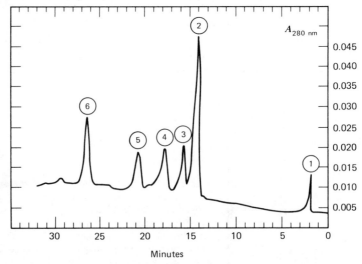

Figure 9.16. Separation of proteins by HPLC on reversed phase, with peaks (1) impurity, (2) insulin, (3) cytochrome c, (4) bovine albumin, (5) catalase, and (6) ovabumin. Reproduced with permission from W. Mönch and W. Dehnen (1978), *J. Chromatogr.* **147**, 415.

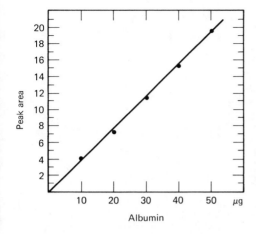

Figure 9.17. Relationship between amount of bovine albumin injected and peak area in planimeter units during separation of proteins on reversed phase. Reproduced with permission from W. Mönch and W. Dehnen, *J. Chromatogr.* 147, 415.

are minimized by adding ethylene glycol to the elution buffer (57). Conversely, the active site of alcohol dehydrogenase from many sources possesses a hydrophobic active site as demonstrated by the increased affinity of the enzyme for increasingly hydrophobic alcohols. Schöpp et al. (58) utilized 10-carboxydecyl–agarose prepared by binding 10-aminodecanoic acid to cyanogen bromide-activated agarose and employed this zwitterionic materix to purify alcohol dehydrogenase from baker's yeast and from *Acinetobacter calcoaceticus*. The crude enzyme was applied in a high ionic strength buffer and was eluted bioselectively by adding ethanol to the buffer. Alternatively, the enzyme could be eluted by lowering the pH and ionic strength (see Figure 9.18).

Massoulié and Bon (59) have demonstrated in an eloquent series of experiments that the active site of acetylcholinesterase isolated from *Electrophorus* is similarly hydrophobic. Many bioselective adsorbents have been made for the purification of this enzyme (see Table 9.4), and each one possesses considerable hydrophobicity (60–70).

Massoulié and Bon (59) synthesized a number of hydrophobic matrices with and without the substrate quaternary nitrogen. Hexyl amidocarboxyphenyl dimethyl ammonium, with a hydrophobic hexyl spacer arm, proved to be a much better affinity ligand than the same compound coupled to a hydrophobic spermine spacer arm.

Further, N-hexylacetamido–agarose, prepared by the cyanogen bromide method, was a very useful hydrophobic chromatographic medium of the "detergent" type, as evidenced by the fact that the enzyme could be eluted by use of either increased ionic strength or by elution with the enzyme inhibitor, tetraethyl ammonium iodide (see Figure 9.19).

There are a large number of other enzymes with hydrophobic active sites where hydrophobic ligands must be employed. For example, lipoamide dehydro-

TABLE 9.4. Some Affinity Systems Which Have Been Described for Acetylcholinesterase Purification

References	Structure of Spacer Arm and Ligand	Enzyme	Wash Buffer	Elution Medium
i, j, k	$-NH-(CH_2)_5-CO-NH-\langle\bigcirc\rangle-\overset{+}{N}(CH_3)_3$	*Electrophorus* trypsin-treated lytic form (G)	0.1 M NaCl, 0.01 M phosphate, pH 7	0.01 M decamethonium bromide, 0.1 M NaCl, 0.01 M phosphate, pH 7
g, h	$-NH-(CH_2)_5-CO-NH-(CH_2)_5-CO-NH-\langle\bigcirc\rangle-\overset{+}{N}(CH_3)_3$	*Electrophorus* autolyzed lytic form (G)	0.02 M phosphate pH 6.9	salt gradient (0.1–0.15 M NaCl or KCl)
b	$-[NH-(CH_2)_3-NH-(CH_2)_3-NH-CO-(CH_2)_2-CO]_2-NH-\langle\bigcirc\rangle-\overset{+}{N}(CH_3)_3$	*Electrophorus*	0.1 M NaCl, 0.04 m MgCl₂ pH 7.8 (CO_3HNa)	0.01 M tensilon, 0.1 M NaCl, 0.04 M MgCl₂, pH 7.8 (CO_3HNa)
c	$-[NH-(CH_2)_3-NH-(CH_2)_3-NH-CO-(CH_2)_2-CO]_2-NH-\langle\bigcirc\rangle-\overset{+}{N}(CH_3)_3$	bovine erythrocyte	0.01 M phosphate, pH 7.5	0.01 M tensilon, 0.1 M NaCl, 0.01 M NaH₂PO₄, pH 7.5
d	$-[NH-(CH_2)_3-NH-(CH_2)_3-NH-CO-(CH_2)_2-CO]_2-NH-\langle\bigcirc\rangle-\overset{+}{N}(CH_3)_3$	bovine brain solubilized without detergent	0.1 M NaCl, 0.03 M phosphate, pH 8	0.01 M tensilon, 0.1 M NaCl, 0.03 M phosphate, pH 8
e	$-NH-(CH_2)_4-NH-CO-\langle\bigcirc\rangle-\overset{+}{N}(CH_3)_3$ and $-NH-(CH_2)_4-NH-CO-\langle\bigcirc\rangle-\overset{+}{N}(CH_3)_3$	guinea-pig brain; extracts contained 0.7% Triton X-100	0.5 M NaCl, 0.01 M phosphate, pH 7.4 or 0.2 M NaCl, 0.01 M phosphate, pH 7.4	salt or choline chloride gradient (≈0.4 M NaCl) 0 to 0.5 M choline chloride gradient

262

m, k					
		Electrophorus asymmetric forms	1 M NaCl, 0.01 M phosphate, pH 7.0	0.02 M decamethonium, 1 M NaCl, 0.01 M phosphate, pH 7.0	
f	$-[NH-(CH_2)_6-NH-CO-(CH_2)_2-CO]_2-NH-(CH_2)_6-\overset{+}{N}-CH_3$ (with CH₃, CH₃)	*Torpedo*	0.5 M NaCl, 0.02 M phosphate, pH 7	0.2 M tetramethylammonium bromide, 0.5 M NaCl, 0.02 M phosphate, pH 7.0	

Source: Reprinted with permission from Massoulie and Bon (1976). *Eur. J. Biochem.* **68**, 531

[a] For the Berman and Young column (*a*), a very extended spacer arm (27 atoms) was thought to be essential for binding, since with one only half as long acetylcholinesterase was not retained (*l*). This seems at variance with other systems but may be due to the ionic strength used, which demanded a greater affinity. Better results were noted with *meta* than *para* isomers. The Yamamura *et al.* (*d*) 'm quat' column was rather similar to that described here except for the length of the aliphatic chain, and this fits well with its lower affinity, as judged from the elution conditions used. In some cases incomplete elution was noted (*m, k, g, c*).

[b] J. D. Berman and M. Young (1971), *Proc. Natl. Acad. Sci. U.S.A.* **68**, 395–398.

[c] J. D. Berman (1973) *Biochemistry* **12**, 1710–1715.

[d] S. L. Chan, D. Y. Shirachi, H. N. Bhargava, E. Gardner, and A. J. Trevor (1972), *J. Neruochem.* **19**, 2747–2758.

[e] H. I. Yamamura, D. W. Reichard, T. L. Gardner, J. D. Morrisett, and C. A. Broomfield (1973), *Biochem. Biophys. Acta* **302**, 305–315.

[f] W. H. Hopff, G. Riggio, P. G. Waser (1973), FEBS Lett. **35**, 220–222.

[g] T. L. Rosenberry, Hai Won Chang and Y. T. Chen (1972), *J. Biol. Chem.* **247**, 1555–1565.

[h] Y. T. Chen, T. L. Rosenberry, and Hai Won Chang (1974), *Arch. Biochem. Biophys.* **161**, 479–487.

[i] N. Kalderon, I. Silman, S. Blumberg, and Y. Dudai (1970), *Biochem. Biophys. Acta* **207**, 560–562.

[j] Y. Dudai, J. Silman, N. Kalderon, and S. Blumberg (1972), *Biochem. Biophys. Acta* **268**, 138–157.

[k] U. Dudai, I. Silman, M. Shinitzky, and S. Blumberg (1972), *Proc. Natl. Acad. Sci. U.S.A.* **69**, 2400–2403.

[l] J. P. Changeux (1966), *J. Mol. Pharmacol.* **2**, 369–392.

[m] S. T. Christian and R. Janetzko (1971), *Arch. Biochem. Biophys.* **145**, 169–178.

263

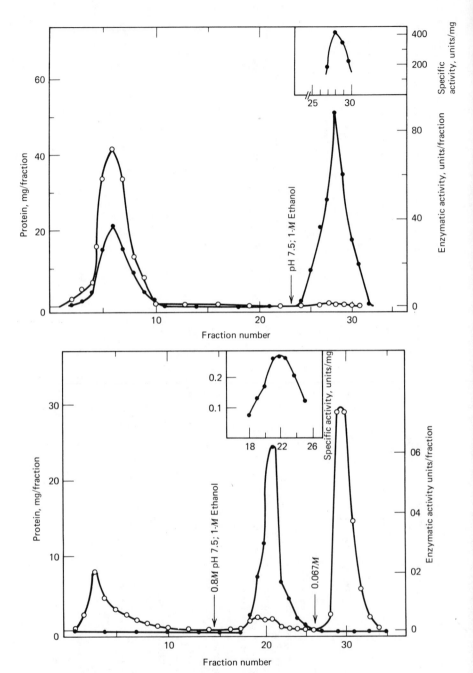

Figure 9.18. Purification of alcohol dehydrogenase from crude enzyme extracts of *Saccharomyces cerevisiae* (*a*) and *Acinetobacter calcoaceticus* (*b*) on 10-carboxydecyl-Sepharose. The extracts were applied to a column (7 cm × 4.6 cm) containing 8 ml of 10-carboxydecyl–Sepharose. Elution was done at 4°C with a flow rate of 6.7 ml/min. The

genase (71–73) has been purified using both lipoamide bound to porous glass and aminoalkyl agarose. Similarly, enzymes dealing with fatty acids, steroids, and the like would all be expected to have a hydrophobic region at the active site and possibly elsewhere on the enzyme surface. An investigator could, therefore, reasonably expect to be able to purify such enzymes by using hydrophobic chromatography.

9.7. DESIGN AND APPLICATION OF HYDROPHOBIC CHROMATOGRAPHY SYSTEMS

These examples can give a general idea as to how a hydrophobic chromatography system is designed. First, does the protein to be purified contain a hydrophobic active site, or is there reason to expect that it is hydrophobic? If not, is there a particular reason for using hydrophobic chromatography? For example, is the protein particularly labile in the absence of salt and/or hydrophobic "stabilizers" such as glycerol? If the answer to these questions is "yes," you may well find the use of hydrophobic chromatography to be useful.

Having decided to use hydrophobic chromatography, you need to next consider the types of hydrophobic media available. There are (1) charged "detergents" bound to agarose, for example, hexylaminoagarose prepared from CNBr–agarose, (2) "imphylytes," hydrophobic zwitterionic media whose charge can be varied from slightly + to slightly – with small pH charges; (3) "purely" hydrophobic media prepared, for example, by coupling alkyl epoxides to agarose or similar media. Finally, you could decide to try agarose or cellulose alone, at high ionic strength, in the fashion of Tiselius's "adsorption by salting out" (27). If your enzyme is very labile to traces of detergent, the first method, detergent chromatography, is obviously risky. On the other hand, if the enzyme is solubilized by detergents or stabilized with detergents, this could be the easiest of the three methods since there are many types of aminoalkyl–agarose commercially available. Conversely, "pure" hydrophobic chromatography, properly designed, would probably be the first method to try if the enzyme is very

---

volume of collected fractions was 3.5 ml or in presence of 1 M ethanol, 2.8 ml (●). Enzymatic activity (0) protein concentration; NAD′-dependent alcohol dehydrogenase from *Saccharomyces cerevisiae*. The extract (2 ml) contained 140 mg of protein in 0.8-M K_2HPO_4/KH_2PO_4, pH 8.5, containing 0.4-M glycinol. Elution was done with a buffer pH 7.5 of the same composition containing 1-M ethanol. The fractions were concentrated before using in electrophoresis; NADP′-dependent alcohol dehydrogenase from *Acinetobacter calcoaceticus*; The extract (2 ml) contained 100 mg of protein in 1-M K_2HPO_4/KH_2PO_4, pH 8.5. Elution was done with 0.8-M K_2HPO_4/KH_2PO_4, pH 7.5, containing 1-M ethanol. Columns were regenerated by washing with 0.1-M NaOH in 80% ethanol. Reproduced with permission from W. Schopp, M. Grunow, T. Tauchert, and H. Aurich (1976), *FEBS Lett.* **68**, 198.

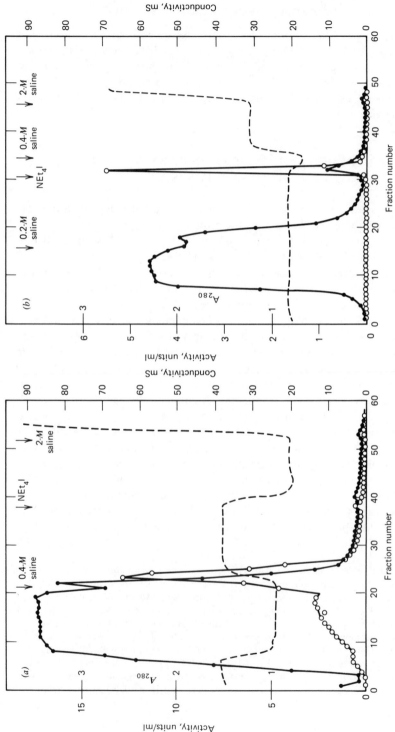

Figure 9.19. Chromatography of acetylcholineesterase on acetamidohexyl-Sepharose. (*a*) Elution with 0.4-*M* saline buffer. The enzyme was and electrophoresis extract (100 ml, 4 units/µl) equilibrated in 0.2-*M* saline buffer. Flow rate 2 ml/hour. (*b*) Elution with inhibitor (tetraethylammonium, NEt₄I). The enzyme (G form) (50 ml, 0.9 unit µl) was applied in 0.2-*M* saline buffer. The flow rate was 50 ml/hour. Reproduced with permission from J. Massoulié and S. Bon (1976), *Eur. J. Biochem.* **68**, 531.

266

labile. Octyl and phenyl–agaroses are commercially available, and other alkyl agaroses can be readily prepared (21,31,32).

Zwitterionic or "imphylyte" hydrophobic chromatography can be thought of as intermediate between these. Whereas "pure" hydrophobic media may bind some proteins irreversibly and "detergent" media may denature other proteins, a zwitterionic medium is, at its isoelectric point, very hydrophobic and yet, because of the equilibrium nature of the proton-exchange procedure, never as hydrophobic as a similar alkylagarose. Further, elution of most proteins will require only very small changes in pH, making it a very useful general technique for proteins that are stable over a very narrow pH range. Moreover, although zwitterionic media are not commercially available, they are easily prepared by coupling the aliphatic α-amino acids or ω-aminocarboxylic acids to cyanogen biomide-activated agarose.

Once the type of hydrophobic medium has been chosen, you next would prepare a series of increasingly hydrophobic matrixes. A kit of alkyl amino agaroses from C_2 to C_{10} is commercially available (Miles, Elkhart, Indiana).

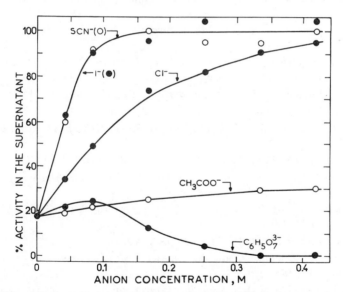

Figure 9.20. Influence of neutral salts on the adsorption, 100 μl of a suspension of propyl-agarose, having adsorbed 0.24 units of β-galactosidase per millileter, were added to 0.5 ml of buffer C supplemented with 0.1-M NaCl and different concentrations of netural salts (pH adjusted to 7.1, if necessary). All the neutral salts employed were potassium salts; only the anion is indicated in the diagram. After 15 minutes of incubation at room temperature, the supernatant obtained by filtration was assayed for enzyme activity. The activities are indicated in percentage of the adsorbed activity. Reproduced with permission from O. A. Hogberg-Raibaud, and M. Goldberg (1975), *FEBS Lett.* **50**, 130.

Small samples of enzyme would be applied to minicolumns of each matrix, and the percent of the enzyme bound would be plotted against carbon length. Application of the sample would be done in the buffer best suited for binding (i.e., low ionic strength for detergent chromatography, isoelectric pH for zwitterionic chromatography, and high ionic strength for hydrophobic chromatography). The least hydrophobic material to retain the enzyme would be the one to employ in the subsequent enzyme purification procedure (7).

It would be useful, perhaps, for us to note once again that the chromatography *system* is much more than the chromatography medium alone. In this case the type of buffer is of extreme importance. In most hydrophobic chromatographic procedures the usefulness of a buffer is related to its lyotropic effect on water. Thus ammonium sulfate, which increases the structure of water, is an excellent component of the enzyme application buffer, whereas thiocyanate could only be used if the enzyme were, in fact, stable in the presence of thiocyanate, an ion that often denatures proteins. The influence of various ions on protein stability and on hydrophobic chromatography have been well studied (74,75). In Figure 9.20 you can see the effect of various neutral salts on the absorption of β-galactosidase on a cyanogen bromide-coupled propylamino agarose column.

References

1. B. H. J. Hofstee (1974), in *Immobilized Biochemicals* and *Affinity Chromatography,* Vol. 42 (R. B. Dunlap, ed.), Plenum, New York, p. 43.

2. B. H. J. Hofstee (1973), *Biochem. Biophys. Res. Commun.* **53**, 1137.

3. B. H. J. Hofstee (1973), *Anal. Biochem.* **52**, 430.

4. B. H. J. Hofstee and N. F. Otillio (1973), *Biochem. Biophys. Res. Commun.* **50**, 751.

5. S. Shaltiel and Z. Er-el (1973), *Proc. Nat. Acad. Sci (USA)* **70**, 778.

6. S. Shaltiel, G. F. Ames, and K. D. Noel (1973), *Arch. Biochem. Biophys.* **159**, 174.

7. S. Shaltiel (1974), in *Methods in Enzymology* (M. Wilchek and W. B. Jakoby, eds.) **34**, 126.

8. Z. Er-el and S. Shaltiel (1974), *FEBS Lett.* **40**, 142.

9. Z. Er-el, Y. Zaidenzaig, and S. Shaltiel (1972), *Biochem. Biophys. Res. Commun.* **49**, 383.

10. S. Shaltiel (1973), in *Metabolic Interconversion of Enzymes* (E. H. Fischer, E. G. Krebs, H. Neurath, and E. R. Stadtman, eds.) Springer-Verlag, Berlin, p. 379.

11. R. Yon (1974), *Biochem. J.* **137**, 127.

12. G. J. Doellgast and W. H. Fishman (1974), *Biochem. J.* **141**, 103.

13. S. Hjertén, J. Rosengren, and S. Påhlman (1974), *J. Chromatogr.* **101**, 281.

14. R. Bigelis and H. E. Umbarger (1974), *J. Biol. Chem.* **250**, 4315.

15. J. Visser, W. Van Dongen, and M. Strating (1978), *FEBS Lett.* **85**, 81.

16. R. Jost, T. Miron, and M. Wilchek (1974), *Biochim. Biophys. Acta* **362**, 75.

17. M. Wilchek and T. Miron (1976), *Biochem. Biophys. Res. Commun.* **72**, 108.

18. H. D. Brown and S. K. Chattopadhyay (1976), in *Methods in Enzymology* (K. Mosbach, ed.) **44**, 288.

19. J. Porath and K. D. Caldwell (1977), *J. Chromatogr.* **133**, 180.

20. J. Porath, L. Sundberg, N. Fornstedt, and I. Olsson (1973), *Nature* **245**, 465.

21. T. Läas (1975), *J. Chromatogr.* **111**, 373.

22. M. Fridkin, P. J. Cashion, K. L. Agarwal, E. Jay, and H. G. Khorana (1974), in *Methods in Enzymology* (M. Wilchek and W. B. Jakoby, eds.) **34**, 645.

23. H. H. Weetall and A. M. Filbert (1924), in *Methods in Enzymology* (W. B. Jakoby and M. Wilchek, eds.) **34**, 59.

24. R. S. Yon and R. S. Simmons (1975), *Biochem. J.* **151**, 281.

25. R. A. Rimerman and G. W. Hutfield (1973), *Science* **182**, 1268.

26. R. S. Yon (1977), *Biochem. J.* **161**, 223.

27. A. Tiselius (1948), *Ark. Kemi Min. Geol.* **B26**, 1.

28. H. L. Brockman, J. H. Law, and F. K. Kézdy (1973), *J. Biol. Chem.* **248**, 4965.

29. J. Monreal (1976), *Biochim. Biophys. Acta* **427**, 15.

30. B. H. S. Hoffstee (1976), *J. Macromol. Sci.* **10A**, 34.

31. S. Hjerten (1973), *J. Chromatogr.* **87**, 325.

32. S. Hjerten (1973), *Ninth Internat. Congr. Biochem.*, Abstracts **9**, 11.

33. M. Wilchek (1977), *J. Macromol. Sci.* **10A**, 15.

34. M. Wilchek and T. Miron (1974), in *Methods in Enzymology* (W. B. Jakoby and M. Wilchek, eds.) **34**, 72.

35. M. Wilchek and R. Lamco (1974), in *Methods in Enzymology* (W. B. Jakoby and M. Wilchek, eds.) **34**, 475.

36. R. Lamed, L. Levin, and M. Wilchek (1973), *Biochim. Biophys. Acta* **304**, 231.

37. B. H. J. Hofstee, *Polym. Preprints,* in press.

38. H. Beikirch, F. v.d. Haas, and F. Cramer (1972), *Eur. J. Biochem.* **26**, 182.

39. M. Robert-Gero and J. P. Waller (1972), *Eur. J. Biochem.* **31**, 315.

40. P. J. Forrester and R. C. Hancoch (1973), *Can. J. Biochem.* **51**, 231.

41. D. D. Nelidova and L. L. Kiselev (1968), *Molec. Biol.* **2**, 60.

42. S. Bartkowiak and J. Pawelkiewicz (1972), *Biochim. Biophys. Acta* **272**, 137.

43. P. Remy, C. Birmele, and J. P. Ebel (1972), *FEBS Lett.* **27**, 134.

44. H. Jakubowski and J. Pawelkiewicz (1973), *FEBS Lett.* **34**, 150.
45. R. Bigelis and H. E. Umbarger (1975), *J. Biol. Chem.* **250**, 4315.
46. G. J. Doellgast and W. H. Fishman (1974), *Biochem. J.* **141**, 103.
47. G. J. Doellgast and G. B. Kohlaw (1972), *Fed. Proc.* **31**, 424A.
48. R. A. Rimerman and G. W. Hatfield (1973), *Science* **182**, 1268.
49. R. J. Simmonds and R. J. Yon (1976), *Biochem. J.* **157**, 153.
50. R. J. Simmonds and R. J. Yon (1977), *Biochem. J.* **163**, 397.
51. L. Liljas, P. Lundahl, and S. Hjerten (1976), *Biochim. Biophys. Acta* **426**, 526.
52. W. Mönch and W. Dehnen (1978), *J. Chromatogr.* **147**, 415.
53. W. Mönch and W. Dehnen (1977), *J. Chromatogr.* **140**, 260.
54. R. W. Frei, L. Michel, and W. Santi (1976), *J. Chromatogr.* **126**, 665.
55. W. S. Hancock, C. A. Bishop, R. L. Prestidge, D. R. K. Harding, and M. T. W. Hearn (1978), *Science* **200**, 1168.
56. A. N. Radhakrishnan, S. Stein, A. Licht, K. A. Gruber, and S. Udenfriend (1977), *J. Chromatogr.* **132**, 552.
57. C. R. Lowe and K. Mosbach (1974), *Eur. J. Biochem.* **52**, 99.
58. W. Schöpp, M. Grunow, H. Tauchert, and H. Aurich (1976), *FEBS Lett.* **68**, 198.
59. J. Massoulié and S. Bon (1976), *Eur. J. Biochem.* **68**, 531.
60. Y. T. Chen, T. L. Rosenberry, and H. W. Chang (1974), *Arch. Biochem. Biophys.* **161**, 479.
61. S. T. Christian and R. Janetako (1971), *Arch. Biochem. Biophys.* **145**, 169.
62. U. Dudai, I. Silman, M. Shinitaky, and S. Blumberg (1972), *Proc. Nat. Acad. Sci. (USA)* **69**, 2400.
63. N. Kalderon, I. Silman, S. Blumberg, and Y. Dudai (1970), *Biochim. Biophys. Acta* **207**, 560.
64. Y. Dudai, J. Silman, N. Kalderon, and S. Blumberg (1972), *Biochim. Biophys. Acta* **268**, 138.
65. T. L. Rosenberry, H. W. Chang, and Y. T. Chen (1972), *J. Biol. Chem.* **247**, 1555.
66. J. D. Berman and M. Young (1971), *Proc. Nat. Acad. Sci. (USA)* **68**, 395.
67. J. D. Berman (1973), *Biochemistry* **12**, 1710.
68. S. L. Chan, D. Y. Shirachi, H. N. Bhargava, E. Gardner, and A. J. Trevor (1972), *J. Neurochem.* **19**, 2747.
69. H. I. Yamamura, D. W. Reichard, T. L. Gardner, J. D. Morrisett, and C. A. Broomfield (1973), *Biochem. Biophys. Acta* **302**, 305.
70. W. H. Hopff, G. Riggio, and P. G. Waser (1973), *FEBS Lett.* **35**, 220.
71. W. H. Scouten, F. Torok, and W. Gitomer (1973), *Biochim. Biophys. Acta* **309**, 521.

72. W. H. Scouten (1974), in *Methods in Enzymology* (W. B. Jakoby and M. Wilchek, eds.) **34**, 288.

73. J. Visser and M. Strating (1975), *Biochim. Biophys. Acta* **384**, 69.

74. O. Raibaud, A. Högberg-Raibaud, and M. E. Goldberg (1975), *FEBS Lett.* **50**, 130.

75. W. Melander and C. Horvath (1977), *Arch. Biochem. Biophys.* **183**, 200.

ANTIBODY–ANTIGEN AND OTHER PROTEIN–PROTEIN INTERACTIONS IN AFFINITY CHROMATOGRAPHY

Chromatography using immobilized antibodies and antigens has developed into one of the most widely applied techniques of immunology. This type of affinity chromatography is often termed *immunosorbent chromatography* to distinguish it from other applications of affinity chromatography. In practice, however, it is really a form of affinity chromatography in which the immobilized ligand is a protein (e.g., lectins (see page 147) or protease inhibitors). In fact, it is easier to use immobilized proteins as ligands than most other types of bioselective adsorbents since the problems of ionic and hydrophilic adsorption and of ligand leakage is minimal when proteins are employed as ligands.

10.1. IMMUNOSORBENTS AS BIOSELECTIVE ADSORBENTS

As early as 1951 Campbell et al. (1) employed antibodies bound to cellulose as bioselective adsorbents. Many of the cellulose–protein coupling procedures now used in immobilized enzyme technology were developed at this time in search for new ways of preparing immunosorbents (2). Later cellulose was replaced by other carriers, for example, agarose, porous glass, and acrylamide. The number of publications in this area have increased exponentially in the past few years. Immunsorbent chromatography specifically has been the subject of several major reviews (3–7).

The basis of immunological adsorbents is the ability of antibodies to react very selectively with specific antigens creating a very strong antibody–antigen complex. This complex in ordinary circumstances forms a precipitate:

$$\text{Antigen + antibody} \rightleftharpoons \text{antigen-antibody complex} \downarrow$$

Classically, a suitable animal (e.g., rabbit or mouse) is injected with a few milligrams of antigen, usually a protein, with repeated injections over several weeks. The injection may include an "adjuvant" (some foreign material; e.g., killed bacteria plus an emulsification agent), the inclusion of which augments the rate of formation of antibodies to the antigen. After antibodies have been formed,

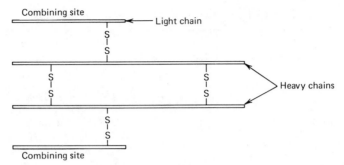

Figure 10.1. Structure of an antibody molecule. Reproduced with permission from C. R. Lowe and P. G. D. Dean (1974), *Affinity Chromatography.* Copyright © John Wiley and Sons, 1974.

blood is withdrawn from the animal periodically, and the red blood cells are removed from the plasma by way of clotting. The amount of antibody, called its *titer,* is determined by testing the resulting serum for its ability to form a precipitate with added antigen. The serum is serially diluted, and the dilution with the lowest concentration of serum that causes precipitate formation is the titer. Thus serum with a 1:2 titer is very low in antibody, whereas a 1:128 titer is much higher. Details of the procedure can be found in any basic text in immunology (e.g., 8–10).

Separation of the antibody from the serum proteins and from other antibodies can be an arduous procedure. Commonly, the antibody is precipitated with added antigen, and the precipitate is collected and dissociated by way of dilution, denaturing agents, and/or other pH extremes. After dissociation the antigen and antibody are separated by physical (10) or enzymatic (11) methods. These methods suffer from rather low yields and low purity of the desired antibody. (See Table 10.1.)

Purification of antibodies by chromatography on immobilized antigens is, in many instances, able to give better yields and/or higher purity. Often yields of over 80% are attained, and essentially homogeneous immunoglobulin may result (12–24). Nevertheless, considerable practical difficulty is observed in immunosorbent chromatography, chiefly because the antibody–antigen complex is so tightly associated that very rigorous conditions are needed for its dissociation. Under conditions where elution is possible, protein denaturation also is likely to occur. Several new techniques for releasing antibodies from immobilized antigens promise to give uniformly higher yields.

For antibody purification, the antigen is immobilized on an inert matrix by one of a number of possible techniques (see Chapter 3). Cyanogen bromide-

TABLE 10.1. Some Eluants for Immunoadsorbents

Hapten	Eluant
	Acids (pH 3)
Insulin	0.01 to 1-M HCl
Insulin	1-M Acetic acid, pH 2.0
Anti-DNP-amino acids	20% Formic acid
	Glycine–HCl, pH 2.8
Ragweed allergens	Glycine–H_2SO_4, pH 2.7
	1-M Propionic acid
	Salts
IgG	Sodium thiocyanate
	2.5-M NaI, pH 7.5
	2.8-M $MgCl_2$
Semliki Forest virus	Carbonate/bicarbonate, pH 11
Serum albumin	1% NaCl (pH 2.0) 37°C
	Protein Denaturants
Human chorionic somatomammotropin	6-M Guanidine–HCl, pH 3.1
β-Lipoprotein	4-M Urea
Galactosidase	8-M Urea
	Haptens
p-Aminophenyl-β-lactoside	0.5-M Lactose
DNP-Lysine	DNP, DNP-lysine
	Others
Human complement, Cl	0.2-M 1,4-Diaminobutane
IgG	Distilled water

Source: Reproduced with permission from C. R. Lowe and P. G. D. Dean, *Affinity Chromatography.* Copyright © John Wiley & Sons, 1974.

activated agarose (24–26) and activated cellulose derivatives (2, 27–35) have been widely employed. Serum is then passed through the column of immobilized antigen, extraneous proteins are washed out of the column with neutral buffer, and finally the antibodies are eluted with a deforming agent. Since the antibody molecule is quite stable (relative to antigens) in denaturing buffers, relatively good yields of antibody are normally obtained.

10.2. HAPTEN ANTIBODIES

A special class of antibodies can be prepared by coupling small organic molecules to proteins and injecting the protein conjugates into experimental animals. The antibodies that result are capable of binding specifically to these small molecules, which are termed *haptens.* Hapten antibodies have proved to be very useful in purifying possible haptens such as dansylated peptides, hormones, monosaccharides, and allergens. Unfortunately, antobodies to haptens are often heterogeneous and contaminated with immunoglobulins to other haptens and/or proteins. Thus affinity chromatography on immobilized haptens is very useful in preparing pure hapten antibodies (36).

Most, if not all, antibodies are heterogeneous, because each antigen has several portions that are recognized by the antibody-forming cells in the body. Figure 10.2 depicts a protein with a covalently bound hapten. The hapten can swing around the protein backbone such that amino acids R_1, R_2, and R_3 may form part of the "antigenic determinant" of the hapten; that is, R_1 and/or R_2 and/or R_3 may, along with the hapten, form part of the site recognized by the cell. Some of the hapten antibodies will recognize only the hapten, others R_1 + hapten; R_2 + hapten, R_1 + R_2 + hapten, and so forth, thus the hapten antibodies will be heterogeneous.

Heterogeneity of antiprotein antibodies is even more readily explained. A protein consisting of only one polypeptide chain may have several areas capable

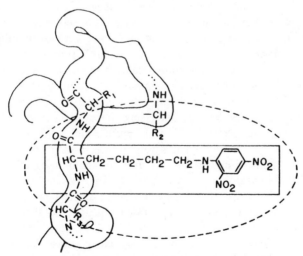

Figure 10.2. Protein with a covalently bound hapten. Reproduced with permission from T. Cooper (1977), *Tools of Biochemistry.* Copyright © John Wiley and Sons, 1977.

of reacting with antibody-forming lyphocytes and eluting the forming of antibodies specific for that region of the protein. Proteins with nonidentical subunits have several such regions, or antigenic determinants, on each peptide subunit. Each antibody to these several antigenic determinants will bind to the protein (or hapten) with a different K_a value. One of the most valuable uses of immunosorbent chromatography is the separation of antibodies with different binding constants.

10.3   EXAMPLES OF ANTIBODY PURIFICATION
BY BIOSELECTIVE ADSORPTION

To date over 100 papers have appeared that deal with the purification of antibodies on immobilized antigen (37). In many instances the purified antibody is, in turn, immobilized and employed as an immunosorbent for the purification of its antigen. In this fashion a small amount of partially purified antigen can be employed to make a much larger amount of antibody that, in turn, can be used repetitively to purify an extremely large amount of very pure antigen.

A very useful example of this procedure is the purification of bovine prothrombin using goat antisera (38). Bovine prothrombin is a major protein involved in blood clotting and thus has been the subject of many investigations using classical purification methods. These were tedious procedures with yields often as low as 15%. To improve upon this, Wallin and Prydz (38) prepared a very small amount of prothrombin by these classical techniques. They used this protein to inject a goat (chosen for maximal yield of serum) twice with 4 mg of prothrombin in Freund's adjuvant. The injections were given subcutaneously 1 month apart. A week after the second injection, 300 ml of blood was drawn from the goat, and the immunoglobulin fraction was separated from it through ammonium sulfate precipitation. The ammonium sulfate fraction was dialyzed, and a portion of it was chromatographed on a column of agarose to which 9 mg of pure prothrombin has been bound using the cyanogen bromide activation method. The column was washed with buffer, and antibodies against prothrombin, which had bound quantatively to the prothrombin-agarose, were eluted with $2\text{-}M$ NaBr–$5\text{-}mM$ EDTA. By using the column repetitively, 178 mg of immunoglobulin was obtained with a total initial expenditure of only 17 mg of the original prothrombin that had been purified by classical methods.

The resulting immunoglobulin was subsequently immobilized to cyanogen bromide activated agarose (80 g of agarose for 178 mg of antibody) and employed to purify more prothrombin in a two-step procedure. First, the prothrombin and other vitamin K-dependent factors from bovine blood were partially purified by adsorption on barium sulfate. [This was a necessary first step since chromatography of crude serum on the antiprothrombin immunosorbent resulted in unspecific adsorption of globulin on the column (39).] Next the proteins adsorbed to the barium sulfate were desorbed with acetate buffer and applied to the antiprothrombin–agarose. After thoroughly washing the column, the prothrombin was eluted with 2-M NaBr-5-mM EDTA. The result (see Figure 10.3) was a 74% yield of electrophoretically homogeneous prothrombin in a procedure that required only 1 day for completion. The yield of 16 mg of prothrombin from 175 ml of plasma could be obtained repeatedly by using the same antiprothrombin column. Further, repeated bleeding of the same goat, followed by subsequent isolation and immobilization of more antiprothrombin globulin, would be easily accomplished, thus permitting even larger-scale isolation.

The importance of having such rapid and reproducible isolation methods should not be overlooked. Often during classical isolation procedures the proteins that are isolated are partially proteolyzed or altered in some fashion. For example, Sjöström et al. (40) and Svensson et al. (41) have prepared two amphilic, highly hydrophobic, peptidases from small intestinal membranes using an immunosorbent approach. The same peptidases, prepared by classical tech-

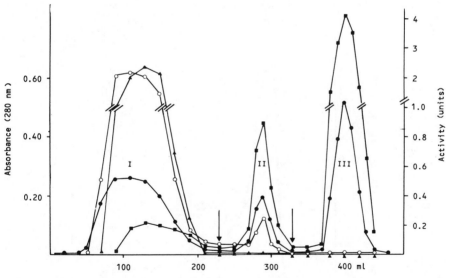

Figure 10.3. Affinity chromatography of barium sulfate eluate on antiprothrombin immunoglobulin–Sepharose column. The column was eluted with 0.05-M Tris-HCl, pH 7.4, containing 5-mM EDTA (step I), 0.5-M NaHCO$_3$ (step II), and 2-M NaBr containing 5-mM EDTA (step III). Changes in elution fluid are indicated by arrows. (o – o – o) Factor X (units/ml); (▲ – ▲ – ▲). Factor VII (units/ml); (■ – ■ – ■). Factor II (units/ml); (● – ● – ●). Optical density 280 nm. Reproduced with permission from R. Wallin and H. Prydz (1975). *FEBS Lett.* **51**, 191.

niques, had very different properties that were demonstrated to be due to proteolytic degradation. These enzymes were prepared by the same two-step procedure, first preparing and purifying antibodies to enzyme isolated by using classical purification techniques, and then using the antibodies to prepare an immunosorbent. Using the immunosorbent, highly purified enzyme could be prepared in a 20% yield, with 1.5 to 3.5 mg total homogeneous peptidase per isolation. The purification was about 500 fold and could be repeated at least 30 times by use of the same imminosorbent column. (See Figure 10.4 and Table 10.2.)

To purify the dipeptidyl peptidase IV, Svensson et al. (41) employed the observation that the dipeptidyl peptidase from pig kidney (which could be purified by classical techniques) was immunologically identical to the intestinal enzyme. The latter, however, could not be easily isolated and always contained at least three active fractions. These were thought to arise in the intestine by proteolytic digestion of a precursor peptidase identical to the renal enzyme. To prove this, these investigators used the renal dipeptidase to immunize several rabbits. Each rabbit received four biweekly injections of 100 μg of the renal enzyme. One week after the last injection 40 ml of blood was removed from

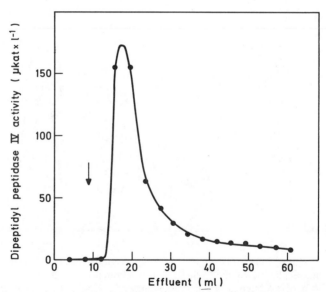

Figure 10.4. Immunoadsorption chromatography of Triton-X-100-solubilized intestinal microvillus fraction on Sepharose conjugated to antibodies against renal dipeptidyl peptidase IV. The arrow indicates the change to 2-m*M* Tris-HCl, pH 8.0, 0.1% in Triton X-100. Reproduced with permission from B. Svensson, H. Sjöström, M. Danielsen, M. Staun, L. Jeppensen, and O. Noren (1978), *Eur. J. Biochem.* 90, 489.

TABLE 10.2. Purification of Dipeptidyl Peptidase IV from 300 g of Intestine

Fraction	Total Activity, μkat	Specific Activity, mkat/kg	Recovery, %	Purification, -Fold
Homogenate	7.2	0.80	100	1
Solubilized microvillus fraction	2.8	8.6	39	11
Purified enzyme	1.6	420	22	525

Source: Reprinted with permission from: B. Svensson, H. Sjöström, M. Danielsen, M. Staun, L. Jeppensen, and O. Noren (1978), *Eur. J. Biochem.* **90**, 489.

each rabbit, and the antisera was fractionated by ammonium sulfate precipitation and ion-exchange chromatography. The immunoglobulin fraction was immobilized to cyanogen bromide-activated agarose to form an immunosorbent for dipeptidyl protease.

To prepare a cell extract from small intestinal membranes that had not been exposed to pancreatic proteases, the pancreatic tubes of several rats were ligated 3 days prior to sacrifice of the animals. After sacrifice, the intestinal membranes were prepared and solubilized with a nonionic detergent. The solubilized fraction was applied to the immunosorbent in a 0.15-M Tris buffer, pH 8.0, containing 0.15-M NaCl. After thorough washing of the column, the pure enzyme could be eluted by lowering the ionic strength of the buffer to 2-mM Tris–HCl, pH 8.0. It is not known why lowering of the ionic strength of the buffer eluted the protein, but it probably is related to the hydrophobic nature of the enzyme. If so, the method might be applicable to many membrane proteins. (Attempts to elute the enzyme with deforming buffers resulted in only inactive protein being eluted.) Using this procedure, a single very hydrophobic peptidase was obtained. On treatment with trypsin a hydrophillic form resulted.

Cuatrecasas (42), in an early application of immunosorbents, used agarose to immobilize porcine insulin, which, in turn, was employed to purify antiinsulin antibodies from guinea pig serum. As can be seen in Figure 10.5, antiinsulin antibodies were only eluted under drastic conditions (1-M HCl) and could be separated into at least four groups of antibodies with different affinities for insulin agarose.

More recently, Sairam and Porath (43) have prepared antibodies to ovine, rat, and human protein hormones by using hormones bound to divinylsulfonyl agarose as an adsorbent for rabbit antisera. Elution was effected with 0.5- or 1.0-M ammonium hydroxide, giving antibodies in high yield. Again, the antibodies could be fractionated into groups with various affinities for the aspective antigen.

Among the other antibodies purified by immunosorbent chromatography are mRNA (44), tRNA (45), and the "Y" base of certain tRNA molecules (46). The antiviral protein interferon has been the subject of several antibody investigations (47,48) and antibodies to brain proteins have been investigated (49–51) recently. The latter are of considerable interest since, under certain pathological circumstances, the body appears to make antibodies to its own neural proteins, often with severe consequences. Among the more recent studies in this area, Reuger et al. (51) have purified the brain-specific ostroglial protein, glial fibrial-acidic protein (GFA), from both human and bovine sources. The immunosorbent step was essential in obtaining highly purified GFA, free from contamination with denatured GFA and from tubulin, both of which are major contaminants in this protein obtained by classical methods. Again, the specificity and speed of isolation by immunosorbent chromatography were responsible for

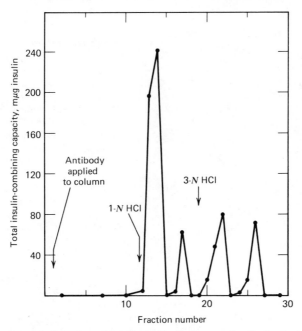

Figure 10.5. Pattern of elution of insulin antibody from an insulin-lys-Sepharose column using strong acid solutions. The sample applied was 3 ml of a 1:100 dilution of antiinsulin serum, containing a total insulin-binding capacity of 467 mμg of insulin. The total amount of antibody recovered was capable of combining with 484 mμg of insulin. Reproduced with permission from P. Cuatrecasas (1969), *Biochem. Biophys. Res. Commun.* **35**, 531.

the successful application (See Figure 10.6.) Moreover, the possibilities of using antibodies against an antigen from one source to obtain the same antigen from a different source are seen in the fact that the antibodies against GFA from human sources could be immobilized to yield an immunosorbent for GFA from either bovine or human white matter. Yields as high as 2 mg/10 ml of immunosorbent were obtained by eluting the column with 1-*M* acetic acid, 5-*M* urea, and 0.8-*M* NaCl at pH 2.8. Since GFA is stable for a short time under these conditions, very little problem with denatured GFA was seen.

10.4. ELUTION METHODS FOR IMMUNOSORBENTS

Until recently, only very stable antigens could be purified by immunosorbent chromatography, since elution required drastic conditions. As antibodies are more stable, they could usually be purified on immobilized antigens either by

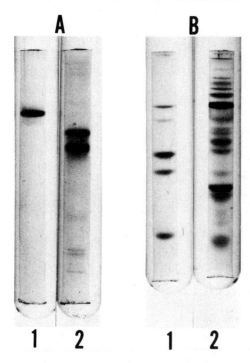

Figure 10.6. (*A*) Comparison of GFA protein (gel 1) isolated by immunoaffinity chromatography with tubulin prepared from calf brain by two cycles of the polymerization procedure on Tris/glycerine–SDS/urea gel electrophoresis. Separation of the proteins was unsatisfactory on phosphate–SDS gel electrophoresis. (*B*) Cyanogen bromide peptides of GFA protein isolated by immunoaffinity chromatography on SDS/urea gel electrophoresis. Gel 1 is an incomplete digest of myoglobin used as the standard. From top to bottom: myoglobin (17,000), peptides I and II (14,900), peptide I (8270), peptide II (6420), and peptide III (2550) Gel 2 is a digest of GFA protein. The bands above ˈ the peptide in the myoglobin range are due to aggregation and not to incomplete cleavage since they were not seen on SDS gel electrophoresis. Reproduced with permission from Reuger et al. (51).

elution with deforming buffers or with elution buffers containing relatively high free antigen concentrations. Various elution conditions have been attempted, for example, chaotropic ions (52) or the application of basic instead of acidic buffers (52). These were only moderately successful. Grenot and Cuilleron (53) and workers in the laboratory of P. D. G. Dean (54–59) have recently introduced an electrophoretic desorption method that may well prove valuable, not only in immunosorbent chromatography, but in other forms of bioselective adsorption. Because of the obvious importance of this newly emerging technique, I shall attempt to describe it in detail.

In concept, the technique depends on the fact that there is always an equilibrium between bound and free protein:

$$\text{Matrix–antibody + antigen}$$
$$\Big\downarrow\Big\uparrow$$
$$\text{[Matrix–antibody - antigen complex]}$$

If the free antigen is rapidly and steadily pulled down the column, it may be eluted before rebinding to the matrix bound antibody. More likely, the antigen will reform a complex several times before it elutes from the gel.

In practice, the electrophoretic elution is done in a special, but simply constructed, electrophoretic cell in which there are discrete chambers for the adsorbent, the desorbed protein, and the electrophoresis buffers. Figure 10.7 depicts one such cell as designed by Iobal et al. (60) for the purification of sex hormone binding globulin and corticosteroid binding globin. In this study, instead of an antibody matrix, immobilized steroids were used as the bioselective adsorbent. The basic problem, however, was the same in that steroid binding proteins cannot be eluted from immobilized hormones, except under the same very rigorous conditions usually employed with immunosorbents.

For the purification of the sex binding hormone proteins, 5 ml of postpartum plasma was stirred for 1 hour with androstandiol–agarose. The gel was washed with 0.05-M Tris, pH 7.4, containing 5-mM CaCl$_2$ and 1-M NaCl, until no protein was detected in the wash. After washing the gel was transferred to the electrophoresis cell, which was then filled with 50-mM Tris-HCl, pH 7.4, and electrophoresis was carried out at 10°C and 10 mA, 110 V. After 5 hours the elution chamber was emptied and its contents analyzed. Essentially homogeneous binding protein was obtained with over a thousandfold purification. (See Table 10.3.)

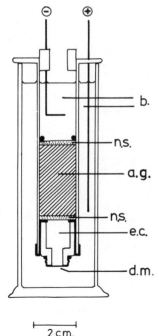

2 cm.

Figure 10.7. Diagram of electrophoresis cell, b., buffer compartment: n.s., porous nylon support; a.g. affinity gel; e.c., elution chamber; d.m., dialysis membrane. Reproduced with permission from M. J. Iobal, P. Ford, and M. W. Johnson (1978), *(FEBS Lett.* 87, 235.

TABLE 10.3. Comparison of Binding Data and Protein Concentrations for Original Plasma and Electrophoretically Desorbed Protein

	Protein Concentration, mg/ml	Binding Capacity, nmole/ml	Total Binding Capacity, nmole/mg	Specific Activity, nmol/mg	ka L/M	Purification Factor	Recovery, %
Original plasma (15 ml)	119	0.238	3.57	0.002	0.68×10^9	–	–
Elution	0.330	0.738 (5 hours)[a]	2.80	2.24	0.47×10^7	1120	78.4
chamber fluid (3.8 ml)	0.330	0.695 (7 hours)	2.64	2.11	0.41×10^7	1053	74.0

Source: Reprinted with permission from Iobal et al. (60).

[a] Represents individual electrophoretic runs; the time for each run is given in parenthesis. Amperage = 10 mA; voltage = 110 V. Affinity gel treated with plasma was washed with 1 l of buffer. Number of DHT binding sites per moles of SHBG for 5-hour run = 0.18 and for 7-hour run = 0.17 (calculated by using a molecular weight of 80,000 for SHBG).

It would seem that ordinary gel electrophoresis apparatus might well be suitable for small-scale purifications by using a method previously employed for concentrating protein bands from several acrylamide gels (61). First, a dialysis membrane is attached to the bottom of a gel tube by a piece of tubing. Next, buffer containing 40 to 60% sucrose is added to the tube to yield the desired elution volume. On top of this is added a very narrow (ca. 1 cm) band of acrylamide gel polymerization mixture (usually a 7% gel), which is then overlayered with water and allowed to polymerize. The water layer is removed, and, if desired, the sucrose elution chamber may be replaced with non-sucrose-containing buffer. The gel is placed in an ordinary gel electrophoresis system and the affinity media with adsorbed protein, prepared as described in the literature (53–60), is added to the top of each tube. Electrophoresis for several hours should yield a small amount of eluate in the elution chamber. Several such chambers may be electrophoresed simultaneously, and, if desired, the optimal time for desorption may be investigated by interrupting the desorption at various times and collecting the elute from one elution chamber at each interruption time. (See Figure 10.8.)

The factors that seem most critical in this desorption technique are temperature, stability of the bioselective adsorbent, pH, and thickness of the layer of bioselective adsorbent (53,54). (Thinner layers of adsorbent yield minimal desorption times). Although the optimal condition will vary from preparation to preparation, Grenot and Cuilleron (53) found optimal elution of antibodies to 5-α-dihydrotestosterone from the immobilized antigen could be obtained at 4°C, pH 4.5, using 30 mA (V ~ 200 V). About 60% of the antibodies were eluted during the first 4 hours. The remaining 40% were eluted much more

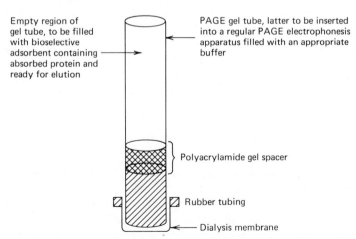

Figure 10.8.   Ordinary gel tube adapted for electrophoretic desorption.

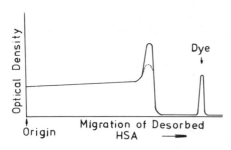

Figure 10.9. The pattern of electrophoretic desorption of HSA from Cibacron Blue 1 3G-A Sepharose as shown by the absorbence of the gel after staining for protein. Dashed line (– – –) shows the effect on pattern of decreasing the time between the washing of the affinity matrix and beginning the desorption. Reproduced with permission from Morgan et al. (54).

slowly (~1%/hour) and represented a class of much tighter binding antibodies. Comparison of this elution method with elution with either free antigen or with deforming buffers (e.g., 1-M NH$_4$OH) revealed that free antigen was not a very effective eluant and required extensive further treatment to separate the excess antigen from the antibody. Ammonium hydroxide was effective in eluting the 60% of looser binding antibody, but none of the higher affinity antibody could be prepared by this technique. Only electrophoretic desorption resulted in recovery of active antibody with both high and low affinity.

Morgan et al. (54) have prepared ferretin by using antiferretin–agarose adsorbents in an electrophoretic desorption system. They have also studied the adsorption of human serum albumin (HSA) on Cibacron Blue agarose by a similar technique. After adsorbing HSA on Cibacron Blue agarose and washing away any unadsorbed protein, they placed a thin layer of the adsorbent–HSA complex on the top of 7% polyacrylamide gels. On electrophoresis and staining, it could be seen that the HSA migrated as a sharp peak followed by a long, declining tail. The peak according to these authors represented the amount of HSA in an unbound state at the beginning of the electrophoresis (assuming, of course, that the layer of HSA–Cibacron Blue–agarose was infinitely thin). Thus the ratio of protein initially desorbed to the total yields an estimate of the K_a of the protein–adsorbent complex. Other, similar, forms of affinity electrophoresis (see page 312) have also been developed to study interactions between proteins and immobilized adsorbents. (See Figure 10.9.)

10.5. HAPTEN PURIFICATION ON IMMUNOSORBENTS

Haptens, or peptides containing hapten groups, are easily purified by using immunosorbents, chiefly because these molecules, having no biological activity and little, if any, secondary structure, are stable to almost any type of deforming buffer, extremes of pH, or high temperatures. This fact is one of the key observations that make hapten-immunosorbent chromatography a valuable tool in protein chemistry. In general, antibodies are prepared by using modified proteins, for example, 2,4 dinitrofluorobenzene-treated proteins (DNP-proteins),

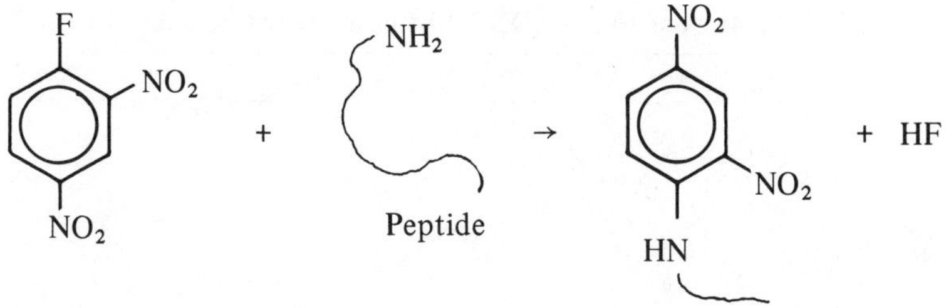

2,4 Dinitrofluorobenzene   Dinitrophenyl (DNP)-peptide

(Sanger reagent)

Figure 10.10. Reaction of peptides with 2,4 dinitrofluorobenzene.

as antigens. Antibodies prepared by injecting rabbits with proteins containing these haptens will now bind any protein or peptide containing the DNP group (or any similar hapten). Thus by immobilizing the hapten antibodies, one obtains an immunosorbent capable of binding to and purifying any proteins or peptide containing the hapten. Haptens that have been employed include the dinitrophenyl (DNP) group (62–71), nitrotyrosyl peptides (72–73), and the arsanilazotyrosyl (Ars) function (62–67). Of these, the DNP derivative have been most widely and successfully employed. By employing various DNP containing protein reagents, DNP derivatives of specific amino acids can easily be prepared (see Table 10.4). In a typical application, Wilchek and Miron (62,69), employed N^{α}-bromo-acetyl-N^{ϵ}-DNP-L-lysine to specifically dinitrophenylate the cysteins (at pH 8.5) or methionine (at pH 4.0) of papain and lysozyme. The labeled protein was digested with trypsin and/or chymotrypsin. The resulting peptides were then applied to an anti-DNP-antibody-agarose column. The DNP labeled peptides became bound to the column, the nonlabeled peptides passed directly through the column. The DNP-peptides were subsequently eluted with 6 M guanidine HCl. About 95% of the DNP peptides were recovered by this technique. The recovered peptide can be sequenced and the sequence employed in deducing the primary structure of the original protein. For example, the DNP-methionine peptide (2) from a given protein will provide the means for ordering cyanogen bromide fragmentation peptides of a protein. (CNBr cleaves the carboxyl side of protein methionine residues).

Tyrosine containing peptides have been purified using arsanilazotyrosine as a hapten (67). First, bovine serum albumin was derivatized with freshly diazo-tized p-arsanilic acid. After 2 hours at 0°, the reaction was stopped and the excess reagent removed via dialysis. The arsanilazo–BSA was injected in goats in complete Freund's adjuvant (2-mg hapten antigen per injection) intradermally. Repeated injections were given under the same conditions if the reaction of a sample of the goat serum with arsanilino–BSA gave little or no precipitate.

The serum was periodically collected and the arsanilazo antibodies purified

TABLE 10.4. Use of Anti-DNP Antibodies for Isolation of Peptides

Amino Acid	Reagent = RX	Conditions	Products
Cysteine	O_2N—⟨NO₂⟩—F O_2N—⟨NO₂⟩—CH₂Cl O_2N—⟨⟩—CH₂Br	pH 5, 1 hour, RT[a]	—NH—CH—CO—NH— | CH₂ | S | R
Methionine	O_2N—⟨NO₂⟩—NH—CH₂—CH₂—NH—C(=O)—CH₂—Br O_2N—⟨NO₂⟩—NH—(CH₂)₄—CH(COOH)—NH—C(=O)—CH₂—I	pH 3, 24 hours, RT 8-M urea	—NH—CH—CO—NH— | CH₂ | CH₂ | (+) S—R | CH₃
Tryptophan	O_2N—⟨NO₂⟩—SCl O_2N—⟨⟩—SCl	50% acetic acid, 1 hour, RT	—NH—CH—CO—NH— | CH₂ | (indole ring) R
Lysine	O_2N—⟨NO₂⟩—SO₃Na	pH 9.5, 16 hours, RT	—NH—CH—CO—NH— | (CH₂)₄ | NH | R
Histidine	O_2N—⟨NO₂⟩—NH—(CH₂)₄—CH(NH—C(=O)—CH₂—I)—COOH	pH 5, 24 hours, RT	—NH—CH—CO—NH— | CH₂ | (imidazole ring N—R)

Source: Reprinted with permission from M. Wilchek et al (1974), *Methods in Enzymology,* **34**, 185.
[a] Room temperature.

by adsorbing the serum to 4-(p-aminophenylazo)-phenylarsonic acid–agarose. The serum was incubated with the biospecific adsorbent for 1 hour, after which the adsorbent was collected by suction filtration and washed with buffer until no further protein appeared in the wash buffer. The adsorbed antibodies were eluted with 0.3-M arsanilic acid, pH 8.0, and the eluate was dialyzed extensively to remove excess free antigen. This was considered complete when the ratio $A_{250}:A_{250}$ was 2.8.

The purified antiarsanilazo antibodies are coupled to agarose and utilized as an adsorbent for peptides prepared by protolytic digestion of arsanlinoazo proteins. The adsorbed arsanilinoazo peptides are eluted with 1-M NH₄OH, which

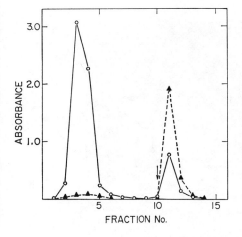

Figure 10.11. Fractionation of Ars-peptides derived from subtilisin digest of Ars-CPD on anti-Ars Sepharose column. A digest of 0.5 μmole of Ars-CPD was applied to a 1-cm $\times$ 3.4-cm anti-Ars Sepharose column (containing 0.44 μmole of antibody). After washing with 0.1-M NH_4HCO_3 elution was started by 1M NH_4Oh (arrow). Fractions of 3.2 ml were collected. (o) Absorbence at 280 nm; (▲) absorbence at 325 nm after neutralization with CO_2. Reproduced with permission from Wilchek et al. (67).

results in a change in their color from yellow at pH $\sim$ 7.0 to orange in 1-M NH_4OH. As soon as the orange peak containing the antigen reaches the end of the column, the column is washed with buffer until the eluate is neutral. The first use of this column results in a loss of 15 to 20% of its initial capacity, probably as a result of both denaturation of some of the antibodies and the release of some of the antibodies due to the nucleophilic attack of the ammonium on the agarose–antibody coupling bonds. Small amounts of antibody often are eluted in this procedure and can be removed from the peptides by gel-permation chromatography on Sephadex G-25.

Figures 10.11 and 10.12 show the results of the chromatography of peptides from arsanilinoazocarboxypeptidase with the use of this method. The peptide isolated was 10 amino acids long and contained the arsaniloazo derivative of tyrosine 248. (See Table 10.4.)

10.6. POTENTIAL MEDICAL USES OF IMMUNOSORBENTS AND OTHER BIOSELECTIVE ADSORBENTS

There have been numerous attempts to produce extracorporeal shunts by means of which blood from animals could be circulated through immunoadsorbents (74–78) and other bioselective adsorbents (79–81). The use of such adsorbents could remove highly toxic compounds, viruses, or cells from serum and thus represent an intriguing potential therapeutic approach to the treatment of a wide variety of illnesses including genetic diseases. Hemoperfusion of patients blood through activated charcoal and other adsorbent resins to remove various toxic compounds has been attempted by many investigators (81),

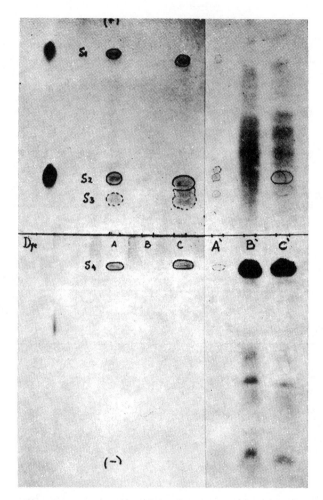

Figure 10.12. Paper electrophoresis of subtilisin digest of Ars-CPD (*A*). Peptides eluted from antibody–Sepharose by 1-*M* NH$_4$OH. (*B*) Peptide mixture unadsorbed by the column. (*C*) Digest before antibody–Sepharose column. Electrophoresis was at pH 6.5 at 4 kV for 20 min. The dye on the left-hand side was a mixture of orange G, xylene cyanol BR, and basic fuchsin that served as visible markers during the run. The right-hand side of the elctropherogram was photographed after staining with ninhydrin, and the left-hand side was photographed without staining. The yellow peptides are marked. Reproduced with permission from Wilchek et al. (67).

although practical problems, due chiefly to the undesired removal of platelets from the blood, have prevented their widespread application.

An alternative approach is to circulate the blood through a bioselective adsorbent contained in an extracorporeal shunt. Terman et al. (74,75) removed DNA antibodies from the plasma of positively immunized rabbits by letting their blood circulate through such an extracorporeal shunt containing immobilized DNA. Terman et al. (78), in another report, used bovine serum albumin immobilized to microcapsules by glutaraldehyde treatment to remove antibodies from the blood of dogs previously injected with anti-BSA antibodies. These were essentially quantitatively removed from the blood stream with few visible side effects for the treated animals. Similar experiments by Schenkein et al. (76) using rabbits as the experimental animals yielded essentially the same results. In this case the anti-BSA antibodies, labeled with ^{125}I isotope, were removed by passing the blood through a column of BSA in agarose or coupled to bromacetyl cellulose.

One of the more serious medical problems that might be amenable to treatment by bioselective adsorbents contained in extracorporeal shunts is the removal of toxic substances released into the body during liver failure. In biliary obstruction, for example, jaundice occurs as a result of the accumulation of toxic levels of bilirubin. Plotz et al. (79,80) have conjugated HSA with agarose using the cyanogen bromide method to prepare an extracorporeal hemoperfusion system as shown in Figure 10.13. Human serum albumin has a high capacity for binding bilirubin and other hydrophobic toxins. Both congenitally

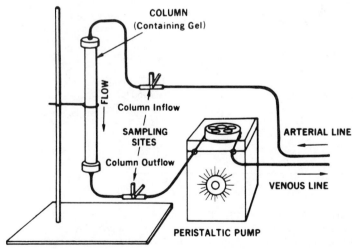

Figure 10.13. Extracorporeal hemoperfusion circuit. Reproduced with permission from the *Journal of Clinical Investigation* (80).

jaundiced rats lacking glucuronyl transferease (and thus accumulate bilirubin) and rats with surgically formed biliary obstructions were hemoperfused by use of this system. The extracorporeal shunt was filled with heparinized saline, and the animals were perfused under pentobarbital for 1 hour. In the study shown in Figure 10.14, the rats were given small doses of ^{3}H-bilirubin as markers to determine the effectiveness of the hemoperfusions. Rats lacking glucuronyl transferase had over 50% of the radioactive bilirubin removed within 1 hour of treatment, whereas rats with surgically created biliary obstructions had an average of over 96% of the plasma bilirubin removed by adsorption onto the bovine serum albumin containing extracorporeal shunt.

After hemoperfusion the adsorbed bilirubin could be eluted from the immobilized serum albumin by using 50% ethanol and the column reused. This method has also been applied to the purification of bilirubin by means of bioselective adsorption from jaundiced plasma as shown in Figure 10.15.

Hughes et al. (81) have reported difficulties in using the same procedure with the use of commercially activated agarose to prepare the serum albumin–agarose adsorbent. The basic problems were the considerable hemolysis observed, the low capacity of their adsorbent, and the water content of the extracorporeal

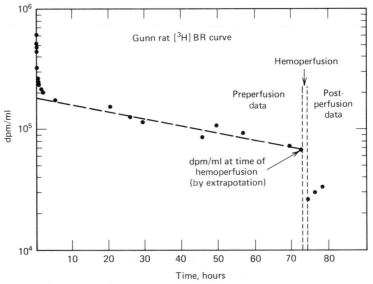

Figure 10.14. (^{3}H) Bilirubin (BR) disappearance curve in a jaundiced rat. The ordinate represents plasma radioactivity present in the lower, nonpolar layer of the Weber Schalm partition. The vertical dashed lines separate data obtained before and after the 1-hour hemoperfusion. Reproduced with permission from the *Journal of Clinical Investigation* (80).

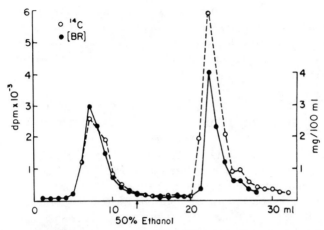

Figure 10.15. Binding of bilirubin (BR) to agarose HSA and elution by ethanol. 1 ml of plasma with 180 μg bilirubin/ml labeled with ^{14}C bilirubin was applied to a column with 3 g of agarose–HSA and eluted with PBS followed by 50% (vol/vol) ethanol in water. The bilirubin was measured by the van den Bergh reaction and by radioactivity. The lower chemical values in the ethanol-eluted fraction are due to the degradation of bilirubin before the chemical assay. Reproduced with permission from the *Journal of Clinical Investigation.*

shunt. These technical problems seem potentially solvable with the use of larger bead sizes (which would permit reasonable flow rates without hemolysis) and by using agarose with higher albumin loadings. Alternatively, serum albumin immobilized by other procedures, such as glutaraldehyde crosslinking, might prove advantageous. Further studies utilizing bioselective adsorbents as extracorporeal shunts would seem worthwhile, since only by this method can one specific toxin be removed from the blood stream while not otherwise altering its normal composition.

10.7. OTHER PURIFICATIONS BASED ON BIOSELECTIVE PROTEIN–PROTEIN INTERACTIONS

Besides immunosorbents, there have been many broad groups of bioselective adsorbents based on protein–protein interactions. Chromatography on immobilized lectins has already been described (page 147). Other groups of protein–protein interactions that have been exploited include proteinase affinities for proteins (82–88) or for broad-spectrum protease inhibitors, for example, soybean trypsin inhibitor (89–104) and blood-clotting enzymes (109–116). Subunits of dissociable multisubunit complexes have also been immobilized to form an adsorbent for other components of the same complex (117–119). Likewise, complimentary peptides, for example, RNAse S-peptide and S-protein (124,125), can be purified in this fashion, as can receptor proteins and protein hormones. A few examples of these types of purification are presented in the following section.

10.8. PROTEASE–PROTEIN INTERACTIONS

Among the peptidases that can be purified by affinity chromatography, collagenase purification is one of the easiest (82,83). Collagen is coupled to cyanogen bromide activated agarose and employed without further treatment. Either crude collagenase or a partially purified fraction (e.g., after an ammonium sulfate fractionation step) is applied to a column containing collagen–agarose previously equilibrated with a neutral buffer. The same buffer is passed through the column after the sample application until the adsorbence at 280 nm is essentially zero. The purified collagenase is then eluted with the same buffer containing 1-M NaCl. Although yields are sometimes lower than that found in other examples of affinity chromatography, electrophoretically homogeneous collagenase is obtained in a very short time (84–86). Various attempts to increase the yield have been made, but only with moderate success. For example, collagenase from tadpole can be eluted from polyacrylamide–collagen in 69% yield by using a pH-5.2 eluant (87), whereas human collagenase loses most of its activity on subjection to these same conditions.

Probably the most versatile protease purification medium as yet employed is immobilized hemoglobin, usually hemoglobin–agarose (88,89). Cathepsin D, one of the autolytic proteases of the cell, was purified by Smith and Turk (89) in several hours from bovine speen and thymus by means of this adsorbent. After a subsequent gel-permeation step, the preparation consisted solely of three electrophoretically distinguishable isoenzymes (see Figure 10.16), each of which possessed proteolytic activity. On SDS electrophoresis a single species of 42,000 molecular weight was seen. This is in contrast to the multiple lower-molecular-weight active protiens seen when cathepsin D was purified by conventional techniques. Moreover, when the tissue was allowed to sit for several hours prior to purification, the cathepsin D produced by affinity chromatography on hemoglobin–agarose also contained these lower-molecular-weight polypeptides, showing that the low-molecular-weight active proteins were fragments of cathepsin D produced by antolysis in the course of the purification.

A fairly large number of proteases have been purified by chromatography on immobilized protease inhbitors. Conversely, protease inhibitors from many sources have been prepared by using immobilized proteinases. Immobilized

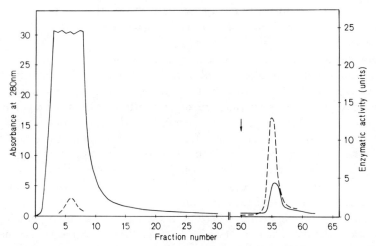

Figure 10.16. Hemoglobin-agarose chromatography of a 40 to 70% acetone-precipitated fraction from thymus. Immediately following sample application the pH of the buffer was changed to 5 from 3.5. Later, at the point marked with an arrow, the pH was changed to 8.6. The absorbence at 280 nm of the effluent was measured with 0.1-cm-pathlength cells. Proteolytic activity (– – – –); absorbence at 280 nm (——). Reproduced with permission from Smith and Turk (89).

trypsin has served as an affinity chromatography media for protease inhibitor from wheat germ (90), soybeans (91), jack bean (91), *E. coli* (19), cow colostrum (92), sea anemone (93), snails (94,95), and urine (96), whereas trypsin has been purified on immobilized trypsin inhibitors from soybean (97–102), lima bean (103), and pancreas (104). An *N*-acylated pentapeptide isolated from actinomycetes (isovaleryl-L-valyl-L-valyl 4 amino-3-hydroxy-6-methylhepanoyl-L-alanyl-4-amino-3-hydroxy-6-methylheptanoic acid), termed "pepstatin," has proved to be a very useful ligand for a broad group of acid proteases (105–108).

Many other proteins are purified by way of bioselective adsorption on small peptides or even single, immobilized, amino acids. Among these, the most important are plasminogen, purified on lysine–agarose (109–112) or lysine–acrylamide (113,114), and urokinase, purified on *p*-aminobenzyamide–agarose (115) or α-benzylsulphonyl-*p*-aminophenylalanine–agarose (116). The purification of urokinase has become very important medically and commercially since its introduction as an agent for dissolving blood clots. Urokinase is specifically proteolytic toward the cross-linked proteins of blood clots, but it must be purified from tissue culture or urine prior to its injection in a patient as a clot-dissolving agent.

10.9. BIOSELECTIVE ADSORPTION THROUGH
SUBUNIT INTERACTIONS

A large segment of proteins, particularly enzymes, with biological importance, can be readily and reversibly dissociated into subunits. This is the basis of bioselective adsorption through subunit interaction. That is, the subunits are immobilized and soluble subunits are subsequently passed into a column containing the immobilized subunits under conditions where reassembly is probable. After allowing an appropriate time for the soluble subunits to reform holoenzyme with the immobilized subunits, the column is washed to remove any contaminating proteins. Finally, the purified enzyme is eluted with a buffer that causes dissociation of the holoenzyme, and the eluted subunits are dialyzed against a buffer that permits reassociation.

An eloquent example of this procedure is the purification of porphobilinogen synthetase by Gurne et al. (117) (see Table 10.5). Prophorobilinogen synthetase, an enzyme of the biosynthesis of heme, is an octameric enzyme that can be dissociated by 4-M urea to form a tetramer. On removal of urea or on dilution, the active octamer reforms. Porphorobilinogen synthetase, isolated by classical methods, was coupled to CNBr-activated agarose, after which the gel was washed extensively to remove any unbound enzyme. Then 5 ml of this enzyme–agarose complex was poured into a column and washed with 20 ml of buffer containing 4-M urea followed by a subsequent wash with buffer alone. The protein eluted by the 4-M urea was diluted with an equal volume of buffer and then freed of urea by gel filtration. The column, now containing only half its original enzyme in the tetrameric form, could be employed to purify more porphorobilinogen synthetase. To do so, a crude extract from bovine liver was made 4 M in urea and kept in the urea containing buffer for 30 minutes at room temperature to dissociate the enzyme present in the extract. To this was added a slurry of the urea-treated enzyme–agarose preparation. After 10 minutes the suspension was diluted fivefold with buffer and the enzyme–agarose was collected by suction filtration. The enzyme agarose was washed extensively with buffer followed by elution of the bound enzyme with 4-M urea as described previously. This cycle of association–dissociation–reassociation could be performed repeatedly. Starting with the crude extract, a 250-fold purification was effected. The preparation yielded an electrophoretically homogenous enzyme with a specific activity equal to that best achievable through the much more laborous classical purification scheme.

Relatively few enzymes have been purified with the use of subunit reassociation techniques, although Kervabon et al. (118) recently reported purification of palmityl CoA-ACP-transacylase 5500 fold using a combination of isoelectric focusing and bioselective adsorption on immobilized acyl carrier protein (ACP). Most of the application of immobilized protein subunits have not been done as a

TABLE 10.5. Assembly of the Synthase: Purification of the Synthase with Use of the Immobilized Preparation [a]

Preparation	Enzyme Activity, Total Units		Protein in Urea Eluate, mg	Protein Recovered by Gel Filtration, mg	Specific Enzyme Activity of Recovered Protein Units/mg
	Immobilized Enzyme	Immobilized Enzyme after Urea Treatment			
Original immobilized enzyme[b]	280	70	11	—	—
Reassembled immobilized enzyme					
First reassembly	230	60	~10	3.0	24
Second reassembly	290	60	~10	5.2	25
Third reassembly	200	40	~10	3.5	20

Source: Reproduced with permission from D. Gurne, J. Chen, and D. Shemin (1977), *Proc. Nat. Acad. Sci.* **74**, 1383.

[a] An enzyme-Sepharose preparation of 280 units of enzyme activity was treated with 4-M urea. The residual immobilized protein was then incubated with 200 ml of a crude bovine liver extract containing 600 units of total enzyme activity (specific activity = 0.1) in urea (4 M). After a five-fold dilution the Sepharose preparation was recovered and thoroughly washed. The reassociated enzyme-Sepharose preparation was treated with urea once again. The procedure was repeated twice more. In each cycle the amount of protein and the enzymic activities in the bound and eluted fractions were determined.

[b] The specific activity of the enzyme used for this preparation was 16.

means of protein purification, but rather as a means of studying the isolated monomeric subunits under conditions where, in free solution, they would normally reassemble in a dimeric or polymeric form (119–123).

In one applicantion Chan (121) immobilized aldolase on CNBr-activated agarose under conditions in which the enzyme was attached to the matrix by only one of the four subunits. The unattached subunits were washed away with 4-M urea. The resulting denatured aldolase–agarose complex (Figure 10.17) was dialyzed in the presence or absence of added aldolase to form a renatured bound subunit (β-subunit) or matrix bound renatured tetramer (MB-aldolase). Using these preparations, he showed that, among other things, the subunit is much more easily denatured than the reassociated tetramer (Figure 10.18).

Other workers have applied this technique to the study of aspartate-4-decarboxylase (119) and to lactate dehydrogenase (120). In each case, denatur-

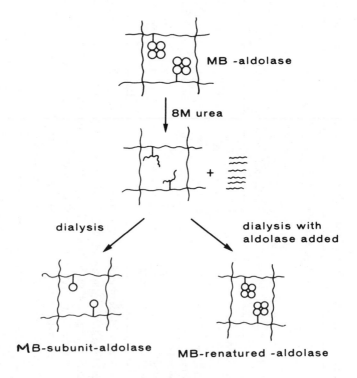

Figure 10.17. Scheme of matrix-bound (MB)-aldolase derivatives. Reproduced with permission from W.–C. Chan, *Biochem. Biophys. Res. Commun.* **41**, 1198 (1970).

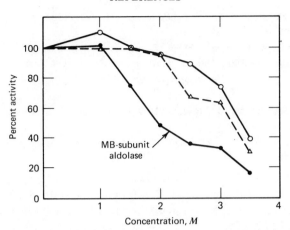

Figure 10.18. Effect of urea concentration on enzymatic activity of matrix-bound derivatives. All three derivatives were diluted to 10 mU/ml with untreated Sepharose 4B and equilibrated with triethanolamine (40 m*M*)-EDTA (10 mM) buffer, pH 7.4, containing 2-mercaptoethanol (5mM) and the appropriate amount of urea. Activity was measured with assay mixtures that contained the same urea concentration as the samples to be tested. Controls show that loss of activity was not due to inactivation of auxiliary enzymes in the coupled assay system. O, MB-aldolase; △, MB-renatured aldolase; ●, MB-subunit aldolase. Reproduced with permission from W.–C. Chan, *Biochem. Biophys. Res. Commun.* **41**, 1198 (1970).

ing and renaturing of the immobilized subunit was much more effective than the same procedure with the soluble enzyme. This observation shows that attachment of the enzyme to a solid surface prevents undesirable chain interactions in the renaturation procedure. Cho and Swaisgood (120) suggest, in fact, that this may be an excellent model for protein subunit folding by newly formed peptides on the surface of the ribosome.

References

1. D. H. Campbell, E. L. Luescher, and L. S. Lerman (1951), *Proc. Nat. Acad. Sci. (USA)* **37**, 575.
2. N. Weliky and H. H. Weetal (1965), *Immunochemistry* **2**, 293.
3. H. H. Weetal (1972), in *Chemistry of Biosurfaces* (M. L. Hair, ed.) Marcell Dekker, New York, p. 597.
4. D. H. Campbell and N. Weliky (1967), in *Methods in Immunology and Immunochemistry* Vol. I, (C. A. Williams and M. W. Chase, eds.), Academic, New York, 1967, p. 365.
5. I. H. Silman and E. Katchalski (1966), *Annu. Rev. Biochem.* **35**, 873.

6. M. Wilchek and D. Givol in *Peptides* (H. Nesvadba, ed.) North-Holland, Amsterdam, 1973, p. 203.

7. B. D. Davis, R. Dulbecco, H. N. Eisen, H. S. Grinsberg, and W. B. Wood, Jr. (1973), *Microbiology*, 2nd ed., Harper and Row, New York.

8. D. M. Weir (1967), *Handbook of Experimental Immunology*, F. A. Davis, Philadelphia.

9. E. A. Kabat and M. M. Mayher (1967), *Experimental Immunology*, Thomas Press, Springfield, Ill.

10. T. Cooper (1977), *The Tools of Biochemistry*, Wiley-Interscience, New York, Chapter 9.

11. C. A. Kabat (1954), *Science* **120**, 782.

12. A. Masseyeff and R. Maoilini (1975), *J. Immunol. Meth.* **8**, 223.

13. P. Cuatrecasas and C. B. Anfinsen (1971), in *Methods in Enzymology* (W. B. Jakoby, ed.) **22**, 345.

14. J. B. Robbins and R. Schneerson (1974), in *Methods in Enzymology* (W. B. Jakoby and M. Wilchek, eds.) **34**, 703.

15. D. M. Livingston (1974), in *Methods in Enzymology* (W. B. Jakoby and M. Wilchek, eds.) **34**, 723.

16. S. F. Schlossman and L. F. Hudson (1973), *J. Immunol.* **110**, 313.

17. C. B. Anfinsen, S. Bose, L. Corely, and D. Gurari-Rotman (1974), *Proc. Nat. Acad. Sci. (USA)* **71**, 3139.

18. R. F. Murphy, A. Iman, A. E. Hughes, M. J. McGlucken, K. D. Buchanan, J. M. Conlon, and D. T. Elmore (1976), *Biochim. Biophys. Acta* **420**, 87.

19. J. V. Ferrante and D. J. D. Nicholas (1976), *FEBS Lett.* **66**, 187.

20. N. Guiso and P. Truffa-Bachi (1973), *Eur. J. Biochem.* **42**, 401.

21. M. L. Ernst-Fonberg, A. W. Schongalla, and T. A. Walker (1977), *Arch. Biochem. Biophys.* **178**, 166.

22. B.-M. Sjöberg and A. Holmgren (1973), *Biochem. Biophys. Acta* **315**, 176.

23. C. A. Ogburn, K. Berg and K. Paucker (1973), *J. Immunol.* **111**, 1206.

24. R. G. C. Gallop, B. T. Tozer, J. Stephen, and H. Smith (1966), *Biochem. J.* **101**, 711.

25. Y. Akanama, T. Kuzuya, M. Hayashi, T. Ide, and N. Kuzuya (1970), *Biochem. Biophys. Res. Commun.* **38**, 947.

26. D. M. Livingston, E. M. Sconick, W. P. Parks, and G. J. Todaro (1972), *Proc. Nat. Acad. Sci. (USA)* **69**, 393.

27. S. H. Polmar, T. H. Waldman, and W. D. Terry (1973), *J. Immunol.* **110**, 1253.

28. A. T. Jagendorf, A. Patchornik, and M Sela (1963), *Biochem. Biophys. Acta.* **78**, 516.

29. J. B. Robbins, J. Haimovich, and M. Sela (1967), *Immunochemistry* **4**, 11.

30. E. Ott, H. M. Spurlin, and M. W. Grafflin (1954), *Cellulose and Cellulose Derivatives*, 2nd ed., Wiley, New York.

31. H. H. Weetal and N. Weliky (1965), *Biochim. Biophys. Acta.* **107**, 150.

32. A. Malley and D. H. Campbell (1963), *J. Am. Chem. Soc.* **85**, 487.

33. A. E. Gurvich, O. B. Kuzuvleva, and A. E. Tumanova (1961), *Biokhimiya* **24**, 129.

34. N. R. Moudgal and R. R. Porter (1963), *Biochim. Biophys. Acta* **71**, 185.

35. M. D. Lily (1976), in *Methods in Enzymology* (K. Mosbach, ed.) **44**, 46.

36. C. R. Lowe and P. D. G. Dean (1974), *Affinity Chromatography*, Wiley, Chichester, UK, p. 152–155.

37. S. Turková (1978), *Affinity Chromatography*, Elsevier, New York, p. 246.

38. R. Wallin and H. Prydz (1975), *FEBS Lett.* **51**, 191.

39. W. C. Palmer (1972), *Biochem. Biophys. Acta* **278**, 299.

40. H. Sjöström, O. Norén, L. Jeppesen, M. Staun, B. Svensson, and L. Christensen (1978), *Eur. J. Biochem.* **88**, 503.

41. B. Svensson, M. Danielsen, M. Staun, L. Jeppesen, O. Norén, and H. Sjöström (1978), *Eur. J. Biochem.* **90**, 489.

42. P. Cuatrecasas (1969), *Biochem. Biophys. Res. Commun.* **35**, 531.

43. M. R. Sairam and J. Porath (1976), *Biochem. Biophys. Res. Commun.* **69**, 190.

44. I. Schechter (1974), *Biochemistry* **13**, 1875.

45. S. Fuchs, A. Aharonov, M. Sela, F. van den Haar, and F. Cramer (1974), *Proc. Nat. Acad. Sci. (USA)* **71**, 2800.

46. J. Stephen, R. G. C. Gallop, and H. Smith (1966), *Biochem. J.* **101**, 717.

47. V. Hajnická, N. Fuchsberger, and L. Borecky (1976), *Acta Virol.* **20**, 326.

48. C. B. Anfinsen, S. Bose, L. Corley, and D. Gurari-Rotman (1974), *Proc. Nat. Acad. Sci. (USA)* **71**, 3139.

49. S. Sirisinha and H. N. Eisen (1971), *Proc. Nat. Acad. Sci. (USA)* **68**, 3130.

50. L. Wofsy and B. Burr (1969), *J. Immunol.* **103**, 380.

51. D. C. Rueger, D. Dahl, and A. Bignami (1978), *Anal. Biochem.* **89**, 360.

52. M. R. Sairam, W. C. Clark, D. Chung, J. Porath, and C. H. Li (1974), *Biochem. Biophys. Res. Commun.* **61**, 355.

53. C. Grenot and C.-Y. Cuilleron (1977), *Biochem. Biophys. Res. Commun.* **79**, 274.

54. M. R. A. Morgan, P. J. Brown, M. J. Leyland, and P. D. G. Dean (1978), *FEBS Lett.* **87**, 239.

55. P. J. Brown, M. J. Leyland, M. P. Keenan, and P. D. G. Dean (1977), *FEBS Lett.* **83**, 256.

56. P. D. G. Dean and D. H. Watson (1978), *First International Symposium on Affinity Chromatography*, Vienna, Pergamon, p. 25.

57. P. D. G. Dean (1978), *Proceedings of Fourth Enzyme Engineering Conference*, Bad Neuenahr, Pergamon, New York.

58. M. R. Morgan, E. J. Kerr, and P. D. G. Dean (1978) *J. Steroid Biochem.*, in press.

59. P. D. G. Dean, P. Brown, M. J. Leyland, D. H. Watson, S. Angal, and M. J. Harvey (1977), *Biochem. Soc. Transact.* **5**, 1111.

60. M. J. Iobal, P. Ford, and M. W. Johnson (1978), *FEBS Lett.* **87**, 235.

61. M. Cohn, L. Wang, W. H. Scouten, and I. R. McManus (1968), *Biochim. Biophys. Acta* **159**, 182.

62. M. Wilchek (1974), in *Methods in Enzymology* (W. B. Jakoby and M. Wilchek, eds.) **34**, 182.

63. Y. Weinstein, M. Wilchek, and D. Givol (1969), *Biochem. Biophys. Res. Commun.* **35**, 694.

64. D. Givol, P. H. Strausbrauch, E. Hurwitz, M. Wilchek, J. Haimovich, and H. N. Eisen (1971), *Biochemistry* **10**, 3461.

65. P. H. Strausbaugh, Y. Weinstein, M. Wilchek, S. Shaltiel, and D. Givol (1971), *Biochemistry* **10**, 4342.

66. J. Haimovich, H. N. Eisen, E. Harwitz, and D. Givol (1972), *Biochemistry* **11**, 2389.

67. M. Wilchek, V. Bocchini, M. Becker, and D. Givol (1971), *Biochemistry* **10**, 2828.

68. D. Givol, Y. Weinstein, M. Gorecki, and M. Wilchek (1970), *Biochem. Biophys. Res. Commun.* **38**, 825.

69. M. Wilchek and T. Miron (1972), *Biochem. Biophys. Acta* **278**, 1.

70. M. Bustin and D. Givol (1972), *Biochem. Biophys. Acta* **263**, 459.

71. D. Givol and M. Rotman (1970), *Israel J. Med. Sci.* **6**, 452.

72. M. Helman and D. Givol (1971), *Biochem. J.* **125**, 971.

73. M. Sokolovsky (1972), *Eur. J. Biochem.* **25**, 267.

74. D. Terman, H. Harbeck, A. Hoffman, I. Steward, J. Robinette, and R. Carr (1975), *Fed. Proc.* **34**, 976.

75. D. Terman, I. Stewart, A. Hoffmann, R. Carr, and R. Harbeck (1974), *Experimentia* **30**, 1493.

76. I. Schenkein, J. C. Byrstryn, and J. W. Uhr (1971), *J. Clin. Invest.* **50**, 1864.

77. S. E. Charm and B. C. Wong (1974), *Biotechnol. Bioeng.* **16**, 593.

78. D. S. Terman, T. Tavel, D. Petty, A. Tavel, R. Harbeck, G. Buffaloe, and R. Carr (1976), *J. Immunol.* **116**, 1337.

79. P. H. Plotz, P. Beck, B. F. Scharschmidt, J. K. Gordon, and J. Vergalla (1974), *J. Clin. Invest.* **53**, 778.

80. B. F. Scharschmidt, P. H. Plotz, P. D. Berk, J. G. Waggoner, and J. Vergalla (1974), *J. Clin. Invest.* **53**, 786.

81. R. D. Hughes, E. H. Dunlop, M. Davis, D. B. A. Silk, and R. Williams (1977), *Biomat. Med. Dev. Art. Org.* **5**, 205.

82. A. Z. Eisen, E. A. Bauer, G. P. Stricklin, and J. J. Jeffrey (1974), in *Methods in Enzymology* (W. B. Jakoby and M. Wilchek, eds.) **34**, 420.

83. E. A. Bauer, J. J. Jeffrey, and A. Z. Eisen (1971), *Biochem. Biophys. Res. Commun.* **44**, 813.

84. E. A. Bauer, A. Z. Eisen, and J. J. Jeffrey (1971), *J. Clin. Invest.* **50**, 2056.

85. A. Z. Eisen, E. A. Bauer, and J. J. Jeffrey (1971), *Proc. Nat. Acad. Sci.* (USA) **68**, 248.

86. E. A. Bauer, A. Z. Eisen, and J. J. Jeffrey (1972), *J. Biol. Chem.* **247**, 6679.

87. Y. Nagi and H. Hori (1972), *Biochim. Biophys. Acta* **263**, 564.

88. G. K. Chua and W. Bushuk (1969), *Biochem. Biophys. Res. Commun.* **31**, 545.

89. R. Smith and V. Turk (1974), *Eur. J. Biochem.* **48**, 245.

90. K. Hochstrasser and E. Werle (1969), *Hoppe-Seyler's Z. Physiol. Chem.* **350**, 249.

91. S. Avrameas and B. Guilbert (1971), *Biochemie* **53**, 603.

92. L. Sundberg, J. Porath, and K. Aspberg (1970), *Biochim. Biophys. Acta* **221**, 394.

93. H. Fritz and B. Föng-Brey (1972), *Hoppe-Seyler's Z. Physiol. Chem.* **353**, 19.

94. H. Tscheche, T. Dietl, R. Marx, and H. Fritz (1972), *Hoppe-Seyler's Z. Physiol. Chem.* **353**, 483.

95. H. Tscheche and T, Dietl (1972), *Eur. J. Biochem.* **30**, 560.

96. P. Casati, M. Grandi, and N. Toccaceli (1975), *Ital. J. Biochem.* **24**, 188.

97. V. V. Mosolov and E. V. Lushnikova (1970), *Biokhimiya* **35**, 440.

98. G. J. Bartling and C. W. Barker (1976), *Biotechnol. Bioeng.* **18**, 1023.

99. J. Porath and L. Sundberg (1972), *Nature New. Biol.* **238**, 261.

100. J. Liepnieks and A. Light (1974), *Anal. Biochem.* **60**, 395.

101. A. Light and J. Liepnieks (1974), in *Methods in Enzymology* (W. B. Jakoby and M. Wilchek, eds.) **34**, 448.

102. M. Wilchek and M. Gorecki (1973), *FEBS Lett.* **31**, 149.

103. E. B. Gillam and G. B. Kitto (1976), *Comp. Biochem. Physiol.* **54B**, 21.

104. J. Chauvet, J. P. Dostal, and R. Archer (1976), *Internat. J. Peptide Protein Res.* **8**, 45.

105. C. Devaux, P. Corvol, C. Auzan, J. Ducloux, and J. Menard (1973), *C. R. Acad. Sci. (Paris)* **277D**, 2561.

106. C. Devaux, J. Menard, P. Sicard, and P. Convol (1976), *Eur. J. Biochem.* **64**, 621.

107. P. Corvol, C. Devaux, and J. Menard (1973), *FEBS Lett.* **34**, 189.

108. K. Murakami and T. Inagami (1975), *Biochem. Biophys. Res. Commun.* **62**, 757.

109. D. G. Deutsch and E. T. Mertz (1970), *Science* **170**, 1095.

110. D. G. Deutsch and E. T. Mertz (1970), *Fed. Proc. Fed. Am. Soc. Exp. Biol.* **29**, 647.

111. L. Summaria, F. Spitz, L. Arzadon, I. G. Boreisha, and K. C. Robbins (1976), *J. Biol. Chem.* **251**, 3693.

112. L. Summaria, L. Arzadon, P. Bernabe, K. C. Robbins, and B. H. Barlow (1973), *J. Biol. Chem.* **248**, 2984.

113. T. H. Liu and E. T. Mertz (1971), *Can. J. Biochem.* **49**, 1055.

114. W. J. Brockway and F. J. Costellino (1972), *Arch. Biochem. Biophys.* **151**, 194.

115. L. Holmberg, B. Bladh, and B. Astedt (1976), *Biochim. Biophys. Acta* **445**, 215.

116. T. Maciag, M. K. Weibel, and E. K. Pye (1974), in *Methods in Enzymology* W. B. Jakoby and M. Wilchek, eds.) **34**, 451.

117. D. Gurne, J. Chen, and D. Shemin (1977), *Proc. Nat. Acad. Sci.* **74**, 1383.

118. A. Kervabon, B. Albert, and A.-H. Etémati (1977), *Biochimie* **59**, 23.

119. S.-I. Ikeda and S. Fukui (1974), *Eur. J. Biochem.* **46**, 553.

120. I. C. Cho and H. E. Swaisgood (1971), *Biochim. Biophys. Acta* **258**, 675.

121. W. W.-C. Chan (1970), *Biochem. Biophys. Res. Commun.* **41**, 1198.

122. W. W.-C. Chan (1973), *Can. J. Biochem.* **51**, 1240.

123. W. W.-C. Chan and H. M. Mawer (1972), *Arch. Biochem. Biophys.* **149**, 136.

124. C. B. Anfinsen (1968), *Pure Appl. Chem.* **17**, 461.

125. J. Kato and C. B. Anfinsen (1969), *J. Biol. Chem.* **244**, 5849.

OTHER APPLICATIONS

Numerous, as yet unmentioned, applications of affinity techniques could be cited. Affinity chromatography have been used in purifying such diverse materials as whole cells or vitamins, resolving enantiomeric mixtures, or determining formation constants for enzyme–substrate complexes. Moreover, affinity chromatography is itself only one many aspects of solid-phase biochemistry, as the application of solid-phase technology to biochemical problems is termed. Among the other areas of solid-phase biochemistry, closely related to affinity chromatography, are solid-phase protein synthesis ("the Merrifield technique"), solid-phase protein sequencing (the "Laursen technique"), immobilized enzymes, and solid-phase radioimmunoassay. In this chapter we briefly outline these other applications of affinity chromatography and some of the other areas of solid-phase biochemistry.

11.1. PURIFICATION OF WHOLE CELLS AND VIRUS

The use of affinity chromatography in the purification of whole cells (1–14) and virus particles (15–20) has received considerable attention in a large number of laboratories. In general, a certain type of cell or virus with a unique surface characteristic, for example, specific surface antigens or receptor sites, can be readily separated from cells that lack these sites with the use of an affinity column. This technique is extremely useful in concentrating dilute solutions of virus or extracting one particular cell type from a mixed lyphocyte population. Unfortunately, cells are very sensitive toward the presence of protein and various other molecules that bind to their surface membranes. Such binding may trigger fundamental changes in their character. This may mean that a cell that is eluted from a matrix that binds specifically to the cell surface proteins may be a very different type of cell from what it was before it was bound to the affinity matrix. This possibility must be borne in mind in any subsequent experiments that might be performed with the isolated cells. With virus such changes are less likely, although they can be ruled out only by comparing the virus before and after purification with reference to those viral properties that are critical for any subsequent investigation.

Romanowska et al. (15) have utilized the lipopolysaccharide of *Shigella sonnei* to prepare an affinity matrix for both *Shigella* antibodies and for the *Shigella* bacterophage, SK VI, in a very typical example of viral purification by way of bioselective adsorption. The lipopolysaccharides were isolated from whole cells and were coupled by their free amine function to CNBr-activated agarose, although epoxide-activated agarose could also be employed. The latter binds considerably more lipopolysaccharide than does CNBr-agarose, but the resulting adsorbent has a lower capacity for bacterophage than does an adsorbent prepared using the CNBr-activated matrix.

Bacterophages were applied to the lipopolysaccharide–agarose column in the presence of Mg^{++} and Ca^{++}. The phage mixture was allowed to sit on the column, since for reasonable binding to occur a contact time at $0°C$ of at least 30 minutes was needed between the column and the phage. As shown in Table 11.1, higher temperatures results in more phage binding to the column, but, conversely, the bound phage could not as readily be removed in an active state. At $0°C$, 60% of the phage applied could be eluted. The same bioselective adsorbent could be used in a similar manner to purify *Shigella* antibodies from rabbit immune sera.

Other viral purifications performed with the use of affinity chromatography include Aleutian mink virus, purified on immunoglobin-agarose (16,17), foot-and-mouth virus (18), tobacco mosaic virus (19), and influenza virus (20). By far the largest use of bioselective adsorption in cell separation has been the isolation of specific cells from the immune system. Most of these separations have been done by using affinity chromatography on agarose or similar materials, but the introduction of nylon fibers and mesh (1) has considerably improved the purity of the final cell fractions. Undesired cells may be trapped between beads when a chromatographic method is used, but they can be readily and clearly washed from nylon mesh. Naturally, the capacity of nylon mesh is far less than that of beaded chromatographic matrices; thus yield is sacrificed for purity when the nylon mesh system is employed.

Figure 11.1 depicts the basic format used in the nylon fiber technique. First, the fiber is derivatized and reacted with an appropriate antibody or ligand to form a bioselective adsorbent. The fiber, held taut in a suitable frame, is slowly agitated in the mixture of cells to be fractionated. The unattached cells are washed away with buffer and the adsorbed cells are subsequently removed by plucking the string, thus shearing the cells free from it, or they can be removed by more classical elution methods. Often the nylon thread is coated with gelatin, thus making it possible to remove the cells by melting the gelatin at $37°C$. This procedure, since it is so very gentle, yields a larger proportion of viable and unaltered cells.

Typical applications of this technique include the separation of erythrocytes and thymocytes on conconavalin A bound-nylon fibers and the purification of DNP-hapten binding lymphocytes on nylon fibers containing either coatings

TABLE 11.1. Elution of Active Phage SKVI Attached to Phase II LPS$_A$-Sepharose 4B[a]

Temperature of Phage Binding, °C	Number of Phage Particles Attached	Effluent	Temperature of Phage Elution, °C	Number of Phage Particles Eluted	Active Phage Eluted, %
0	$1 \cdot 10^5$	2 M NaCl in 0.5-M borate buffer, pH 6.0	25	$6.2 \cdot 10^4$	62
0	$2.5 \cdot 10^{5\,b}$	0.05 EDTA in 0.01-M phosphate buffer, pH 7.3	25	$1.28 \cdot 10^5$	52
0	$1.06 \cdot 10^{6\,b}$	0.05-M EDTA in 0.5-M borate buffer, pH 7.3	20	$4 \cdot 10^5$	40
8 to 10	$1.56 \cdot 10^6$	2.5-M NaCl in 0.2-M glycine buffer, pH 4	10	$2.4 \cdot 10^4$	1.5
20	$1 \cdot 10^6$ – $1.7 \cdot 10^6$	2.5-M NaCl in 0.01-M phosphate buffer, pH 7.3 2.5-M NaCl in 0.2-M glycine buffer, pH 4 0.05-M EDTA in 0.01-M phosphate buffer, pH 7.3 0.9-M borate buffer, pH 6.0	20	$<2 \cdot 10^3$	0

Source: Reprinted with permission from Romanowska *et al.* (15).

[a]The gel (10 ml) was eluted with approximately 60 ml of effluent added in three portions during 45 to 60 minutes. The eluants were dialyzed 3 times against 0.9% NaCl in 0.001-M phosphate buffer, pH 7.3, before the determination of phage activity.

[b]The phage binding was carried out in the presence of 0.01-M MgCl$_2$, 0.003-M CaCl$_2$, and 0.003-M MnCl$_2$.

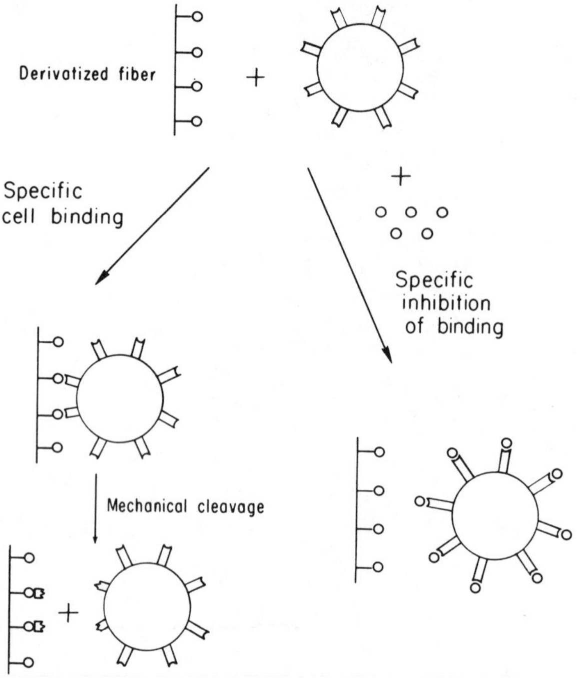

Figure 11.1. General scheme of fiber fractionation. Reprinted with permission from Edelman and Ruitishauer (1).

of DNP, DNP–albumin, or DNP–gelatin. As can be seen from Figures 11.2 and 11.3, this technique is readily applicable to the microscopic study of the isolated cells whereas, on the other hand, large-scale purifications of cells are feasible by using stacks of nylon mesh held in a suitable frame as shown in Figure 11.4. Fractionation of cells by bioselective *column* chromatography has been very successful despite the lack of a suitable support matrix. Agarose, when activated by CNBr, contains fissures that trap whole cells. Conversely, glass and plastics tend to nonspecifically adsorb many types of cell. Polyacrylamide beads seem to be the most suitable for cell purification. The usual problem in applying polyacrylamide to bioselective adsorption is the very small pore diameter of the derivatized bead. However, since only the bead surface is actually employed in cell purification, this objection doesn't apply to cell-chromatography. Truffa-Bachi and Wofsy (4) purified antibody-producing lymphocytes from azido-phenyl-β-lactoside immunized mice spleen cells on a column of acrylamide by using histamine as a spacer arm. The cells (108) were applied in a phosphate-saline buffer to a column containing this adsorbent. After the unabsorbed cells are washed away, the specific hapten binding fraction is eluted with a solution

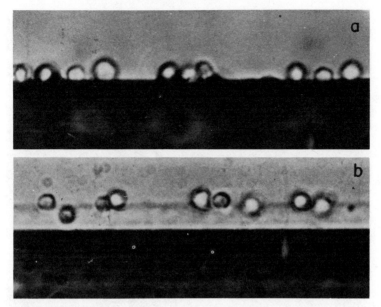

Figure 11.2. Mouse spleen cells bound to (*a*) dinitrophenylated (Dnp) bovine serum albumin fibers and (*b*) Dnp-gelatin fibers. Reprinted with permission from Edelman and Ruitishauer (1).

of the azidophenyl-β-lactoside hapten. Recovery of 99% of the cells was possible with a 500-fold enrichment in antilac lympocytes.

11.2. AFFINITY DENSITY PERTURBATION

Cell membranes or cell membrane fractions can be purified by methods similar to those described for whole cells. Another and even gentler bioselective method has been developed by Wallach (21–23). This technique, which has been termed *affinity density perturbation,* is based on bioselective adsorbents that alter the density of the desired membrane fragment. (See Figure 11.5.)

In the model system developed by Wallach, conconavalin A is labeled with ^{125}I to serve as a marker. The labeled conconavalin A was covalently coupled to bacteriophage by use of the glutaraldehyde method, although the choice of coupling method does not seem to be critical. The resulting conconavalin A phage is much denser than either membrane protein or unreacted conconavalin A and is purified by cesium chloride isopycnic centrifugation. The purified complex has a density of about 1.6 because of the presence of large amounts

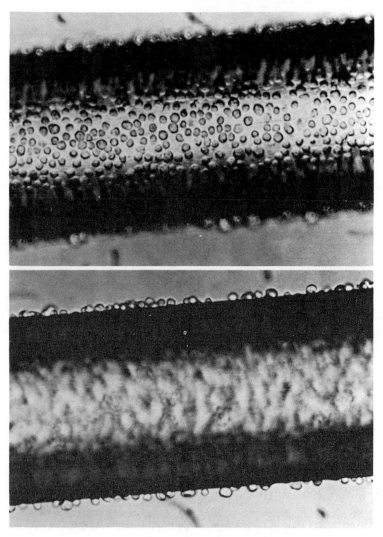

Figure 11.3. Mouse thymocytes bound to a concanavalin A-derivatized nylon fiber. Top: the field was focused on the face of the fiber by use of bright-field microscopy. Bottom: the field was focused on the edge of the fiber using bright-field microscopy. The fibers have a diameter of 250 μm. Reprinted with permission from Edelman and Ruitishauer (1).

Figure 11.4. Petri dish containing polyethylene frame strung with nylon monofilament. The length of the longest fiber segment is 2.5 cm. Reprinted with permission from Edelman and Ruitishauer (1).

of nucleic acid in the phage. In contrast, pure membrane proteins have a very sharp density distribution (1.18) in a cesuim chloride gradient.

A membrane fraction is subsequently prepared and mixed with the conconavalin A phage. Those proteins with glycosidic groups that bind conconavalin A will form a membrane protein–conconavalin A–phage complex with a broad density range of 1.30 to 1.34. This complex is separated from those proteins that do not bind conconavalin A by isopycnic certifugation in a cesuim chloride gradient. The isolated complex is then mixed with an excess of a sugar that competes with the membrane glycoproteins for the conconavalin A sugar binding site, after which the mixture is again centrifuged on a cesuim chloride gradient. The isolated glycoproteins now band well above the conconavalin A–virus complex.

This technique has not been widely employed but in theory could be used with almost any bioselective adsorbent by preparing antibody-linked phage or phage-containing immobilized bioligands (NAD, ATP, etc.). Perhaps one reason why affinity partitioning has been so seldom employed is that the biochemist is much more accustomed to purifying proteins by chromatographic procedures than by any other procedure and relegates density methods to the purification of nucleic acids and cell organelles. One real difficulty in using affinity partitioning is the fact that isopycnic density gradient materials, such as high cesuim chloride concentrations, must be employed, thus creating condi-

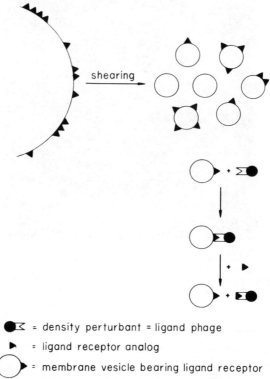

shearing

●◁ = density perturbant = ligand phage

▶ = ligand receptor analog

○▶ = membrane vesicle bearing ligand receptor

Figure 11.5. Schematic of affinity density perturbation applied to membrane subfractionation. A plasma membrane bearing multiple receptors is sheared into membrane fragments carrying different numbers of receptors in varying distributions. These are allowed to react with the ligand (◁) coupled to the density perturbant, that is, K_{29} phage (●), producing a membrane–receptor–ligand phase complex of higher density than the membrane itself and of lower density than the density perturbant. Addition of a low-molecular-weight dissociating agent (△) returns the membrane and density perturbant to their original densities. From D. F. H. Wallach, B. Kranz, E. Ferber, and H. Fischer, *FEBS Lett.* **21**, 29 (1972). Reprinted with permission from Wallach (21).

tions that may weaken bioselective interactions between proteins and bioligands. Conversely, the same high ionic strength would be favorable for hydrophobic bond formation. Therefore, phages with hydrophobic coatings might be adaptable to development of a hydrophobic density perturbation system.

11.3. AFFINITY ELECTROPHORESIS

Affinity electrophoresis is a very convenient analytical tool for determining binding constants between enzymes and their substrates or inhibitors as well as for predicting the probable results of affinity chromatography (24-26) experiments. Basically the technique involves electrophoresis in or through a matrix of an immobilized bioligand. The bioligand can be either a polymer entrapped in the acrylamide matrix, (for example, starch granules or Cibacron Blue–dextran), or the ligand can be present as a substituent on the gel monomer, as, for example, copolymers of alkenyl-O-glycosides with acrylamide.

Horesji and Kocourek (24) have described the copolymerization of allyl-O-α-D-galactose and allyl-O-α-D-mannose in a polyacrylamide mixture to analyze the presence of lectins in a phytohemmaglutinin preparation from pea seeds (*Pisum sativum*). First, a small pore separation gel was prepared, followed by an "affinity gel" containing the allyl–glycoside, and finally on the top is a large pore "stacking gel." The gel is next washed thoroughly for about 8 hours, after which the sample is applied and electrophoresis is performed in the usual fashion. If the concentration of allyl glycoside in the gel is large enough, the lectins will be totally adsorbed into the affinity layer. With lower allyl glycoside concentrations, the electrophoretic migration of the lectin is retarded relative to its migration in gels that do not possess a glycoside affinity layer.

Bøg-Hansen (27) has added the element of affinity electrophoresis to immuno-electrophoresis to form a crossed immunoaffinielectrophoresis system. Four different gels are used in this two-dimensional method. The gels are arranged as shown in Figure 11.6. Gel 1 is an ordinary agarose gel in which the protein components of the sample are separated. Gel 2 contains conconavalin A coupled to the agarose using CNBr, gel 3 is another simple agarose gel, and gel 4 is an antibody-containing agarose. By performing electrophoresis of human serum in such a system, Bøg-Hansen was able to ascertain which serum proteins were

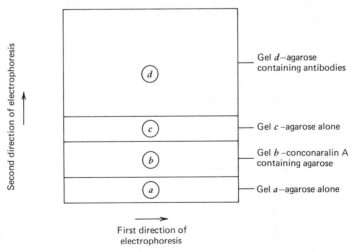

Figure 11.6. Crossed immunoaffinity electrophoresis.

glycoproteins. Prediction of affinity chromatography experiments can also be readily made using this procedure.

A simplified type of crossed affinity electrophoresis or rocket electrophoresis has also been employed to determine the glycoprotein content of serum samples (28). In this system only two gels are used: (1) a gel containing no adsorbent and then (2) electrophoresis in a second dimension through a gel containing conconavalin A or a similar lectin:

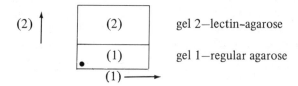

The result is that each glycoprotein forms a precipitate by complexing with the lectin. The distance from the origin of these precipitin lines or "rockets" is directly related to the amount of glycoprotein. Horesji et al. (32) have performed affinity electrophoresis to determine both the binding constants for lectins with free sugars and with immobilized saccharide-acrylamide complexes. The technique is quite different from that described previously in that the saccharide-acrylamide complexes are included in the entire acrylamide system. First, allyl glycoside (0.1 to 2 g), acrylamide (1 g), and ammonium persulfate are dissolved in 20 ml of water, and the mixture is boiled for 5 minutes. The mixture is then added to 20 ml of water and dialyzed extensively against water. The products are polymers of $2.5S$ to $3.5S$ (where S is the sedimentation rate) containing 3 to 30% sugar monomer as determinal spectrophotometrically. Polyacrylamide gels are now prepared by incorporating a defined amount of

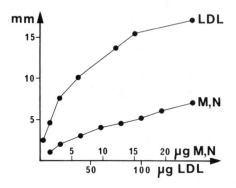

Figure 11.7. Correlation of rocket height and amount of purified protein in rocket-affinity electrophoresis; purified human serum LDL into conconavalin A (30 μg cm^2) and purified human erythrocyte M, N-glycoprotein into wheat germ agglutinin (50 μg cm^2). Reproduced with permission from Bϕg-Hansen (27).

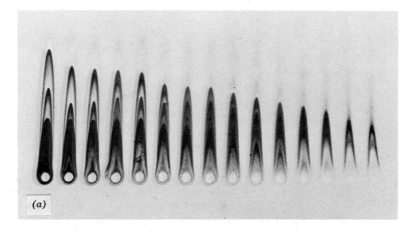

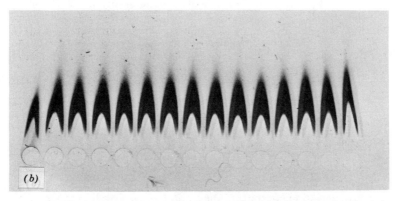

Figure 11.8. Analysis of heat-denatured concanavalin A-binding proteins from human serum (*a*) rocket immunoelectrophoresis of 3-μl samples in a gel containing antibodies against human serum proteins (μl/cm^1), (*b*) rocket-affinity electrophoresis of 15-μl samples in a gel containing concanavalin A (24 μl/cm^2). Purified concanavalin A-binding proteins were denatured at 85°C for 0, 1, 2, 4, 8, 16, 24, 32, 40, 48, 60, 90, 120, 180, and 245 minutes, respectively, in dilute Tris-veronal, pH 8.6. Reprinted with permission from Bϕg-Hansen (27).

the saccharide–acrylamide copolymer. In addition, various amounts of a competitive free saccharide is included in various gels. As seen in Figure 11.9, the lectin is strongly retarded in the allyl-mannosyl copolymer-containing gel, but not retarded at all in a control gel. The addition of free D-mannose to the gels increases the mobility of the lectin. This increase is dependent on the affinity of the lectin for both free and immobilized saccharide, the dependence

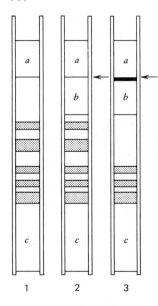

Figure 11.9. Electrophoresis of the protein fraction with hemagglutinating activity from peak seeds: (a) large-pore gel; (b) affinity gel; (c) small-pore gel; (1) control standard gel; (2) affinity gel (b), formed by 0-α-p-galactopyranosyl polyacrylamide; (3) affinity gel (b) formed by 0-α-p-mannosyl polyacrylamide. Arrows in 2 and 3 indicate boundary between large-pore gel and affinity gel. The large-pore gel layer has been removed. Reprinted with permission from Horejsi (33).

following the equation:

$$\frac{l}{l_0 - l} = \frac{K_i}{c_i}\left(1 + \frac{c}{K}\right)$$

where K_i = dissociation constant of the lectin–saccharide–copolymer

K = dissociation constant of the lectin-free saccharide complex

c_i = concentration of immobilized saccharide

c = concentration of free sugars

l = mobility of lectin at any given c and c_i

l_0 = mobility of lectin under identical conditions in the absence of immobilized saccharide

This equation comes directly from the derivation by Dunn and Chaiken (34) for the method of determining equilibrium constants in column affinity chromatography experiments. The elution volumes of the equation of Dunn and Chaiken were replaced by the electrophoretic migration distance.

If ($l/l_0 - l$) is plotted against c, a straight line results with a y intercept equal to K_i/c_i. Thus K_i can be easily determined at any known c_i. If c is 0, then $K_i = (l/l_0 - l)(c_i)$.

The x intercept, on the other hand, yields K directly: $-c = K$, when l/l_0 $- l = 0$.

Figure 11.9 depicts the determination of K and K_i for various sugar–lectin interactions, using $L.$ $sativin$ lectin as an example. Using similar gels, Horesji et al. have calculated the K values for a large number of lectin–saccharide complexes. These measurements are performed very easily with much greater precision than possible with the use of any of the column affinity chromatographic techniques described to date. These authors have also (33) used this technique to study the alterations effected on the binding constants through chemical modification or demetalization of various lectins. A similar technique (35) has been employed by using immobilized Blue Dextrin in the polyacrylamide gels to study the interaction between various dehydrogenases and the dextran per se. (See Table 11.2.)

Although affinity electrophoresis is more convenient and often more accurate than other affinity techniques that have been used to determine equilibrium constants, there does exist the possibility that the equilibrium constant can be altered by the electric field. Caron et al. (29–31, 36–39) have noticed this effect in their studies of various antibody–antigen interactions. This effect is minimized when a competitive free legand is employed, as is done in the technique employed by Horejsi; thus this technique is to be preferred over the systems of Caron (36), Bøg-Hansen (40), and Morgan (41) when large electric field effects are suspected to be present. This effect is not normally great, with the K' between serum albumin antibodies and serum albumin ranging from 3.3×10^{-4} M^{-1} to 0.7×10^{-4} M^{-1} over an approximately three-fold span in electric field.

One could even consider an electric field to be a mild deforming medium that can be readily varied to give the desired effect. Certainly more work is needed to establish firmly the basis of this effect and to apply it to such uses as electrophoretic elution from affinity matrices (see page 283).

11.4. AFFINITY PARTITIONING

Aqueous solutions of hydrophilic polymers can form two-phase, water-rich systems in which proteins, nucleic acids, and other biomolecules can be partitioned without danger of denaturation (42). In fact, such polymeric solutions often stabilize otherwise labile enzymes. One such system is composed of aqueous solutions of polyethylene glycol and either dextran or dextran sulfate. When protein is mixed in such a two-phase system, some proteins will favor one phase and some will favor the other phase, depending on such factors as pH, ionic strength, and the molecular size of the polymer components of the system.

TABLE 11.2. Various Modifications of Affinity Electrophoresis with Interacting Components and Some Applications[a]

No.	Modification	Interacting Component that is Electrophoresed	Interacting Component in the Medium	Purpose	Reference[b]
1.	Crossing diagrams	Enzyme Glycoproteins	Substrate in-inhibitor Lectin	Identification of interacting components	2
2.	Affinity electrophoresis	Lectin Enzyme	Carbohydrate Substrates	Identification of lectins Determination of dissociation constants	3 4
3.	One-dimensional affinity electrophoresis	Glycoprotein enzymes	Immobilized lectin	Identification of glycoprotein enzymes	5
4.	Rocket-affinity electrophoresis	Glycoproteins Lectin	Lectin Glycoproteins	Quantification of glycoproteins Quantification of lectins	1
5.	Fused rocket-affinity electrophoresis	Glycoproteins	Lectins	Analysis of Fractionations Progressive changes Treatments	1
6.	Crossed-affinity electrophoresis	Lectin Glycoproteins	Glycoproteins Lectin	Identification of lectins Identification and quantification of lectin-binding glycoproteins	1
7.	Crossed immuno-electrophoresis with ligand in first dimension	Glycoproteins Lectins	Lectin Glycoproteins	Identification of interacting components Determination of binding specificity	6 6

318

8.	Crossed immuno-electrophoresis with ligand in an intermediate gel	Glycoproteins	Lectins	Analysis of microheterogeneity	Determination of dissociation constants	Identification of interacting components	Partial characterization of (number of binding sites)
		Lectins	Glycoproteins	6	1	8	7

Source: Reprinted with permission from T. C. Bøg-Hansen, O. J. Bjerram, and C. H. Brogren (1977), *Anal. Biochem.* **81**, 78.

[a] Agarose gel was used as supporting medium except in 1 (paper) and 2 (polyacrylamide gel).

1. T. C. Bøg-Hansen, O. J. Bjerram, and C. H. Brogren (1977), *Anal. Biochem.* **81**, 78.
2. S. Nakamura (1966), *Cross Electrophoresis. Its Principles and Applications*, Igaku Shoin, Tokyo and Elsevier, Amsterdam.
3. V. Horeŝji and J. Kocourek (1974), *Biochim. Biophys. Acta* **336**, 338–343.
4. K. Takeo and S. Nakamura (1972), *Arch. Biochem. Biophys.* **153**, 1.
5. T. C. Bøg-Hansen, C. H. Brogren, and I. McMurrough (1974), *J. Inst. Brew.* **80**, 443–446.
6. T. C. Bøg-Hansen, O. J. Bjerram, and J. Ramlau (1975), *Scand. Immunol.* **4**, Supplement 2, 141–147.
7. T. C. Bøg-Hansen and C. H. Brogren (1975), *Scand. J. Immunol.* **4**, Supplement 2, 135–139.
8. T. C. Bøg-Hansen (1973), *Anal. Biochem.* **56**, 48.

To form an affinity partitioning system (43–45), a bioligand is covalently attached to one of the polymers forming the two-phase system. Since under very mild conditions many proteins favor the polyethylene glycol phase, the bioligand may be bound to the polydextran. Chemically this is the easiest of the two polymers to derivatize, and polydextran with pre-activated epoxy residues is commercially available. Other derivatives can be prepared using the same techniques that have been applied to agarose.

Flanagan and Baronaes (45) demonstrated that conconavalin A, which has receptor sites that bind to dextrin, partitioned almost exclusively in the dextrin phase of a polyethylene glycol–polydextrin system. However, when a soluble competitive inhibitor of conconavalin A was added to the system (as, e.g., D-mannose) then the partitioning effect was nullified with a K_d approaching 1 as the concentration of free mannose approached saturation. The same investigators also used dinitrophenol (DNP)-substituted polyethylene glycol to separate normal IgB (which does not bind DNP) from the S-23 mycloma protein, which has a strong tendancing to bind DNP. In this system (DNP-polyethylene glycol: polydextrin), the S-23 protein normally favors the polyethylene glycol phase. However, when DNP lysine is added to the system, the partitioning effect is reversed. As shown in Figure 11.9, countercurrent distribution of these proteins in this system separated the two proteins.

Affinity partitioning, like other partitioning systems, probably will be of limited use until a better method of applying it becomes available. Currently it remains a tedious process unless the partitioning coefficients are very large. One example of such a system is the purification of serum albumin using a palmitic acid derivative of polyethylene glycol (P-PEG). With no P-PEG present in the polyethylene glycol phase, serum albumin is found completely in the polydextrin phase. Conversely, it is 74% in the polyethylene glycol phase if P-PEG replaces 10% of the PEG. Even larger amounts will distribute into the PEG phase if K_2SO_4 is added and the pH is chosen carefully.

The chief advantage of affinity partitioning is the gentleness of the method. Since whole cells and organelles can be separated by similar partitioning methods, affinity partitioning may be a superior tool when applied to whole cells and fragile cell organelles.

11.5. IMMOBILIZED PROTEIN-MODIFICATION REAGENTS

Certain reactive functional groups of proteins can be readily modified by using selectively reactive reagents (46). For example, cysteine sulfhydryls add, through a Michael-type reaction to N-ethyl maleimide or other alkyl maleimides. Sulfhydryls can also displace halides from various α-keto halides.

Normally these reagents are employed in aqueous solution. The rate of re-

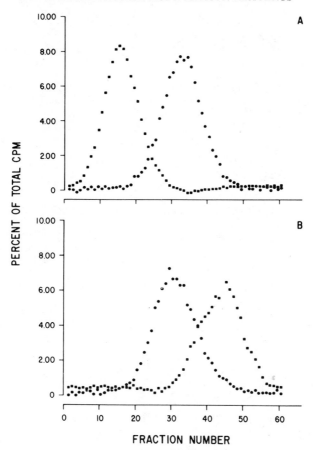

Figure 11.10. Countercurrent distribution of ^{125}I-labeled *S*-23 myeloma protein (■) and ^{131}I-labeled γ-globulin (●): (*A*) in the presence of 5.4×10^{-4} moles of ε-*N*-Dnp-lysine per kilogram of phase system; (*B*) in the presence of 2.6×10^{-4} moles of DNP-poly (EO) per kilogram of phase system. The phase system contained 55 g of dextran T-500, 40 g of poly (EO), 6000, 80.9 mmoles of NaCl, and 8.7 mmoles of sodium phosphate, pH 6.8 per kilogram of phase system; 10^{-9} moles of each protein was added to the first chamber of the apparatus. Reprinted with permission from Flanagan and Barondes (45).

action is dependent on such factors as the pH of the solution, the pK of the reactive group (which varies widely with the nature of its microenvironment) and the accessibility of the reactive amino acid. (See Table 11.3.) Often the rate of its reaction is taken as a measure solely of the accessibility of a particular amino acid on a protein surface but this isn't necessarily true; a very accessible amino acid may be unreactive toward one or more reagents as the result of the microenvironment of the amino acid whereas a partially buried amino acid

TABLE 11.3. Side-Chain Reactivities[a] of Various Reactive Amino Acids

Reagent	$-NH_2$	$-SH$	$-OH$ (phenol)	$-NH$ (imidazole)	$-NH-C(=\overset{+}{N}H_2)NH_2$	$-COOH$	$N-H$ (indole)	$-S-S-$	$-SCH_3$
Acetic anhydride	+++	+++[b]	+++[c]	+++[b]	—	—	—	—	—
N-Acetylimidazole	±±	+++[b]	+++[c]	+++[b]	—	—	—	—	—
Acrylonitrile	±±	+++	—	—	—	—	—	—	—
Aldehyde–NaBH4	+++	—	—	—	—	—	—	—	—
N-Bromosuccinimide	—	+++	++	—	—	—	+++	—	—
N-Carboxyanhydrides	+++	—	—	+[b]	—	—	—	—	—
Cyanate	+++	+++[b]	+[b]	—	—	+[b]	—	—	—
Cyanogen bromide	—	+	—	—	—	—	—	—	+++
1,2-Cyclohexanedione	±	—	—	—	+++	—	—	—	—
Diacetyl trimer	+	—	—	—	+++	—	—	—	—
Diazoacetates	—	++	+++	+++	—	+++	—	—	—
Diazonium salts	+++	+	++	+++	+	—	+	—	—
Diketene	+++[c]	—	+	—	—	—	—	—	—
Dinitrofluorobenzene	+++	+++	++	++	—	—	—	—	—
5,5'-Dithiobis(2-nitrobenzoic acid)	—	+++[c]	—	—	—	—	—	—	—
Ethoxyformic anhydride	+++	—	—	+++[c]	—	—	—	—	—

Reagent								
Ethylenimine	–	+++	–	–	–	–	–	+
N-Ethylmaleimide	±±	+++	–	–	–	–	–	–
Ethyl thiotrifluoro-acetate	+++^c	–	–	–	–	–	–	–
Formaldehyde	+++	+++	+++	+	–	+	–	–
Glyoxal	++	–	–	–	+++	–	–	–
Haloacetates	+	+++	–	–	++	–	+	+
Hydrogen peroxide	–	++	–	–	–	+	+	+++
2-Hydroxy-5-nitro-benzyl bromide	–	++	–	–	–	+++	–	–
Iodine	–	+++	+++	+++	–	–	–	–
o-Iodosobenzoate	–	+++	–	–	–	–	–	–
Maleic anhydride	+++^c	++^c	++^c	++^c	–	–	–	–
p-Mercuribenzoate	–	+++	–	–	–	–	–	–
Methanol–HCl	–	–	–	–	+++	–	–	–
2-Methoxy-5-nitrotrpone	+++^c	–	–	–	–	–	–	–
Methyl acetimidate	+++	–	–	–	–	–	–	–
O-Methylisourea	+++	–	±	–	–	–	+	–
Nitrous acid	+++	+++	+	–	–	–	+	+
Performic acid	–	+++	–	–	–	++	+++	+++
Phenylglyoxal	++	–	–	–	+++	–	–	–
Photooxidation	–	+++	++	+++	–	+++	±	+++
Sodium borohydride	–	–	–	–	–	–	+++	+++
Succinic anhydride	+++^c	++^c	++^c	+++	–	++^c	–	–
Sulfenyl halides	–	+++	–	–	–	+++	–	–
Sulfite	–	+++^c	–	–	–	–	+++^c	–
Sulfonyl halides	+++	+++	+++	+++	–	–	–	–
Tetranitromethane	–	+++	+++	–	–	–	–	+
Tetrathionate	–	+++	–	–	–	–	–	–

323

(*Continued*)

TABLE 11.3. (*Continued*)

Thiols	—	—	—	—	—	—	+++	—
Trinitrobenzene-sulfonic acid	+++	++[c]	—	—	—	—	—	—
Water-soluble car-bodiimide and nucleophile	±	±	±	—	+++	—	—	—

Source: Reprinted with permission from G. E. Means and R. E. Feeney (1971), *Chemical Modification of Protein*, Holden Day, San Francisco.

[a]—, +, ++, and +++ indicate relative reactivities; ±, ±±, and ±±± likewise indicate relative reactivities that may or may not be attained, depending on the conditions employed.

[b]Spontaneously reversible under the reaction conditions or on dilution, regenerating original group.

[c]Easily reversible, regenerating original group.

324

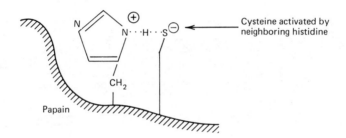

$$R-SH + I-CH_2-\overset{O}{\overset{\|}{C}}-R' \rightarrow R-S-CH_2-\overset{O}{\overset{\|}{C}}-R' + HI$$

may be very reactive as the result of an "activating" microenvironment. For example, a histidine adjacent to the active site sulfhydryl of papain greatly accelerates its reactivity:

Cysteine activated by neighboring histidine

Papain

In the author's laboratory (47–49) and in those of Patchornick (50–52) and Neckers (53,54), protein-modification reagents have been immobilized on inert supports. The resulting immobilized protein modification reagents possesses two inherent advantages over the use of soluble reagents, namely (1) they are readily removed by simple filtration and (2) they may be so designed as to sterically restrict their reactivity to the surface of simple proteins, macromolecular complexes, membranes, and even whole cells. (74,75)

Probably the first example of an immobilized protein modification reagent was the finding by Wilchek and Topper (55,56) that insulin, which had been immobilized on cyanogen bromide-activated agarose, could be released from the agarose by incubating it with nucleophiles such as primary amines, sulfhydryls, or even other proteins. The released insulin exhibited a transient degree of increased reactivity (particularly toward brown fat cells) and thus was termed "super insulin." The nature of the transient increased activity has not yet been fully determined, but it has been shown (57–59) that isourea functions (through which proteins are attached to CNBr-activated agarose) will undergo nucleophilic cleavage with ⁻OH, RNH_2 or R–SH groups, yielding a protein with a urea, guanidine, or thiourea function, respectively. The end result is the change of a protein amine group into one of these functions. Since in most cases only surface amines react with CNBr-activated agarose, this can be thought of as a stereoselective modification of lysyl residues on the protein surface.

Kamido and Nikaido (60) have used a reverse form of this procedure by adding CNBr activated soluble polydextrin to *Salmonella typhimurium* cells and to invented vesicles from the cellular membrane. The protein on the outside of the outer cell membrane (or vesicle membrane) reacted with the polydextran to form very large aggregates that could not enter polyacrylamide gels. Comparison of the electrophoretic pattern of membrane proteins before and after treatment with CNBr-activated polydextrin revealed the loss of several proteins on treatment with the activated dextrin. Using this method, they were able to deduce which proteins were exposed on the outside of the membrane, which were on the inside of the membrane, and which proteins were completely shielded by the membrane.

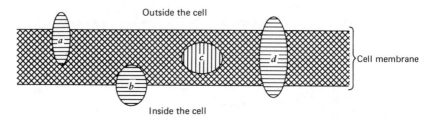

The four types of cell membrane protein are (1) exposed to the outside, (2) exposed to the inside, (3) shielded and buried, or (4) extending through both sides of the membrane. Conceivably the same type of information could also be obtained by using CNBr-activated agarose beads, which would immobilize only the exposed membrane proteins. The immobilized protein could then be released by using $^{14}CH_3NH_2$, thus radioactively labeling only the exposed membrane proteins. Such experiments have not yet been reported in the literature.

Rubenstein et al. (52) have isolated tryptophan-containing peptides by using aryl sulfenyl chlorides $(Ar-S_xCl)$ immobilized on polyacrylamide matrices. The synthesis and use of these polymeric reagents as shown in Figures 11.11 and 11.12 is extremely simple. When either reagent B or D is added to tryptophan or tryptophan-containing peptides in dilute acetic acid (15 to 100%), tryptophan reacts very rapidly, with 90% removed in less than 2 minutes. The only competing reaction under these conditions is with sulfhydryl groups of the peptide. To prevent these groups from reacting, they can be blocked with DTNB prior to the reaction. Unfortunately, sulfenyl chlorides have so far

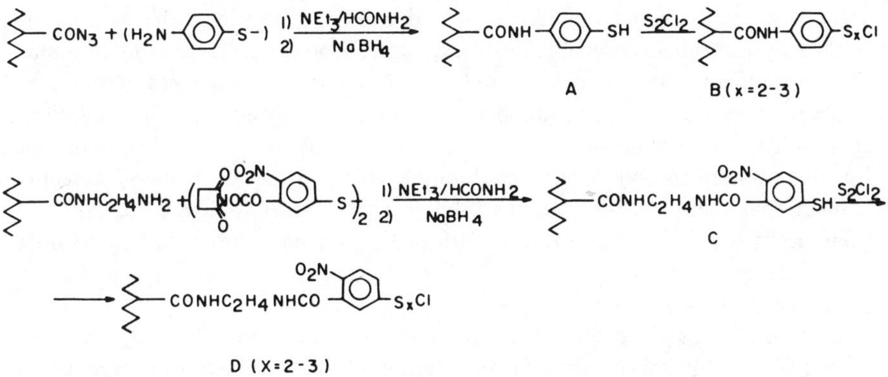

Figure 11.11. Preparation of the polymeric reagents B and D. Reprinted with permission from M. Rubenstein, Y. Schechter, and A. Patchornik (1970), *Biochem. Biophys. Res. Commun.* **70**, 1257.

found application only with small peptides, for example, proteolytic hydrolysates, chiefly because of the harsh acidic conditions needed for the reaction. Rubenstein et al. (52) did show that the immobilized sulfenyl chloride would be readily employed to purify tryptophan peptides from hydrolysates from proteolysis of serum albumin and glucagon. The resulting peptides were homogeneous and were isolated in 49 to 70% yield.

Shechter et al. (53) have also prepared a chloracetamide–acrylamide reagent that specifically reacts at $\sim$ pH 3.5 with methionyl residues. Peptides that contain methinione can be isolated from proteolytic digests in a fashion similar to that used with acyl sulfenyl halides for the isolation of tryptophanyl peptides. More importantly, not only can the peptides be isolated by using reducing

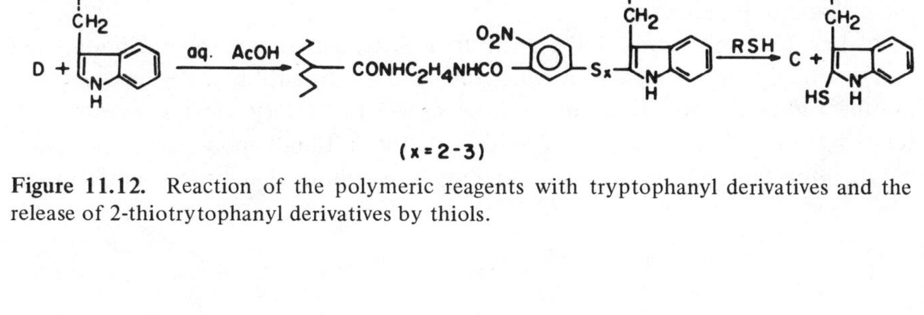

Figure 11.12. Reaction of the polymeric reagents with tryptophanyl derivatives and the release of 2-thiotrytophanyl derivatives by thiols.

agents, but the isolated peptides can, if desired, be cleaved at the methionine residue by a mechanism analogous to cyanogen bromide cleavage at methionine. Further, the cleavage is much simpler than ordinarily employed with CNBr cleavage in that all that is required is that the immobilized methionyl-peptide-acrylamide complex be heated in water for 4 hours at 110°C. The conditions for the immobilization of the methionine (pH 3.3, 37°C, 6 days) are mild enough that many of the proteins used in this study retained 40 to 100% of their activity after immobilization, although many proteins, other than those used here, might well be denatured on prolonged exposure to such acidic pH. (See Figure 11.13.)

Scouten (48) has immobilized a large variety of protein-modification reagents through cleavable spacer arms in an attempt to prepare insoluble reagents for protein modification that could be utilized under mild, nondenaturing, conditions. A general scheme for the preparation and use of such immobilized reagents is shown in Figure 11.14. A wide variety of cleavable spacer arms have been reported in the literature and are commercially available. Those used in the author's laboratory include compounds containing vicinal hydroxyls (cleaved with metaperiodate), these containing disulfides or diazolinkages (cleaved by reduction with dithionite, borohydride, or 2-mercaptoethanol), and those containing a labile ester (cleaved by way of hydroxylamine or other mucleophiles). Anilinonaphthalene maleimide has, as one example, been employed to label the reactive sulfhydryl of papain and the nonessential cysteine from lipoamide dehydrogenase (61).

11.6. AFFINITY DRUGS

Polymeric affinity drugs, drugs that are directed toward specific cells, for example, tumor cells, have been developed independently in several laboratories, most notably those of Wilchek (62) and Goldberg (63). The basic principle of polymeric affinity drugs is to attach a drug to a polymer that in turn is also attached to a molecule, for example, an antibody, which will direct the polymer-bound drug to a particular area or cell type in the body. The polymeric drug may be effective by itself, but in many instances the drug, once it finds the cell toward which it is directed, needs to be released from the polymer to penetrate the cell and to be effective.

Wilchek (62) immobilized the antitumor drugs adriamycin and dannomycin either directly or through a polydextrin support, to antibodies against various murine tumor strains. These drugs were shown to be very specifically directed toward tumor cells when conjugated with the antibody-polymer. This could be an important finding in chemotherapy since most chemotherapeutic drugs

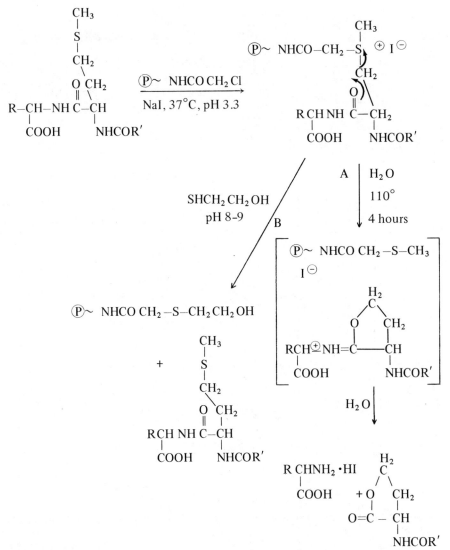

Figure 11.13. Covalent Chromatography of Methionine. Reprinted with permission from Schechter et al., *Biochemistry* **16**, 1424 (1977).

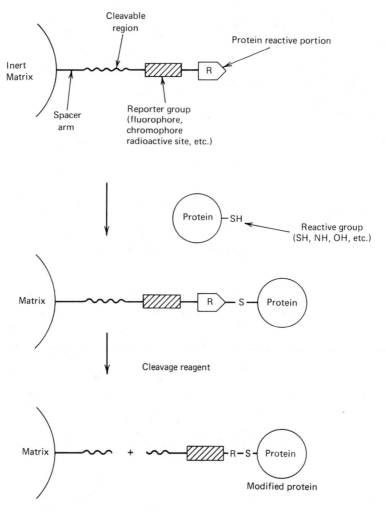

Figure 11.14. Generalized scheme for the application of immobilized protein modification reagents.

also exert a cytotoxic effect against normal cells, limiting the maximum usable dosage. If antibody-directed drugs are used, it should be possible to employ lower dosages, with virtually all of the drug directed to the affected cells. Other types of bioaffinity ligands (lectins, plasma protein, hormones, etc.) could be employed instead of antibodies to direct drugs to various other types of cells or organs.

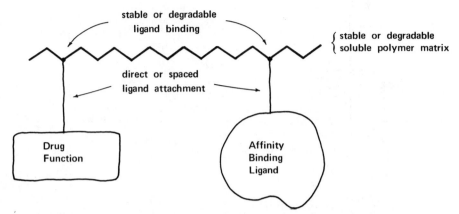

Figure 11.15. Model polymeric affinity drug (PAD). Reprinted with permission from E. P. Goldberg (1978), *Polymeric Drugs* (L. G. Donaruma and O. Vogl, eds.). Copyright © Academic Press, 1978.

11.7. SEPARATION OF ENANTIOMERS

Separation of enantiomeric mixtures by chemical and mechanical methods is tedious and often requires a time-consuming trial-and-error approach. Resolution of enantiomers by biological methods usually entails the modification and/or destruction of one of the isomers and is normally no less difficult than the chemical methods. Recently, however, approaches based on biological specificity of immobilized enzymes or other materials with bioselected adsorption characteristics have yielded rapid methods for efficient and non-destructive isolation of a variety of biologically important enantiomeric compounds. Stuart and Doherty (64) employed bovine serum albumin–agarose column for the resolution of DL-tryptophan. Serum albumin has a strong tendency to bind L-tryptophan, but not D-tryptophan, and thus D-tryptophan passes directly through the serum albumin–agarose whereas L-tryptophan can be eluted only with a deforming buffer of $0.1\text{-}M$ acetic acid. If the serum albumin is bound to agarose by a suitable spacer arm, the products of D,L-tryptophan resolution are essentially 100% optically pure, with yields of 67 to 100%. Serum albumin bound directly to agaraose was not suitable for the separation of tryptophan racemates, possibly because of small allerations in the protein that occur in the immobilization process. (See Figure 11.16.)

Several interesting reports have been presented in which the biospecificity of enzymes has been built into nonbiological support materials. Becket and Ander-

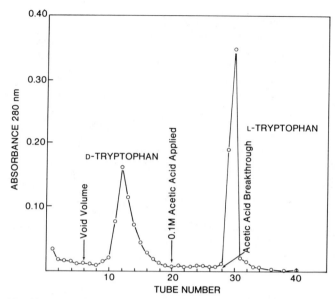

Figure 11.16. Chromatography of DL-trytophan on defatted bovine serum albumin–succinoylaminoethyl–agarose. DL-Tryptophan (500 nmole) dissolved in 0.1 ml of 0.1-M borate buffer (pH 9.2) containing 1% (v/v) $(CH_3)_2SO$ was applied to a 0.9-cm × 25-cm column of defatted bovine serum albumin–succinoylaminoethyl–agarose. The column contained a total of 630 nmole of bovine serum albumin. The column was eluted at 30 ml/hour with the borate buffer [without $(CH_3)_2SO$] for 20 tubes and then with 0.1-N acetic acid. The void volume was determined from the elution volume of $(CH_3)_2SO$. Reprinted with permission from Stewart and Doherty (64).

son (65–67) have dried silica gel preparations in the presence of optically pure quinine, levophanol, dextrophan, and similar biologically important compounds. After drying, the organic material was washed out of the dry silica by reflex in methanol. This supposedly leaves holes or "footprints" in the silica into which one of the enantiomeric compounds could fit. Separations of dextrophan and its enantioner could be performed by chromatography on this material, although the purity of the product did not match that achievable with the immobilized serum albumin–D,L-tryptophan system.

More recently Wulff et al. (68–70) have prepared what they termed "enzyme-analogue built polymers" by copolymerizing styrenyl borate derivatives of various optically pure enantiomers of biological importance followed by hydrolyzing the enantiomer, again leaving a sterically designed holes in the matrix, as for example, with D-glyceric acid:

E.   F.

The resulting polymer binds D-glyceric acid in preference to L-glyceric acid, although chromatography of a D,L-glyceric acid mixture on this medium did not yield products with optical purity above 87% in any of the systems used and often lower. Nonetheless, the synthesis of nonbiological polymers that have binding specificities this great suggests that the synthesis of such polymers, in an attempt to make an analogue of an enzyme active site, holds much promise. The great advantage of such systems, once developed, is that such polymers are not denatured by conditions that are ordinarily deleterious to enzymes and other proteins.

11.8. SOLID-PHASE RADIOIMMUNOASSAY
AND ENZYME-LINKED IMMUNOASSAY

Radioimmunoassay, developed by Yallow and Berson in 1960 (71) has become such an important tool in clinical chemistry that Yallow in 1977 received the Nobel prize for her part·in its development. In its simplest form, radioimmunoassay is based on the ability of an antibody to bind radioactive antigen and the competitive inhibition of this binding by unlabeled antigen. The bound antigen–antibody may form a precipitate that can be separated from the antigen fraction, and one or both of these is measured by scintillation counting.

The processes of precipitate formation, which may require 30 minutes to 4 hours, and separation of the bound and unbound antigen are greatly facilitated if the antibody is covalently bonded to a solid phase. After an appropriate incubation period, the antigen–antibody matrix is removed by centrifugation, washed, and its radioactive content determined. The procedure is a competitive one in which an unknown quantity of unlabeled antigen in serum, urine, cell extracts, or similar samples is competing with a known quantity of radioactive antigen. Then the result is an inverse relationship between the radioactivity in the antigen–antibody complex and the antigen present in the original sample. The amount of antigen present in the sample is determined with the aid of a standard curve created with various known concentrations of unlabeled antigen in place of the sample.

Solid-phase radioimmunosassay was first introduced by Catt and Tregar (72) in 1966, 6 years after the first radioimmunoassays. They used the iso-thiocyanate derivative of a copolymer of styrene and tetrafluoroethylene as a support matrix to which they bound antibodies against the human growth hormone, somatotropin. Since then, many other different types of support matrix, including porous glass nylon tubes and magnetic particles, have been used as immunosorbents in solid-phase radioimmunoassays. A wide variety of antigens have been determined by this technique, including digitoxin, thyroxine, and triiodothyronine. Many of these procedures have been automated.

The most significant advantage of solid-phase over nonsolid phase radio-immunoassay is that in the former the formation of the antibody–antigen complex is essentially irreversible since the immobilized complex is removed from the normal thermal motion that makes precipitation in ordinary immuno-assay a reversible process. This fact offers several advantages; first, the sample can be added, incubated, and after incubation, the excess sample can be removed by filtration. The radioactive antigen is then added, and, instead of competing with the antigen from the sample, the radioactive antigen simply binds to the unoccupied antibody-binding sites. This confers greater sensitivity to the procedure and also increases the rate of the binding process, thus decreasing the time needed per assay. Finally, the bound antibodies are more stable than the free antibody (73).

References

1. G. M. Edelman and U. Ruitishauer (1974), in *Methods in Enzymology* (W. B. Jakoby and M. Wilchek, eds.) 34, 195.
2. U. Ruitishauer, P. d'Eustachio, and G. M. Edelman (1973), *Proc. Nat. Acad. Sci. (USA)* 70, 3894.
3. U. Ruitishauer and G. M. Edelman (1972), *Proc. Nat. Acad. Sci. (USA)* 69, 3774.
4. P. Truffa-Bachi and L. Wofsy (1970), *Proc. Nat. Acad. Sci. (USA)* 66, 685.

5. H. Wigzell and B. Anderson (1969), *J. Exp. Med.* **192**, 23.

6. J. B. Robins and R. Scheerson (1974), in *Methods in Enzymology* (W. B. Jakoby and M. Wilchek, eds.) **34**, 703.

7. Z. Eshhar, T. Waks, and M. Bustin (1974), in *Methods in Enzymology* (W. B. Jakoby and M. Wilchek, eds.) **34**, 750.

8. Z. Eshhar (1973), *Eur. J. Immunol.* **3**, 668.

9. M. Bustin, Z. Eshhar, and M. Sela (1972), *Eur. J. Biochem.* **31**, 541.

10. D. D. Soderman, J. Germershausen, and H. M. Katzen (1973), *Proc. Nat. Acad. Sci. (USA)* **170**, 792.

11. G. N. Trump (1973), *Biochem. Biophys. Res. Commun.* **54**, 544.

12. N. S. Abdon and M. Richter (1969), *J. Exp. Med.* **130**, 141.

13. B. R. Venter, J. C. Venter, and N. O. Kaplan (1976), *Proc. Nat. Acad. Sci. (USA)* **73**, 2013.

14. W. H. Evans, M. G. Mage, and E. A. Peterson (1969), *J. Immunol.* **102**, 899.

15. E. Romanowska, C. Lugowski, and M. Mulczyk (1976), *FEBS Lett.* **66**, 82.

16. A. S. Kenyon, J. E. Gander, O. Lopez, and R. A. Good (1973), *Science* **179**, 187.

17. J. W. Yoon, A. S. Kenyon and R. A. Good (1973), *Nature New Biol.* **245**, 205.

18. H. D. Matheka and M. Mussgay (1969), *Arch. Ges. Virusforsch.* **27**, 13.

19. G. E. Galvez (1966), *Virology* **28**, 171.

20. C. Sweet, J. Stephen, and H. Smith (1974), *Immunichemistry* **11**, 295.

21. D. F. H. Wallach (1974), in *Methods in Enzymology* (W. B. Jakoby and M. Wilchek, eds.) **34**, 171.

22. D. F. H. Wallach, B. Kranz, E. Ferber, and H. Fischer (1977), *FEBS Lett.* **21**, 29.

23. D. F. H. Wallach and P. S. Lin (1973), *Biochim. Biophys. Acta* **300**, 211.

24. V. Horesji and J. Kocourek (1974), in *Methods in Enzymology* (W. B. Jakoby and M. Wilchek, eds.) **34**, 178.

25. V. Horesji and J. Kocourek (1974), *Biochim. Biophys. Acta* **336**, 338.

26. K. Takeo and S. Nakamura (1972), *Arch. Biochem. Biophys.* **153**, 1.

27. T. C. Bøg-Hansen (1973), *Anal. Biochem.* **56**, 480.

28. T. C. Bøg-Hansen, O. J. Bjerrum, and J. Ramlau (1975), *Scand. J. Immunol.* **4**, Supplement 2, 141.

29. M. Caron, A. Faure, and P. Cornillot (1975), *J. Chromatogr.* **103**, 160.

30. M. Caron, A. Faure, and P. Cornillot (1976), *Anal. Biochem.* **70**, 295.

31. M. Caron and R. J. Lefkowitz (1976), *Biochem. Biophys. Acta* **444**, 472.

32. V. Horejsi, M. Ticha, and J. Kourcourek (1977), *Biochem. Biophys. Acta* **499**, 290.

33. V. Horesji, M. Ticha, and J. Kocourek (1977), *Biochem. Biophys. Acta* **499**, 301.

34. B. M. Dunn and I. M. Chaiken (1975), *Biochemistry* **14**, 2343.

35. M. Ticha (1978), *Biochim. Biophys. Acta* **534**, 58.

36. M. Caron, A. Faure, R. V. Keros, and P. Cornillot (1977), *Biochim. Biophys. Acta* **491**, 558.

37. M. Caron, A. Faure, and P. Cornillot (1975), *L'actual Chim.* **4**, 43.

38. M. Caron, A. Faure, P. Keros, and P. Cornillot (1976), *C. R. Acad. Sci. (Paris)* **283(D)**, 1253.

39. A. Faure, M. Caron, and P. Cornillot (1975), *Ann. Immunol. (Inst. Pasteur)* **127(c)**, 122.

40. T. C. Bøg-Hansen (1976), Abstr. *Tenth International Congress of Biochemistry*, Hamburg, p. 188.

41. M. R. A. Morgan, P. J. Brown, M. J. Leyland, and P. D. G. Dean (1978), *FEBS Lett.* **87**, 239.

42. P. Å. Albertsson (1974), in *Methods in Enzymology* (S. Fleischer and L. Packer, eds.) **31**, 761.

43. V. P. Shanbhag and V. Johansson (1974), *Biochem. Biophys. Rec. Commun.* **61**, 1141.

44. S. D. Flanagan, P. Taylor, and S. H. Barondes (1975), *Croat. Chem. Acta* **47**, 449.

45. S. D. Flanagan and S. H. Barondes (1975), *J. Biol. Chem.* **250**, 1484.

46. G. E. Means and R. E. Feeney (1971), *Chemical Modification of Protein*, Holden Day, San Francisco.

47. W. H. Scouten (1974), in *Methods in Enzymology* (W. B. Jakoby and M. Wilchek, eds.) **34**, 288.

48. W. H. Scouten (1978), in *Enzyme Engineering* (G. B. Broun, G. Manecke, and L. B. Wnigard, Jr., eds.) **4**, 391.

49. C. Lewis and W. Scouten (1976), *Biochim. Biophys. Acta* **444**, 326.

50. C. Lewis and W. Scouten (1976), *J. Chem. Ed.* **53**, 395.

51. W. H. Scouten and G. L. Firestone (1976), *Biochim. Biophys. Acta* **453** 277.

52. M. Rubenstein, Y. Schechter, and A. Patchornik (1976), *Biochem. Biophys. Res. Commun.* **70**, 1257.

53. Y. Shechter, M. Rubenstein, and A. Patchornik (1977), *Biochemistry* **16**, 1424.

54. D. C. Neckers (1975), *J. Chem. Ed.* **52**, 695.

55. M. Wilchek, T. Oka, and Y. J. Topper (1975), *Proc. Nat. Acad. Sci. (USA)* **72**, 1055.

56. Y. S. Topper, T. Oka, B. K. Vonderhaar, and M. Wilchek (1975), *Biochem. Biophys. Res. Commun.* **66**, 793.

57. T. Oka and Y. S. Topper (1974), *Proc. Nat. Acad. Sci. (USA)* **71**, 1630.

58. G. I. Tesser, H.-U. Fisch, and R. Schwyzer (1972), *FEBS Lett.* **23**, 56.

59. G. I. Tesser, H.-U. Fisch, and R. Schwyzer (1974), *Helv. Chim. Acta* **57**, 1718.

60. Y. Kamido and H. Nikaido (1977), *Biochim. Biophys. Acta* **464**, 589.

61. W. H. Scouten, W. Iobst, and C. Lewis (1978), unpublished results.

62. M. Wilchek (1978), in *Enzyme Engineering* (B. G. Broun, G. Manecke, and L. B. Wirgard, Jr. eds.) **4**, 435.

63. E. P. Goldberg (1978), in *Polymeric Drugs* (L. G. Donaruma and O. Vogl, eds.) Academic, New York, p. 239.

64. K. K. Stewart and R. F. Doherty (1973), *Proc. Nat. Acad. Sci.* (*USA*) **70**, 2850.

65. A. H. Beckett and P. Anderson (1957), *Nature* **179**, 1074.

66. A. H. Beckett and P. Anderson (1960), *J. Pharm. Pharmacol.* **12**, 228T.

67. A. H. Beckett and P. Anderson (1959), *J. Pharm. Pharmacol.* **11**, Supplement, 258T.

68. G. Wulff, W. Vesper, R. Grobe-Einsler, and A. Sarhan (1977), *Makromol. Chem.* **178**, 2799.

69. G. Wulff, R. Grobe-Einster, W. Vesper, and A. Sarhan (1977), *Makromol. Chem.* **178**, 2817.

70. G. Wulff, A. Sarhan, and K. Zabrocki (1973), *Tetrahed. Lett.* **44**, 4329.

71. R. S. Yallow and S. A. Berson (1960), *J. Clin. Invest.* **39**, 1157.

72. K. Catt and G. W. Tregar (1966), *Biochem. J.* **100**, 31c.

73. S. Siegel, W. Line, N. Yang, A. Kwong, and C. Frank (1973), *Clin. Endocrinol. Metab.* **37**, 526.

74. Eshdat, Y. and Prujansky–Jakobovits. (1979), *FEBS Lett.* **101**, 43.

75. Singh, P., Lewis, S. D. and Shafer, J. A. (1980) *Arch. Biochem. Biophys.* **203**, 774.

INDEX

339

Cyanogen bromide:
   activation, 42, 44
   dangers of, 42
1-Cyclohexyl-3-(2 morpholinoethyl)
   carbodiimide, 79, 211
Cysteic acid, 49
Cysteine, 164
Cysteine-containing peptides, 173

Dannomycin, 328
DEAE cellulose, 8, 184, 188, 191, 214, 251
Deforming buffers, 286
Dehydrogenases, 109
Denaturants, 51
Denaturation, 10
Denaturing buffers, 274
5'-Deoxyadenosylcobalamin, 140
Deoxynucleotide kinases, 130
Deoxynucleotidyltransferase, 219
Desorption technique, 285
Detergent chromatography, 106, 129, 244,
   259
Detergents, 265
Dextorphan, 332
Dextran, 317
Dextran sulfate, 216
Dextrins, cross-linked, 37
Diacetylchitobiose, 150
1,4-Diaminobutane, 274
4-Diazobenzoic acid, 230
2,3-Dibromopropanol, 51
Dicyclohexyl carbodiimide, 64, 75,
   76, 78, 113, 218, 223
Dicyclohexylurea, 75
Diethylaminoethyl (DEAE) cellulose,
   8, 184, 188, 191, 214, 251
Digitoxin, 334
Dihydrofolate reductase, 34, 136
Diisopropylfluorophosphate, 163
1-(3-Dimethylaminopropyl)-3-
   ethylcarbodiimide, 56
Dimethyl formamide, 71, 79
Dimethyl sulfoxide, 24
2,4-Dinitrofluoroaniline, 64
2,4-Dinitrofluorobenzene, 178
Dinitrophenol (DNP), 287, 308, 320
Dipeptidyl peptidase, 278, 279
2,2'-Dipyridyldisulfide, 166
Discrete-staged transfer and equilibrium
   model, 88

Dissociation constants, 85, 316, 318
Dithiothreitol, 164, 171
Divinyl sulfone, 50, 52, 280
DNA, 13
   immobilized, 211
   melting, 13
   removal of endogenous, 214
DNA agarose, 37
DNA-binding proteins, 10, 214
   denatured, 216
   half-life, 210
c-DNA-cellulose, 221, 222
DNA glass, 37
DNA polymerase, 37, 208, 211, 219, 227
DNA regulatory proteins, 208
DNAse, 216, 247
DNP lysine, 274, 320
DNP-methionine peptide, 287
Dyes, reactive, 54

Egg white, 145
Electric field, effects on binding
   constants, 317
Electrophoresis:
   affinity, 312, 317, 318
   crossed affinity, 314
   immunoaffinity, 313
   rocket-affinity, 318
   SDS, 294
Electrophoretic desorption, 281
Elution:
   bioselective, 14
   with soluble ligand, 97
Enantiomeric mixtures, separation, 331
Enantiomers, 331
Enolase, 194, 201
Enzacryl, 20, 34, 64
"Enzyme-analogue polymers," 332
Enzyme recovery, 93, 94
Enzymes, immobilized, 4, 305
Epichlorohydrin, 23, 50, 170
   cross-linking with, 50
Epoxieds, 49, 265
Equilibrium constant, 317
Equilibrium model, 89
Erythrocyte membrane, 256
Estradiol, 31
17-β-Estradiol dehydrogenase,
   178
Ethanedithiol, 49

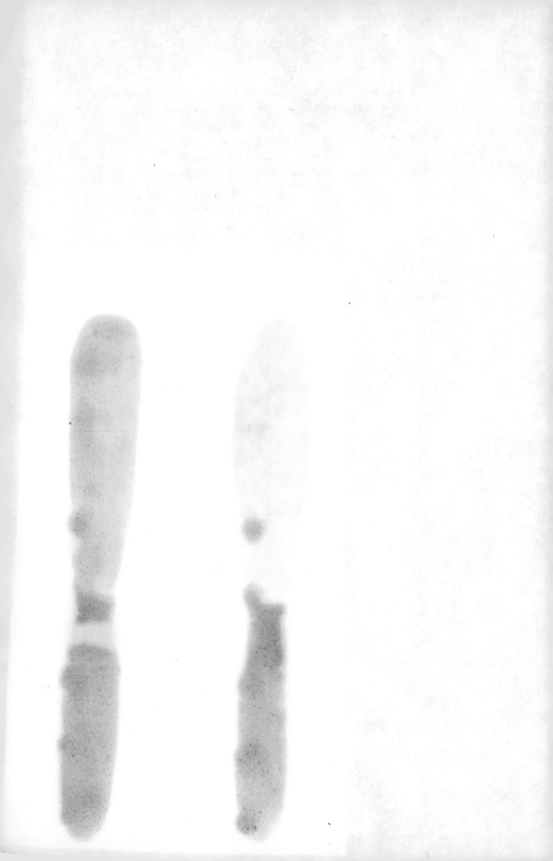